DISCRETE AND INTEGRATED ELECTRONICS

DISCRETE AND INTEGRATED ELECTRONICS

ERVINE M. RIPS
New Jersey Institute of Technology

Prentice-Hall, Englewood Cliffs, New Jersey 07632

Library of Congress Cataloging-in-Publication Data

RIPS, ERVINE M. (date)
  Discrete and integrated electronics.

  Includes index.
    1. Analog electronic systems. 2. Integrated
circuits. 3. Digital electronics. I. Title.
TK7870.R53 1986   621.381   85-16971
ISBN 0-13-215153-7

Editorial/production supervision and interior design: **Diana Drew**
Cover design: **Whitman Studio, Inc.**
Manufacturing buyer: **Rhett Conklin**

The author and publisher of this book have used their best efforts in preparing this book. These efforts include the development, research, and testing of the theories and programs to determine their effectiveness. The author and publisher make no warranty of any kind, expressed or implied, with regard to these programs or the documentation contained in this book. The author and publisher shall not be liable in any event for incidental or consequential damages in connection with, or arising out of, the furnishing, performance, or use of these programs.

Printed in the United States of America

10 9 8 7 6 5 4 3 2 1

ISBN 0-13-215153-7 025

Prentice-Hall International, Inc., *London*
Prentice-Hall of Australia Pty. Limited, *Sydney*
Editora Prentice-Hall do Brasil, Ltda., *Rio de Janeiro*
Prentice-Hall Canada Inc., *Toronto*
Prentice-Hall Hispanoamericana, S. A., *Mexico*
Prentice-Hall of India Private Limited, *New Delhi*
Prentice-Hall of Japan, Inc., *Tokyo*
Prentice-Hall of Southeast Asia Pte. Ltd., *Singapore*
Whitehall Books Limited, *Wellington, New Zealand*

**To the memory of
LAWRENCE BAKER ARGUIMBAU
Teacher, Engineer, Citizen**

Contents

5 Dynamic Stability of Feedback Amplifiers   161

6 Oscillators   189

7 Characteristics and Applications of Operational Amplifiers   209

11 Digital Memory Devices   343

12 Other Integrated-Circuit Devices   365

Appendix A: Classification of Feedback Circuits   398

Contents

Preface

Electronic systems today are based more upon integrated circuits than on discrete elements. Nevertheless, discrete transistors and diodes are still widely used, especially in high-power applications. Moreover, to make effective use of the ubiquitous IC building blocks, it is necessary to have some idea of how their internal, discrete elements work. We have, therefore, two reasons to study the electronics of discrete elements—first, to allow us to use discrete elements as such, and second, to understand how they function in integrated circuits.

Necessary as such a micro approach may be, it is not sufficient. We must also work on a macro level to understand the operation of larger scale systems whose elements are integrated circuits. In this book I have tried to give adequate treatment to both points of view.

Emphasis is placed on analog circuits. However, since we seem to be living in a world that is becoming increasingly digital, I have devoted three chapters to topics from that area. Chapter 2 deals with the internal structure and operation of basic logic gates. Chatper 11 treats memory devices—from basic flip-flops to RAMs and ROMs. Chapter 12 is devoted largely to those devices that operate at the analog-digital interface—analog comparators, D/A and A/D converters. In the discussion of digital devices I have chosen not to introduce such topics as Boolean algebra or Karnaugh maps; these matters are better left to books that concentrate on digital circuits.

You will find certain departures from what is usually found in a book of this sort. The first of these is the treatment of feedback, which stresses the asymptotic gain formula and Blackman's impedance formula. This approach obviates the troublesome and often confusing classification of input and output circuits as shunt- or series-connected. For those who want to connect the two approaches, conventional feedback-circuit classification is reviewed in Appendix A.

Other departures from standard textbook treatments are the discussion of the differential amplifier and its related current-source circuitry and the detailed discussion of direct-coupled structures such as those used in the internal configuration of operational amplifiers. The applications of operational amplifiers that are considered include inductance and capacitance simulation, band-pass and notch filters, and power supply regulators.

Listings of eleven computer programs are given at appropriate points in the text. These allow computation of such things as the gain-bandwidth product for a common-emitter amplifier, gain and phase crossover points when DC gain and the real poles of an amplifier are given, and pole locations for low-pass and band-pass Butterworth amplifiers. The programs are written in BASIC and are suitable for use on a personal computer.

The book is intended for use by fourth-year students in electrical engineering technology. It may also be useful for third-year electrical engineering students. The chief distinction between students in these two programs is the higher level of mathematical proficiency expected of those in engineering. However, in either curriculum the study of electronic circuits requires very little mathematical sophistication. For those students who are interested, I have tried to provide adequate mathematical explanations and derivations where these are appropriate and to do so in a direct manner. The instructor may, of course, elect to omit some of the more abstract derivations and proceed directly to the application of the principles involved.

For technology students, there is more material here than can be covered in one semester. If the students come to the course well prepared, the instructor might select Chaps. 3, 4, 5, 6, 7, 11, and 12. Another possible selection is Chaps. 1, 3, 4, 7, 9, and 10.

Every section includes worked-out, numerical examples. Problems at the ends of chapters rarely require derivations but can usually be solved using only simple arithmetic in a two- or three-step logical process. Plug-in problems have been avoided and the few that are included serve only to illustrate the magnitude of the quantities involved. An Instructor's Manual is available.

I am pleased to acknowledge the helpful suggestions of my colleagues, particularly Solomon Rosenstark, who brought to my attention the utility of the asymptotic gain formula and Blackman's equation. Discussions with Joseph Frank helped to clarify a number of ideas.

Thanks are also due to Nancy Bogen who provided encouragement and support during early phases of the writing.

—Ervine M. Rips

DISCRETE
AND INTEGRATED
ELECTRONICS

CHAPTER 1

Review of Basic Principles of Semiconductor Devices

1-1 JUNCTION TRANSISTOR OPERATION

A fundamental element of many semiconductor devices is the *pn* junction. Such junctions are usually formed by diffusing an *n*-type doping material through some depth of a *p*-type base material.

We can gain some understanding of the action of *pn* junctions by recalling that the *p* region contains a certain population of mobile holes, whereas the *n* region contains mobile electrons. At the junction, some holes diffuse into the *n* material and some electrons diffuse into the *p* region. Bear in mind that the holes were produced by *p*-doping elements, whose nuclei have one *less* positive charge than the surrounding silicon (or germanium) nuclei. Likewise, the electrons originated from *n*-doping elements, whose nuclei have one *more* positive charge than the neighboring silicon nuclei. When the mobile electrons and holes diffuse across the junction, they leave behind these nuclei, which—of course—are locked, immovable, in the crystal structure. Thus the region on the *p* side is deficient in positive charge, but the region on the *n* side has excess positive charge. The resulting situation is illustrated in Fig. 1-1.

A depletion region is thus formed in the immediate vicinity of the junction, which results in a built-in potential. If we connect an external battery-resistor circuit to the *pn* terminals, we have the situation shown in Fig. 1-2.

Looking at the circuit analogs, we see that in (a) and (b), the external battery and the built-in potential are *aiding*, while in (c) and (d), they are *in opposition*. From this we conclude that if the external source is polarized to direct current *into*

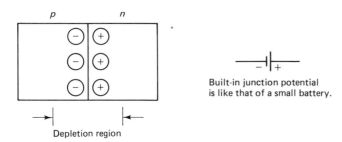

Built-in junction potential
is like that of a small battery.

Fig. 1-1 *pn* junction showing excess positive and negative charges resulting from stripped nuclei.

Depletion region

the *p side* of the junction, current can be made to flow *easily*. But if we attempt to direct current *into* the *n side*, current flows only with difficulty.

Thus a *pn* junction has *rectifying* properties. The current direction from *p* to *n* is called the *forward* direction. The direction from *n* to *p* is called the *reverse* direction. The relation between *p* and *n* regions and the conventional circuit symbol for a diode is illustrated in Fig. 1-3.

If we connect a semiconductor diode as shown in Fig. 1-4(a) and vary the voltage *V* applied across it, the current *I* will behave as shown in the curve of Fig. 1-4(b).

The equation that relates *I* to *V* is

$$I = I_s(e^{qV/kT} - 1) \tag{1-1}$$

where *q* and *k* are constants, *T* is the junction temperature in degrees Kelvin, and I_s is a quantity called the *reverse saturation current*. For *V* positive (forward) and equal to only a few tenths of a vote, *I* will be positive and will increase rapidly. But when *V* is negative (reverse), the term $e^{qV/kT}$ becomes negligible, so that *I* is approximately $-I_s$.

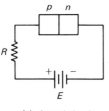

(a) Actual circuit

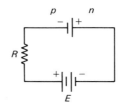

(b) Analog of circuit (a)

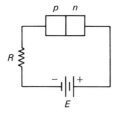

(c) External battery reversed

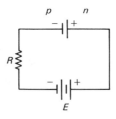

(d) Analog of circuit (c)

Fig. 1-2 An external battery-resistor circuit connected to the *pn* junction.

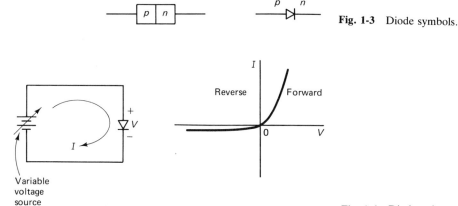

Fig. 1-3 Diode symbols.

(a)

Variable
voltage
source

(b)

Fig. 1-4 Diode volt-ampere character-istics.

Junction transistors are formed by successive diffusions of *n* and *p* doping materials to yield structures like those shown in Fig. 1-5. Since both types of material are used to form these transistors, they are often called bipolar junction transistors or BJTs.

To understand transistor action, we use the schematic structure shown in Fig. 1-6. The batteries shown in Fig. 1-6 are meant to suggest bias polarities for the two internal junctions. We can see that the left-hand *np* junction is *forward* biased, whereas the *pn* junction on the right is *reverse* biased. Thus current flows readily through the left-hand junction, in the direction shown. This indicates that electrons are injected from the left-hand *n* body through the junction into the *p* region. These electrons diffuse through the *p* material until they encounter the strong electric field that exists at the reverse-biased junction on the right. At that point they are pulled through the junction, exiting to the right. Thus a current must enter the right-hand *n* body as shown.

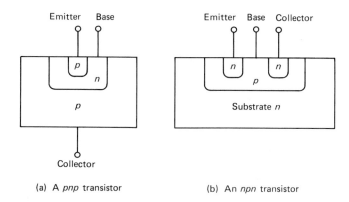

(a) A *pnp* transistor

(b) An *npn* transistor

Fig. 1-5 Transistor structures.

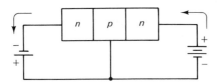

Fig. 1-6 Schematic transistor structure showing normal directions of operating current.

Not all the electrons injected at the left junction arrive at the one on the right because a certain number of them recombine with the holes that are present in the *p* region. (Remember, a hole is the absence of an electron.) For this reason only a fraction of electrons *emitted* at the left are *collected* at the right. (You will recall that the transistor structure consists of an emitter, a base, and a collector.) Thus we see that the collector current is related to the emitter current by the equation

$$I_C = \alpha I_E \tag{1-2}$$

By Kirchhoff's current law, we can see that the current into the base must be given by

$$I_B = I_E - I_C = (1 - \alpha)I_E \tag{1-3}$$

Typical values of α are 0.98 and 0.99. This number varies from one device to another.

1-2 JUNCTION TRANSISTOR BIASING

Before considering circuits that amplify or control signals, we first have to establish DC operating conditions. Perhaps the best way to review the procedure for doing so is by means of an example.

Example 1.1

A typical *npn* transistor circuit is shown in Fig. 1-7. (Note that the arrow on the emitter symbol points in the direction of normal-operating, emitter-current flow.) The problem here is to determine R_B so that I_E will be 1 mA. To solve this problem we need to know how much voltage will appear across the base-emitter junction. One way to find out would be to include in our equations a solution for V_{BE} taken from Eq. (1-1). But to do so would be needlessly complicated and not very useful. Instead, we simply make use of a fact, based on many observations, that in normal, forward bias, the voltage V_{BE} for a silicon transistor is usually about 0.7 V.

For $I_E = 1$ mA,

$$I_B = (1 - \alpha) \times 1 \text{ mA}$$

$$\alpha = 0.99$$

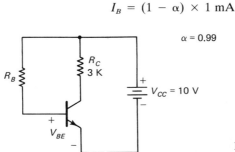

Fig. 1-7 Transistor circuit.

or

$$I_B = (1 - 0.99) \cdot 10^{-3} = 10 \ \mu A$$

To find R_B we observe that

$$V_{CC} = V_{RB} + V_{BE}$$

Hence V_{RB}, the voltage across R_B, must be

$$V_{RB} = 10 - 0.7 = 9.3 \ V$$

Then $R_B = V_{RB}/I_B$, or $R_B = 9.3/10^{-5} = 9.3 \times 10^5 = 930 \ K\Omega$.

Sometimes, when V_{CC} is large compared to V_{BE}, we can neglect V_{BE} altogether with little effect on the results.

It is interesting to observe that, in Example 1-1, the voltage from collector to emitter is $V_{CE} = V_{CC} - I_C R_C$. Since $I_C \approx I_E$ we can calculate V_{CE} quite simply.

$$V_{CE} = 10 - 10^{-3} \times 3 \times 10^3 = 10 - 3 = 7 \ V$$

It is convenient to define a parameter that relates base current to collector current directly in dealing with circuits like the one in Fig. 1-7. Since $I_B = (1 - \alpha)I_E$ and $I_C = \alpha I_E$, then

$$I_B = \left(\frac{1 - \alpha}{\alpha}\right)I_C \quad \text{or} \quad I_C = \frac{\alpha}{1 - \alpha} I_B$$

We can define the quantity h_{FE}, sometimes taken as β, to be

$$h_{FE} = \beta = \frac{\alpha}{1 - \alpha} \tag{1-4}$$

Then we can write

$$I_C = h_{FE}I_B \tag{1-5}$$

A word of caution: Eq. (1-5) takes no account of the presence of a reverse-bias current I_{CBO} through the collector-base junction. When the transistor is connected as shown in Fig. 1-7, this current is magnified by a factor $1 + h_{FE}$. The correct equation should therefore be

$$I_C = h_{FE}I_B + (1 + h_{FE})I_{CBO}$$

To make matters even more complicated, I_{CBO} is temperature sensitive. However, its value for silicon transistors is usually extremely small. Consequently, since our purpose here is only to review a few basic principles, we shall ignore it. You are advised to review material from previous courses that deal with this matter.

A commonly used biasing circuit is shown in Fig. 1-8. We analyze it in Example 1.2.

Example 1.2

The problem is to find the collector current and the collector-emitter voltage. We begin by assuming that the base current will be small compared to the current through R_2 and R_1, which are connected in series across V_{CC}. We shall check this

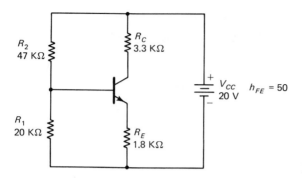

Fig. 1-8 Circuit example.

assumption later. The voltage at the base terminal is, by the voltage-divider rule, 20 × (20 KΩ)/(20 KΩ + 47 KΩ) = 20 × 20/67 ≈ 6 V. Since the base-emitter voltage V_{BE} is 0.7 V, the emitter voltage is 6 − 0.7 = 5.3 V. This must be the voltage *across* R_E, so that I_E = 5.3/1.8 K = 2.94 mA ≈ 3 mA. The voltage *drop across* R_c must therefore be very nearly 3 mA × 3.3 K = 9.9 V. To find the collector-emitter voltage, we use Kirchhoff's voltage law and write

$$V_{CC} = V_{RC} + V_{CE} + V_{RE}$$

or

$$20 = 9.9 + V_{CE} + 5.3.$$

$$V_{CE} = 20 - 9.9 - 5.3 = 4.8 \text{ V}$$

What about our assumption that I_B is small compared to the current through R_2 and R_1? First of all,

$$I_B = \frac{I_E}{1 + h_{FE}} = \frac{3 \times 10^{-3}}{51} \approx 0.06 \times 10^{-3}$$

$$= 60 \text{ μA}$$

The current through R_1 and R_2 is 20/(20KΩ + 47 KΩ) = 20/67 mA, or about 0.3 mA = 300 μA. Thus the assumption appears to be of doubtful validity since the base current is 20% of the current through the voltage divider. Ordinarily, we consider that our assumption is reasonable when I_B is no more than 10% of the voltage-divider current.

What can be done about this situation? To answer this question, we begin by calculating the resistance that is seen looking into the base terminal. The current into the base is, of course, I_B. The voltage *due to* R_E is

$$I_E R_E = (1 + h_{FE})I_B R_E$$

Then the DC resistance looking into the base must be given by

$$R_{\text{base-to-ground}} = R_{B_{in}} = \frac{\text{base-to-ground voltage}}{\text{current into base}} = \frac{(1 + h_{FE})I_B R_E}{I_B}$$

or

$$R_{B_{in}} = (1 + h_{FE})R_E$$

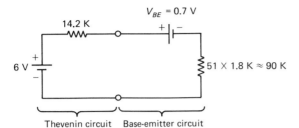

$V_{BE} = 0.7$ V

14.2 K

6 V

51 X 1.8 K ≈ 90 K

Thevenin circuit   Base-emitter circuit

Fig. 1-9 Equivalent circuit.

If we replace the voltage source V_{CC} and the voltage divider by their Thevenin equivalent circuit, we have the circuit shown in Fig. 1-9.

From this circuit we can readily calculate I_B:

$$I_B = \frac{6 - 0.7}{104.2 \text{ K}} = \frac{5.3}{104 \text{ K}} = 51 \text{ μA}$$

The emitter current must then be 51 × 51 μA or 2.60 mA. These values should be compared with the approximate values calculated earlier, which are 60 μA and 3 mA, respectively.

You should review the procedure for locating the quiescent operating point of a transistor on the transistor collector characteristic curves.

1-3 SMALL-SIGNAL TRANSISTOR OPERATION

In small-signal operation, excursions of base and collector currents are such that we can assume (1) neither cutoff nor saturation occurs, and (2) operation is entirely linear, that is, transistor parameters are constant.

Typical collector and base characteristics for an *npn* transistor are shown in Fig. 1-10. It is assumed that the reference terminal is the emitter (common emitter characteristics).

If we assume a constant value of I_B, then collector operation is confined entirely to the single curve appropriate to that value of I_B. The slope of this curve is $\Delta I_C / \Delta V_{CE}$. This leads to the definition

$$h_{oe} = \left. \frac{\Delta I_C}{\Delta V_{CE}} \right|_{I_B = \text{constant}}$$

If V_{CE} is held constant, the change in I_C (the effect) is related to the change in I_B (the cause) by

$$h_{fe} = \left. \frac{\Delta I_C}{\Delta I_B} \right|_{V_{CE} = \text{constant}}$$

Similarly, if V_{CE} is held constant, a change ΔI_B occurs as a result of a change

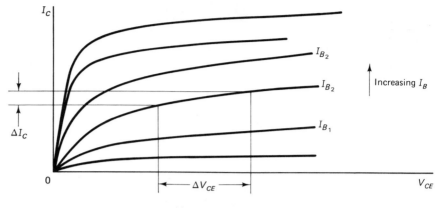

(a) Collector characteristics

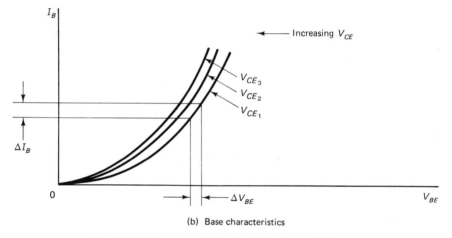

(b) Base characteristics

Fig. 1-10 Common-emitter transistor characteristics.

ΔV_{BE}. Hence we can define

$$h_{ie} = \frac{\Delta V_{BE}}{\Delta I_B}\bigg|_{V_{CE} = \text{constant}}$$

If I_B is held constant, a change ΔV_{BE} is caused by a change ΔV_{CE}. This leads to

$$h_{re} = \frac{\Delta V_{BE}}{\Delta V_{CE}}\bigg|_{I_B = \text{constant}}$$

It should be noted that these ratios have different dimensions; h_{fe} and h_{re} are dimensionless current and voltage ratios, respectively, and h_{oe} is a conductance, whereas h_{ie} is a resistance.

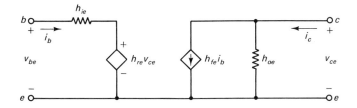

Fig. 1-11 h-parameter, common-emitter circuit.

1-3.1 Small-Signal Equivalent Circuits

These hybrid, or h, parameters can be used to define a small-signal equivalent circuit for a transistor, as shown in Fig. 1-11. It should be remembered that the small-signal h parameters and the corresponding equivalent circuit are the same for both *npn* and *pnp* transistors.

In Fig. 1-11, observe that the capitalized variables ΔI_B, ΔV_{BE}, ΔI_C, and ΔV_{CE} have been replaced by the lowercase variables i_b, v_{be}, i_c, and v_{ce}, respectively. The new variables represent *only* those small-current and voltage components that *vary*, while I_B, V_{BE}, I_C, and V_{CE} represent *constant*, or DC, values. In general, separate solutions are required for the DC quantities and the varying quantities. The equivalent circuits used to find varying quantities have little, if any, significance in finding DC quantities.

Typical values for the h parameters are:

$$h_{fe} = 100$$

$$h_{ie} = 2600 \ \Omega$$

$$h_{re} = 5 \times 10^{-3}$$

$$h_{oe} = 10^{-5} \ \text{mho}$$

Often h_{re} and h_{oe} may be neglected. *Whether this is permissible has to be investigated. Particular care must be used with regard to h_{oe}.* Assuming that these two parameters may be neglected, we are led to the simplified equivalent circuit in Fig. 1-12.

1-3.2 Determination of h_{ie}

It is useful to understand the physical basis for the parameter h_{ie}. The base characteristic shown in Fig. 1-10(b) resembles the forward characteristic of a semiconductor diode. This is not surprising, since the emitter-base junction is forward biased in normal operation. We would therefore expect the dynamic resistance of the emitter-base junction to have the same value as that found in a forward-biased diode and that is, indeed, the case. The emitter current in terms of base-emitter voltage is given by a modified form of Eq. (1-1),

$$I_E = I_{EO}(e^{qV_{BE}/kT} - 1)$$

where I_{EO} is the saturation value of the emitter-base diode current. What is the

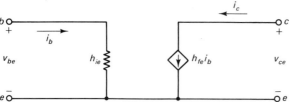

Fig. 1-12 Simplified equivalent circuit.

change ΔI_E resulting from the change ΔV_{BE}? To find out, we differentiate the above equation to obtain

$$\frac{dI_E}{dV_{BE}} = \frac{q}{kT} I_{EO} e^{qV_{BE}/kT} \approx \frac{\Delta I_E}{\Delta V_{BE}}$$

If V_{BE} is more than a few tenths of a volt, then $e^{qV_{BE}/kT} \gg 1$. Consequently,

$$I_E \approx I_{EO} e^{qV_{BE}/kT}$$

Hence

$$\frac{\Delta I_E}{\Delta V_{BE}} \approx \frac{q}{kT} I_E$$

The reciprocal of this ratio is

$$\frac{\Delta V_{BE}}{\Delta I_E} = \frac{1}{(q/kT)I_E} = \frac{kT/q}{I_E}.$$

At room temperature, kT/q has the value of about 0.026 V. We know

$$\frac{\Delta V_{BE}}{\Delta I_E} = \frac{\Delta V_{BE}}{(1 + h_{fe})\Delta I_B}$$

since $\Delta I_E = (1 + h_{fe})\Delta I_B$. Hence

$$\frac{\Delta V_{BE}}{\Delta I_B} = (1 + h_{fe})\frac{0.026}{I_E}$$

But $\Delta V_{BE}/\Delta I_B = h_{ie}$. Therefore, we have the very useful result that

$$h_{ie} \approx h_{fe} \times \frac{0.026}{I_E} \tag{1-6}$$

As before, I_E is capitalized to emphasize the point that this is the DC value of emitter current. A simple rule can be stated that if I_E is the *number of milliamperes* of DC emitter current, then h_{ie} is h_{fe} times 26 Ω per milliampere of I_E. Thus if I_E (DC) is 2 mA and $h_{fe} = 50$,

$$h_{ie} = 50 \times \frac{26}{2} = 650 \ \Omega$$

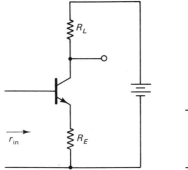
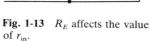

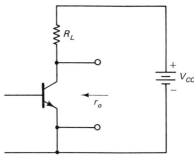

Fig. 1-13 R_E affects the value of r_{in}.

Fig. 1-14 To determine r_o.

1-3.3 Input and Output Resistances

When the transistor is connected with the emitter terminal at reference potential (common emitter), the small-signal input resistance to the base is, of course, about equal to h_{ie}. But when a resistor is connected between the emitter and the reference terminal as in Fig. 1-13, the situation changes.

Recall that the emitter current is $(1 + h_{fe})i_b$, where i_b now refers to signal, or varying, current. The voltage between base and reference is v_{in}, given by $v_{in} = i_b h_{ie} + (1 + h_{fe})i_b R_E$. Then

$$r_{in} = \frac{v_{in}}{i_b} = h_{ie} + (1 + h_{fe})R_E$$

For the transistor we have just been considering, if $R_E = 220\ \Omega$, r_{in} would be

$$r_{in} = 650 + 50 \times 220 = 11{,}650\ \Omega$$

We now wish to find the output resistance of a common-emitter-connected amplifier. The circuit is shown in Fig. 1-14.
We remember that the battery V_{CC} has an AC impedance equal to zero. Consequently it acts as a short circuit between the top of resistor R_L and the reference terminal. Replacing the transistor with its approximate equivalent circuit, we get Fig. 1-15. We see from Fig. 1-15 that if we short-circuit the base terminal, the current source is inactive, $i_b = 0$, and the current source $h_{fe}i_b$ is also inactive. Its internal impedance is infinite, so that the resistance looking into the collector terminal is

$$r_o = \left(\frac{1}{h_{oe}}\right) \parallel R_L.$$

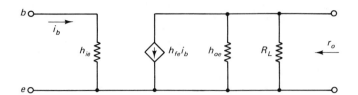

Fig. 1-15 Approximate equivalent circuit to find r_o.

If R_L is small compared to $1/h_{oe}$, as is often the case, then $r_o \approx R_L$. One must be careful in calculating r_o to understand what resistance is being sought. For example, the load resistor R_L sees, as the output resistance of the transistor, $r_o = 1/h_{oe}$, while an external circuit connected across r_o would see

$$\left(\frac{1}{h_{oe}}\right) \| R_L$$

All the foregoing calculations have neglected the effect of parameter h_{re}. Its inclusion complicates the equations, although the basic approach is the same. To simplify matters hereafter, we shall simply refer to Table 1-1, which summarizes exact and approximate formulas for various quantities for the common-emitter circuit. In Table 1-1, notice that r_0 does *not* include Z_L and that an input generator resistance R_s is assumed in calculating output resistance.

TABLE 1-1

Quantity	Exact	Approximate
Current Gain, A_i	$\dfrac{h_{fe}}{1 + h_{oe}Z_L}$	h_{fe}
Voltage Gain, A_v	$\dfrac{-h_{fe}Z_L}{h_{ie} + (h_{ie}h_{oe} - h_{fe}h_{re})Z_L}$	$\dfrac{-h_{fe}Z_L}{h_{ie}}$
Input Resistance, r_i	$h_{ie} - \dfrac{h_{fe}h_{re}Z_L}{1 + h_{oe}Z_L}$	h_{ie}
Output Resistance, r_o	$\dfrac{1}{h_{oe} - \dfrac{h_{fe}h_{re}}{h_{ie} + R_s}}$	$\dfrac{1}{h_{oe}}$

Program 1-1 allows us to calculate easily the four quantities A_i, A_v, r_i, and r_o.

```
10   PRINT "COMPUTATION OF CURRENT GAIN,   VOLTAGE GAIN, INPUT RESISTANCE"
20   PRINT "AND OUTPUT RESISTANCE "
30   PRINT "FOR A COMMON-EMITTER, SINGLE-STAGE AMPLIFIER"
31   PRINT
40   INPUT "ENTER Hfe.";HFE
50   INPUT "ENTER Hoe. (ESTIMATE, IF NECESSARY.)";HOE
60   INPUT "ENTER Hre. (ENTER 0 IF VALUE OF Hre IS NOT KNOWN.)";HRE
70   INPUT "ENTER Hie.";HIE
80   INPUT "ENTER SOURCE RESISTANCE.";RS
90   INPUT "ENTER LOAD RESISTANCE.";RL
```

Prog. 1-1.

```
100   PRINT "SELECT DESIRED FUNCTION BY RESPONDING (Y/N)."
101   PRINT
110   INPUT "DO YOU WANT CURRENT GAIN, Ai?";A$
120   IF A$ = "Y" THEN 140
130   GOTO 160
140 AI = HFE / (1 + HOE * RL)
150   PRINT "Ai = ";AI
151   PRINT
160   INPUT "DO YOU WANT VOLTAGE GAIN, Av?";A$
170   IF A$ = "Y" THEN 190
180   GOTO 210
190 AV =  - HFE * RL / (HIE + (HIE * HOE - HFE * HRE) * RL)
200   PRINT "Av=   ";AV
201   PRINT
210   INPUT "DO YOU WANT INPUT RESISTANCE , Ri?";A$
220   IF A$ = "Y" THEN 240
230   GOTO 260
240 RI = HIE - (HFE * HRE * RL) / (1 + HOE * RL)
250   PRINT "Ri = ";RI;" OHMS"
251   PRINT
260   INPUT "DO YOU WANT OUTPUT RESISTANCE, Ro?";A$
270   IF A$ = "Y" THEN 290
280   GOTO 310
290 RO = 1 / (HOE - (HFE * HRE) / (HIE + RS))
300   PRINT "Ro = ";RO;" OHMS"
301   PRINT
310   INPUT "MORE?";A$
320   IF A$ = "Y" THEN 40
330   END
```

Prog. 1-1 (cont.).

Let us apply the information we have accumulated to the study of the circuit shown in Fig. 1-16.

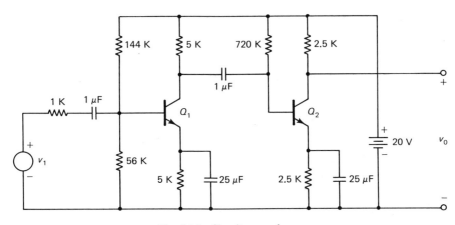

Fig. 1-16 Circuit example.

Example 1.3

In this example, we wish to calculate the gain, which is the ratio v_0/v_1. To put the question more precisely, we wish to find the ratio of the phasors $\mathbf{V}_0/\mathbf{V}_1$ at midband frequencies.

First of all, we must undertake a DC analysis. Transistor Q_1 is biased with a voltage divider in the base circuit. The voltage at the base terminal is $[56/(56 + 144)]20 = 5.6$ V, providing that the base current is small enough. Current through the divider resistors alone is 20/(56 + 144) = 0.1 mA. Allowing 0.7 V for V_{BE}, the emitter voltage is 5.6 − 0.7 = 4.9 V. Hence I_E is 4.9/5 K ≈ 1 mA. Since h_{fe} = 100 (assuming h_{FE} ≈ h_{fe}), I_B is about 10 μA, which is small enough to have negligible effect on the voltage divider.

Transistor Q_2 is biased differently from Q_1. Considering the battery, base resistor, V_{BE}, and emitter resistor as a loop, we can write

$$20 = I_B \times 720 \text{ K} + V_{BE} + h_{FE}I_B \times 2.5 \text{ K}$$

Then I_B is equal to

$$[20 - 0.7]/[(720 + 100 \times 2.5) \times 10^3] = 19.3/970 \times 10^3 = 20 \text{ μA}$$

Therefore, I_E is about 100 × 20 μA = 2 mA.

Next we turn to the AC analysis of the circuit. With 1 mA of DC emitter current, h_{ie} for Q_1 is about 100(0.026/0.001) = 2600 Ω. At ordinary audio frequencies, such as 1 kHz, the reactances of the 25-μF capacitors are $1/(2\pi \times 10^3 \times 25 \times 10^{-6})$, or about 6 Ω. This value is so small compared to the 5-KΩ and 2.5-KΩ emitter resistances that it can be neglected. In effect, the emitter resistors are short-circuited. We can therefore use the formulas of Table 1-1 without modification for emitter impedance.

Since the resistance looking into the base of Q_1 is about equal to h_{ie} = 2600 Ω, the 56-KΩ and 144-KΩ resistors are "swamped" and can be neglected. The reactance of the 1-μF capacitor is $1/(2\pi \times 10^3 \times 1 \times 10^{-6})$ = 159 Ω, which can be neglected in comparison to h_{ie} and to the 1-KΩ source resistance.

Before we can calculate voltage gains, we must consider the effective value of load impedance seen by Q_1. Evidently, this must consist of the combined effect of the 5-KΩ load resistor and the resistance looking into the base of Q_2. Strictly speaking, the input resistance of Q_2 depends to some extent upon the collector load of Q_2. In this example, such dependence is unimportant, but, in general, it is usually a good procedure to begin the calculation of gain starting at the output stage, since its loading is easy to determine. Then we work back, stage by stage, to the input.

In this example, the approximate voltage gain of Q_2 is $-h_{fe}R_L/h_{ie}$, as given in Table 1-1. Substituting numerical values, we get

$$A_{v2} \approx \frac{-100 \times 2.5 \text{ K}}{h_{ie}}$$

As before, we estimate h_{ie} to be 100 × 26/2 = 1300 Ω. Hence,

$$A_{v2} \approx \frac{-2.5 \times 10^5}{1.3 \times 10^3} = -192$$

The input resistance of Q_2 is (720 K)‖h_{ie}. Since h_{ie} is 1300 Ω, we can neglect the 720 KΩ. However, the effective load resistance seen by the collector of Q_1 is (5 K)‖1300 ≈ 1030 Ω.

Before calculating A_v for Q_1, we must investigate its base circuit. This investigation is necessary because the voltage gain desired for the entire amplifier is to be calculated with respect to the input voltage V_1. Because of the 1-KΩ source resistance, the voltage at the base of Q_1 will be somewhat different from V_1. We begin the analysis by finding the effect of the voltage divider. Looking into the voltage-divider-base node, we see the resistance 144 K‖56 K‖h_{ie}. This amounts to 144 K‖56 K‖2.6 K ≈ 2.45 KΩ. Thus, for approximate purposes, we can neglect the effect of the voltage divider. Then the proportion of V_1 that appears at the base of Q_1 will be $V_1 \times h_{ie}/(h_{ie} + 1 \text{ K})$, or 0.722 V_1.

For Q_1, A_v is therefore $-0.722 \times h_{fe} R_{Leff}/h_{ie}$. This value is $-0.722 \times 100 \times$ (1.03 K)/(2.6 K) ≈ -28.6. The overall voltage gain is therefore $-28.6 \times -192 =$ 5491.

1-4 COMMON-COLLECTOR AND COMMON-BASE CIRCUITS

Up to this point we have considered only transistors connected in the common-emitter configuration. Among the many fortunate attributes of the transistor is the ease with which it may be used in any configuration—common emitter, common collector, or common base. As you no doubt know, h parameters are defined for all three configurations. For simplicity our treatment here relies almost exclusively on the common-emitter h parameters; we shall calculate necessary gains and resistances for the other configurations in terms of this one set of parameters.

1-4.1 The Common-Collector Amplifier

The circuit of Fig. 1-17 shows a *common-collector amplifier*, often called an *emitter follower*. At frequencies of interest, we can assume that the capacitor C looks like a short circuit. Then, short-circuiting the input terminals, we can imagine a current source I_0 pumping in current at the output terminals. If we solve for the resulting voltage V_0 and divide the resulting expression by I_0, we get the output resistance r_o. This yields the result

$$r_o = \cfrac{1}{\cfrac{1}{h_{ie}/(1 + h_{fe})} + \cfrac{1}{R_E} + h_{oe}}$$

The significance of this result is that the output resistance r_o is the parallel combination of R_E, $1/h_{oe}$, and $h_{ie}/(1 + h_{fe})$. This last term is simply the inherent input resistance of the transistor itself, h_{ie}, divided by the familiar current-multiplication factor $1 + h_{fe}$.

We can apply a similar procedure to finding the input resistance. In this case,

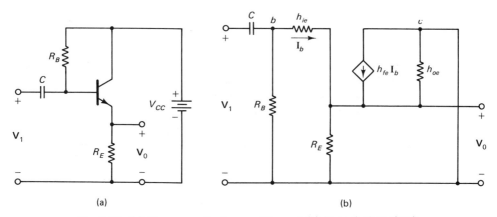

(a)   (b)

Fig. 1-17 (a) Common-collector amplifier and (b) its equivalent circuit.

we assume that $\mathbf{V}_0$ sees, looking to the right, an open circuit. If an input voltage $\mathbf{V}_1$ is applied to the input terminals, the resulting $\mathbf{V}_0$ is found to be

$$\mathbf{V}_0 = \mathbf{V}_1 \frac{\left(\dfrac{1 + h_{fe}}{h_{ie}}\right)\left(\dfrac{1}{h_{oe}} \parallel R_E\right)}{1 + \left(\dfrac{1 + h_{fe}}{h_{ie}}\right)\left(\dfrac{1}{h_{oe}} \parallel R_E\right)}$$

After slight rearrangement this expression yields the voltage-gain formula

$$A_V = \frac{\left(\dfrac{1}{h_{oe}} \parallel R_E\right)}{\left(\dfrac{h_{ie}}{1 + h_{fe}}\right) + \left(\dfrac{1}{h_{oe}} \parallel R_E\right)}$$

Clearly, A_v must always be a number less than 1.

It is not difficult to show that the input resistance is

$$r_{\text{in}} = \frac{1}{1/R_B + (1 - A_v)/h_{ie}} \quad \text{or} \quad R_B \parallel \frac{h_{ie}}{1 - A_v}$$

This suggests that if R_B is large and A_v is made to be close to 1, very large input resistances can be achieved, though they are never larger than R_B.

It is often possible to use approximate variations of the formulas just given. These variations are summarized below. Which formulas to use depends upon circuit values in the case at hand.

$$r_o \approx R_E \parallel h_{ie}/(1 + h_{fe})$$

$$A_v \approx 1$$

$$r_{\text{in}} \approx R_B$$

1-4.2 The Common-Base Amplifier

The *common-base configuration* is shown in Fig. 1-18. As far as DC considerations apply, the circuit does not differ from a common-emitter circuit with an emitter resistor added. However, both capacitors C_1 and C_2 are taken to be short circuits at frequencies of interest. Thus the base is connected to the reference terminal. One might at first wonder why resistor R_E is necessary. Without this resistor, the emitter terminal would be short-circuited to the reference terminal, leaving no way to apply the input voltage to the emitter.

Using the same procedures as in the cases of the common-emitter and common-collector amplifiers, we obtain the following results for the common-base amplifier.

$$A_v = \frac{h_{oe}R_L + h_{fe}R_L/h_{ie}}{1 + h_{oe}R_L} \tag{1-7}$$

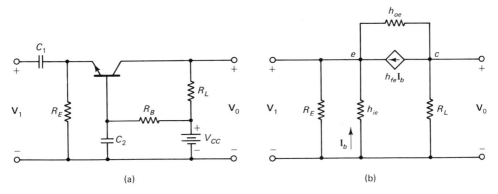

Fig. 1-18 (a) Common-base amplifier and (b) its equivalent circuit.

When h_{oe} can be neglected, $A_v \approx h_{fe}R_L/h_{ie}$. This is the same formula that applies for the common-emitter amplifier, except that there is no minus sign. A little rearrangement of Eq. (1-7) yields the following formula:

$$A_v = \left(\frac{1}{h_{oe}} \parallel R_L\right)\left(h_{oe} + \frac{h_{fe}}{h_{ie}}\right)$$

Short-circuiting the input eliminates $\mathbf{V}_1$, reducing $\mathbf{I}_b$ to zero, thus also eliminating the current source $h_{fe}\,\mathbf{I}_b$. Then r_o is simply

$$r_o = \frac{1}{h_{oe}} \parallel R_L$$

The input resistance turns out to be

$$r_{in} = \frac{1}{1/R_E + (1 + h_{fe})/h_{ie} + (1 - A_v)h_{oe}}$$

This can also be expressed as

$$r_{in} = R_E \parallel \frac{1/h_{oe}}{1 - A_v} \parallel \frac{h_{ie}}{1 + h_{fe}}$$

In many cases, $[1/h_{oe}]/(1 - A_v)$ is large compared to $h_{ie}/(1 + h_{fe})$. If we wish to find the input resistance to the emitter, exclusive of the emitter resistance R_E, we obtain $r_{in} \approx h_{ie}/(1 + h_{fe})$, a formula that is often useful.

Summarizing approximate results we have

$$A_v \approx \frac{h_{fe}R_L}{h_{ie}}$$

$$r_o \approx R_L$$

$$r_{in} \approx \frac{h_{ie}}{1 + h_{fe}}$$

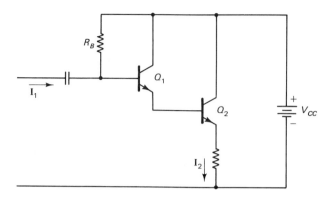

Fig. 1-19 Darlington connection.

1-5 THE DARLINGTON CONNECTION

Certain special circuits deserve attention before we leave the area of junction transistors. One of these is the Darlington circuit shown in Fig. 1-19.

This circuit is essentially a cascade arrangement of emitter followers. Recognition of that fact allows us to state at once that $A_v < 1$. But what is of interest in this circuit is the current gain. It is its large current gain that makes the Darlington circuit popular.

To find the overall current gain, we shall find the current gain for the equivalent circuit of a single emitter-follower stage, such as the one shown in Fig. 1-20.

Clearly the current into the emitter node is $(1 + h_{fe})\mathbf{I}_{in}$. This current splits between R_E and h_{oe}, the ratio of $\mathbf{I}_o$ to $(1 + h_{fe})\mathbf{I}_{in}$ being $(1/R_E)/(h_{oe} + 1/R_E)$. Hence the current gain of this stage is

$$A_i = \frac{(1 + h_{fe2})/R_E}{h_{oe2} + 1/R_E}$$

This expression can also be written

$$A_{i2} = \frac{1 + h_{fe2}}{1 + h_{oe2}R_E}$$

Evidently, when h_{oe2} is small, $A_{i2} \approx h_{fe2}$, as we would expect.

In the circuit of Fig. 1-19, the output transistor is the emitter load seen by the input transistor. This suggests that we can use the last formula to find the current gain of Q_2 if we substitute the input resistance of Q_1 in place of R_E. This

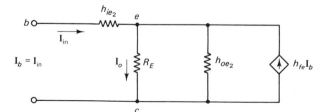

Fig. 1-20 Equivalent circuit of a single emitter follower.

resistance is

$$r_{in_2} = h_{ie_2} + (1 + h_{fe_2}) \left(\frac{1}{h_{oe_2}} \parallel R_E \right)$$

Then A_{i_1} is given by

$$A_{i_1} = \frac{1 + h_{fe_1}}{1 + h_{oe_1} r_{in_2}}$$

The overall current gain is the product of $A_{i_1} \times A_{i_2}$, or

$$A_i = A_{i_1} \times A_{i_2} = \frac{(1 + h_{fe_2})(1 + h_{fe_1})}{(1 + h_{oe_2} R_E)(1 + h_{oe_1} r_{in_2})}$$

If, as is usually the case, $h_{oe_2} R_E \ll 1$, then we can consider the first factor in the denominator to be equal to 1. The influence of the second factor requires more study. In the product $h_{oe_1}[h_{ie_2} + (1 + h_{fe_2})(1/h_{oe_2} \parallel R_E)]$, $h_{oe_1} h_{ie_2}$ is likely to be small compared to $h_{oe_1}(1 + h_{fe_2})[(1/h_{oe_2}) \parallel R_E]$ because of the multiplying factor $1 + h_{fe_2}$. Again, $(1/h_{oe_2}) \parallel R_E$ is likely to be approximately equal to R_E. Thus we are left, for approximate results, with $h_{oe_1}(1 + h_{fe_2})R_E$. The expression for A_i now reduces to

$$A_i \approx \frac{(1 + h_{fe_1})(1 + h_{fe_2})}{1 + h_{oe_1}(1 + h_{fe_2})R_E}$$

A computer implementation of this formula is given in Prog. 1-2. Rather than use the equation we have developed, we are tempted to take the current gain of the Darlington connection to be simply $(1 + h_{fe_1})(1 + h_{fe_2})$, or approximately $h_{fe_1} h_{fe_2}$. Sometimes this is correct, but we cannot decide the matter a priori without looking into the sizes of the quantities involved. The following example is intended to illustrate this point.

Example 1-4

Approximate values for h_{ie_1} and h_{ie_2} are:

$$h_{ie_1} = 2170 \ \Omega \qquad h_{ie_2} \approx 44 \ \Omega$$

First we examine the factor $(1 + h_{oe_2} R_E)$. This is seen to be

$$1 + 2 \times 10^{-5} \times 500 \approx 1 + 10 \times 10^{-3} = 1.01$$

Clearly the approximation that $1 + h_{oe_2} R_E = 1$ is excellent. Next we calculate $h_{oe_1} h_{ie_2}$, which is $2 \times 10^{-5} \times 44 = 8.8 \times 10^{-4}$. Now we find $h_{oe_1}(1 +$

```
10   PRINT "APPROXIMATE COMPUTATION OF CURRENT GAIN FOR DARLINGTON-CONNECTE
     D TRANSISTORS."
20   INPUT "ENTER VALUES OF Hfe1,Hfe2,Hoe1, AND EMITTER RESISTOR, RE.";HFE1
     ,HFE2,HOE1,RE
30 AI = (1 + HFE1) * (1 + HFE2) / (1 + HOE1 * (1 + HFE2) * RE)
40   PRINT "CURRENT GAIN Ai = ";AI
50   INPUT "MORE?";A$
60   IF A$ = "YES" THEN 20
70   END
```

Prog. 1-2.

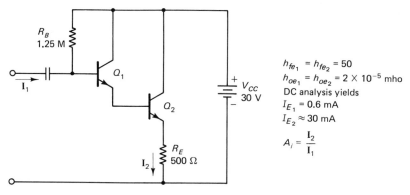

Fig. 1-21 Darlington circuit example.

$h_{fe2})(1/h_{oe2} \parallel R_E)$. The parallel combination of $1/h_{oe2} = 50$ KΩ and $R_E = 500$ Ω is obviously very nearly equal to R_E, or 500 Ω. Consequently, the product becomes $(2 \times 10^{-5})(1 + 50)(500) \approx 2 \times 10^{-5} \times 2.5 \times 10^4 = 0.5$. Hence A_i is equal to about $(50)^2/(1 + 0.5) = 1670$. There is a considerable difference between this result and the hasty and erroneous guess that

$$A_i \approx (1 + h_{fe1})(1 + h_{fe2}) \approx 50^2 = 2500$$

1-6 OUTPUT RESISTANCE OF THE COMMON-BASE CONNECTION

There is another special circuit that must be considered. This is the *common-base circuit* shown in Fig. 1-22. It is perhaps surprising that our interest in this circuit rests exclusively on the small-signal resistance that is seen looking into the collector and reference terminals. The reason for this interest will become apparent when we study differential amplifiers in Chap. 3.

V_{BB} is a constant bias voltage that establishes whatever quiescent operating conditions we choose. It also places the base at signal ground so that we are justified in using the small-signal equivalent circuit shown in Fig. 1-23. This is one of the rare cases in which all four h parameters are used. The necessity for so doing is one of the things we wish to examine.

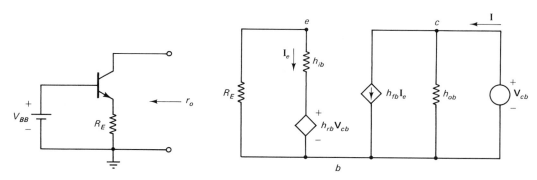

Fig. 1-22 Common-base circuit.　　　　**Fig. 1-23** Equivalent circuit.

At first glance we might be tempted to guess that r_o is simply equal to $1/h_{ob}$. Instead, let us imagine that an external phasor voltage $\mathbf{V}_{cb}$ is applied to the circuit as shown. We wish to determine the resulting phasor current $\mathbf{I}$. Then r_o will be equal to $\mathbf{V}_{cb}/\mathbf{I}$. Clearly,

$$\mathbf{I}_e = \frac{-h_{rb}\mathbf{V}_{cb}}{h_{ib} + R_E}$$

Then

$$\mathbf{I} = h_{fb}\mathbf{I}_e + \mathbf{V}_{cb}h_{ob}$$

$$= \mathbf{V}_{cb}\left(h_{ob} - \frac{h_{fb}h_{rb}}{h_{ib} + R_E}\right)$$

Since $h_{fb} \approx -1$,

$$g_o = 1/r_o \approx h_{ob} + h_{rb}/(h_{ib} + R_E) \qquad (1\text{-}8)$$

The result given by Eq. (1-8) is stated for convenience in terms of conductance g_o rather than r_o. It is a simple matter to invert g_o when the value of r_o is needed.

Example 1.5

Let us compute r_o for a transistor connected as shown in Fig. 1-22 with the following parameters: $I_E = 1.3$ mA, $R_E = 300\ \Omega$, $h_{ob} = 0.49 \times 10^{-6}$ mho, $h_{rb} = 2.9 \times 10^{-4}$.

We know that $h_{ie} \approx (h_{fe} \times 0.026)/I_E$ and $h_{ib} = h_{ie}/(1 + h_{fe}) \approx h_{ie}/h_{fe}$. Hence $h_{ib} \approx 0.026/I_E$. Therefore, $h_{ib} = 0.026/0.0013 = 20\ \Omega$.

Using Eq. (1-7) we find

$$g_o = 0.49 \times 10^{-6} + \frac{2.9 \times 10^{-4}}{20 + 300} = 0.49 \times 10^{-6} + 0.906 \times 10^{-6}$$

$$= 1.396 \times 10^{-6}\ \text{mho}$$

The resistance $r_o = 1/g_o = 716.3$ KΩ. This is quite a different result from that obtained by assuming that $r_o = 1/h_{ob}$, which yields $r_o = 2.04$ MΩ.

It is also interesting to note the great effect of the resistor R_E. If we short-circuit this resistor, making $R_E = 0$, then g_o becomes

$$g_o = 0.49 \times 10^{-6} + \frac{2.9 \times 10^{-4}}{20}$$

which leads to $r_o = 149.9$ KΩ. In this last case, also note the very great effect of h_{rb}.

1-7 FIELD-EFFECT TRANSISTORS (FETs)

A somewhat different mechanism of conduction and control is embodied in the field-effect transistor. A simplified representation of the structure of a junction FET is shown in Fig. 1-24.

The two p regions, called the *gate*, are connected to each other and are reverse biased with respect to the interior n region. As in the case of a collector-base

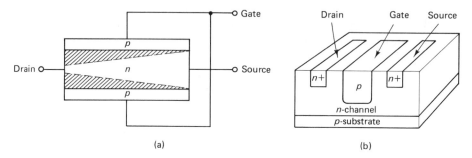

Fig. 1-24 (a) Simplified structural representation of junction FET (the shaded area is the depletion region) and (b) an actual FET.

junction in a junction transistor, this reverse biasing gives rise to a region in which mobile carriers, in this case electrons, are depleted. Since conduction of electrons occurs through the *n* material from source to drain, the larger the depletion regions become, the narrower is the *channel* available for conduction. Thus the depletion regions act like a throttle to control the conduction current. Since reverse bias is applied to the gate-channel junction, the gate current is extremely small and the resistance looking into the gate is very high. This is in contrast to the situation in the junction transistor, where base-emitter currents are higher and the input resistances are much lower than for the FET.

It is interesting to note, in passing, that conduction of electrons in the FET shown in Fig. 1-24 is through the mechanism of *drift*. This is the same conduction mechanism that occurs in a piece of copper wire. In an *npn* junction transistor, conduction across the base region takes place mainly through the *diffusion* of electrons.

The schematic symbols for *n* channel and *p* channel junction FETs (or JFETs) are shown in Fig. 1-25. Also shown are letter symbols, with their defining directions, for the gate-source voltage V_{GS}, the drain-source voltage V_{DS}, and the drain-source current I_D. Although the actual, physical currents and voltages for the *p* channel JFET are opposite to those of the *n* channel type, the *reference* directions for the two types are the same.

For any given reverse-bias voltage V_{GS}, there is a limiting value of current I_{DS} that the FET will pass. This means that there is a set of fairly well defined

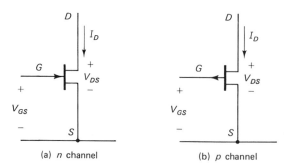

(a) *n* channel   (b) *p* channel   **Fig. 1-25** JFET symbols.

values of V_{DS} beyond which I_D can be increased only slightly. The curves of I_D versus V_{DS} have "knees" at these values.

JFET operation is usually at values of V_{DS} beyond these knees. The situation is illustrated by the curves in Fig. 1-26. You should note that the spacing between adjacent curves is decidedly nonuniform, in contrast to the fairly uniform spacing of junction transistor collector characteristics. Thus nonuniformity occurs because in an FET, I_D varies with the *square* of V_{GS}.

The value of V_{GS} that reduces I_D to zero for all values of V_{DS} is called the *pinch-off voltage*. When that value of reverse bias is applied to the gate-channel junction, the channel width is almost completely depleted of current carriers and is said to be pinched off. The value of this voltage is a parameter that depends upon the structure of a particular FET and is given in the manufacturer's data as a quantity designated by V_P.

Another parameter that varies from one FET type to another is the amount of channel current that flows (for $V_{DS} > |V_P|$) when $V_{GS} = 0$. This current is designated by I_{DSS}.

If the parameters V_P and I_{DSS} are known, the drain characteristics and the transconductance g_m can readily be found. It can be shown analytically and verified by measurement that for values of V_{DS} greater than V_P, the drain current is

$$I_D = I_{DSS} \left(1 - \frac{V_{GS}}{V_P} \right)^2 \tag{1-9}$$

Thus, using this equation, the value of I_D *in the flat part of the curve* for a given value of V_{GS} can be calculated. For example, if $I_{DSS} = 8$ mA, $V_P = -4$ V and $V_{GS} = -2$ V,

$$I_D = 8 \left(1 - \frac{-2}{-4} \right)^2 = 8 \left(1 - \frac{1}{2} \right)^2 = 2 \text{ mA}$$

The fact that I_D is controlled almost exclusively by the value of V_{GS} suggests that an FET can be modeled by a voltage-controlled current source. The voltage V_{DS} has only a slight influence on I_D in the pinch-off region. This suggests that the changes in I_D that result from changes in V_{DS} can be modeled by simply including a large resistor across the drain-source terminals. Because the gate-channel junction

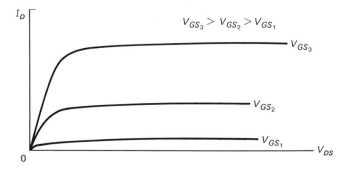

Fig. 1-26 Drain characteristics for an *n*-channel JFET. V_{GS} values are all negative.

is always reverse biased and thus passes extremely small gate currents, we can infer that the gate-source mechanism can be approximated by an open circuit. A small-signal equivalent circuit embodying all these ideas is shown in Fig. 1-27.

The parameter g_m is a *transconductance* defined by

$$g_m = \frac{\Delta I_D}{\Delta V_{GS}}\bigg|_{V_{DS} = \text{constant}}$$

Since g_m is the rate at which I_D changes with variations in V_{GS}, a little calculus exercise immediately gives us a formula for g_m:

$$g_m = \frac{\partial I_D}{\partial V_{GS}} = 2I_{DSS}\left(\frac{-1}{V_P}\right)\left(1 - \frac{V_{GS}}{V_P}\right)$$

or

$$g_m = -\frac{2I_{DSS}}{V_P}\left(1 - \frac{V_{GS}}{V_P}\right) \tag{1-10}$$

For the FET considered above, g_m is given by

$$g_m = -\frac{2 \times 8 \times 10^{-3}}{-4}\left(1 - \frac{-2}{-4}\right) = 4 \times 10^{-3} \times \frac{1}{2}$$

$$= 2 \times 10^{-3} \text{ mho}$$

The drain resistance r_d is given by

$$r_d = \frac{\Delta V_{DS}}{\Delta I_D}\bigg|_{V_{GS} = \text{constant}}$$

Another class of FETs exists, in which control of the channel conduction is accomplished without using a gate-channel semiconductor junction. These FETs are sometimes called *insulated-gate FETs*, or IGFETs. The insulation of the gate is accomplished by oxidizing the silicon of the channel. Thus a thin layer of SiO_2 (glass) separates the gate electrode from the channel. These devices are usually referred to as *metal-oxide field-effect transistors* (MOSFETs).

Two different structures are used in making MOSFETs. These are shown in smiplified form in Fig. 1-28.

When a negative potential is applied to the gate of the depletion-mode MOSFET, electrons in the *n* material of the channel are driven away, so a depletion region is formed. This narrows the remaining *n* channel so that the conduction and

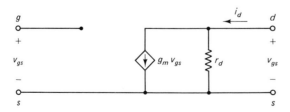

Fig. 1-27 Small-signal equivalent circuit for an FET.

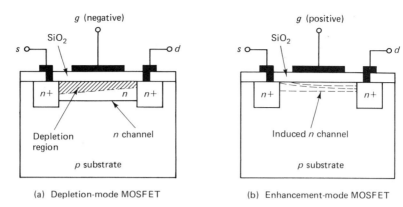

(a) Depletion-mode MOSFET   (b) Enhancement-mode MOSFET

Fig. 1-28 Basic MOSFET structures.

control mechanisms of this type of MOSFET are quite similar to those of the corresponding n-channel JFET.

In the enhancement-mode MOSFET, application of a positive gate-source potential produces an electric field beneath the gate electrode whose lines must terminate on negative charges. For complex reasons that are beyond the scope of this book, this electric field actually converts a thin layer (about 100 Å thick) of p material to n-type material. This takes place just beneath the gate electrode at the Si-SiO$_2$ interface. Electrons for this n-type surface layer may be of thermal origin or they may be supplied by the $n+$ diffusion wells.

As the positive gate potential varies, the concentration of mobile electrons in this induced n channel also varies, causing the conductivity of the channel to vary accordingly. Thus, although the internal control mechanism of the enhancement-mode MOSFET is entirely different from that of the JFET, the terminal behavior of the devices in the saturation region (where normal operation takes place) is quite similar.

For either type of MOSFET, Eq. (1-9) is modified to read

$$I_D = \left(\frac{K}{V_T^2}\right)\left(1 - \frac{V_{GS}}{V_T}\right)^2 \tag{1-11}$$

Here, V_T is called the threshold voltage and plays the same role as V_P does in the JFET. Also, I_{DSS} no longer appears in the equation; its role is taken by the quantity K/V_T^2 (K depends upon the physical dimensions and electrical constants of the MOSFET).

For the circuit analyses that we will consider here, there is little difference between MOSFETs and JFETs. The characteristic curves of Fig. 1-26 and the small-signal equivalent circuit of Fig. 1-27 will be assumed to apply to both devices. Schematic circuit representations for various types of MOSFETs are shown in Fig. 1-29.

A convenient way to bias the JFET is by the self-biasing scheme illustrated in Fig. 1-30. The current I_D passing through resistor R_S produces a voltage drop

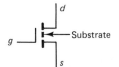

(a) n-channel enhancement MOSFET

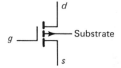

(b) p-channel enhancement MOSFET

(c) *n*-channel depletion MOSFET

(d) *p*-channel depletion MOSFET

Fig. 1-29 Schematic representations of MOSFETs.

$I_D R_S$. The ground-return resistor R_G brings the gate nearly to zero potential. Since the source is now positive, the *relative* voltage from gate to source is negative.

If we apply Kirchhoff's voltage law to the loop formed by R_G, the gate-to-source voltage V_{GS}, and the voltage across R_s (which is $I_D R_s$), we have

$$-V_{RG} + V_{GS} + I_D R_s = 0.$$

Since, as discussed above, $V_{RG} = 0$, then

$$V_{GS} = -I_D R_S$$

and we see that V_{GS} is indeed negative. The resistor R_G must be in the circuit even though it carries negligible current because this resistor establishes the gate potential at the reference (ground) level.

Example 1-5

As an example, consider the circuit shown in Fig. 1-31. The FET for this example has the drain characteristics shown in Fig. 1-32.

Since the total resistance in series with the FET is $667 + 1333 = 2$ KΩ and the supply voltage V_{DD} is 20 V, the DC load line is established as shown. Intelligent use of a trial-and-error procedure will establish the operating point. For example, suppose we assume that $V_{GS} = -1$ V. The intersection of the load line with the I_D versus V_{DS} characteristic for $V_{GS} = -1$ V yields a drain current $I_D = 6$ mA. The voltage across the 667-Ω source resistor will thus be $6 \times 10^{-3} \times 667 = 4$ V, which is too large and, of course, is inconsistent with the assumed value of V_{GS}. If we assume $V_{GS} = -3$ V,

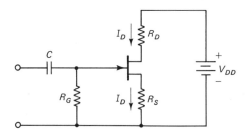

Fig. 1-30 Resistor R_S provides self-biasing for the FET.

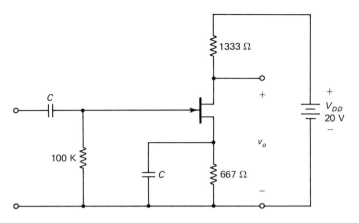

Fig. 1-31 FET circuit example.

we find $0.5 \times 10^{-3} \times 667 = 0.33$ V across the 667-Ω resistor, which is too small. But the assumption that $V_{GS} = -2$ V yields $I_D = 3$ mA and $3 \times 10^{-3} \times 667 = 2$ V, which is in exact agreement. The voltage V_{DS} is now seen to be $20 - 3 \times 10^{-3} \times 2000 = 14$ V.

We might well wish for a more systematic way to determine the operating point of a self-biased JFET amplifier than the procedure just described. For this purpose we need to make use of the JFET transfer characteristic. This is simply a curve that relates I_D to V_{GS} and is readily obtained either from the flat region of the idealized drain characteristics, such as the curves shown in Fig. 1-32, or from Eq. (1-9).

Figure 1-33 shows the transfer curve for a JFET having $I_{DSS} = 12$ mA and $V_P = -4$ V.

If the JFET is connected in the circuit shown in Fig. 1-34, we can replace the gate-bias circuit by its Thevenin equivalent, which—as you can verify—consists simply of a 4-V source in series with a 16-KΩ resistor.

A KVL equation for the gate-source loop yields

$$-V_{TH} + I_G R_{TH} + V_{GS} + I_D R_S = 0$$

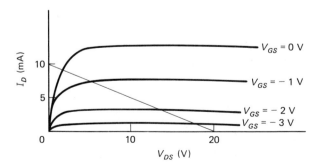

Fig. 1-32 Drain characteristics for FET.

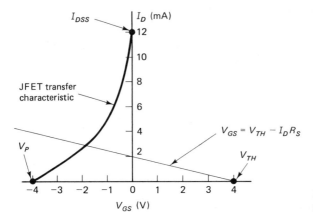

Fig. 1-33 JFET transfer characteristic and gate-source line.

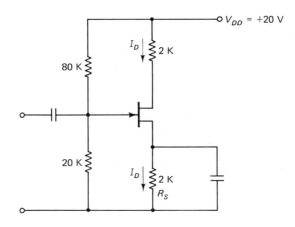

Fig. 1-34 JFET circuit.

```
10  PRINT "DETERMINATION OF DC OPERATING POINT OF JFET
20  PRINT "WITH SOURCE RESISTOR AND GATE BIAS NETWORK
21  PRINT "AS SHOWN IN FIG. 1-34."
22  PRINT
30  INPUT "DRAIN SUPPLY VOLTAGE = ?";VDD
40  INPUT "UPPER GATE BIAS CIRCUIT RESISTOR = ?";R1
50  INPUT "LOWER GATE BIAS CIRCUIT RESISTOR = ?";R2
60  INPUT "ENTER IDSS IN MILLIAMPERES.";IDSS
70  INPUT "ENTER PINCH-OFF VOLTAGE MAGNITUDE.";VP
80 VP =   - 1 * VP
90  INPUT "ENTER SOURCE RESISTANCE IN KILOHMS.";RS
100  INPUT "ENTER DRAIN RESISTANCE IN KILOHMS.";RD
105 VTH = VDD * R2 / (R1 + R2)
110 A = IDSS * (RS / VP) ^ 2
120 B = 2 * IDSS * RS * (1 / VP - VTH / VP ^ 2) - 1
130 C = IDSS * (1 - VTH / VP) ^ 2
140 ID = ( - B -   SQR (B ^ 2 - 4 * A * C)) / (2 * A)
150 VD = VDD - ID * (RS + RD)
160  PRINT "OPERATING POINT IS VD = ";VD;", ID = ";ID
170  INPUT "DO YOU WANT ANOTHER COMPUTATION? (Y/N)";A$
180  IF A$ = "Y" THEN 30
190  END
```

Prog. 1-3.

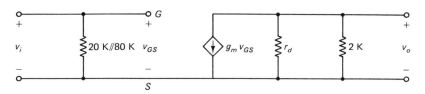

Fig. 1-35 Small-signal equivalent circuit.

I_G is, of course, zero. Hence the solution for V_{GS} is

$$V_{GS} = V_{TH} - I_D R_S$$

where R_{TH} is the Thevenin resistance and V_{TH} is the Thevenin voltage, in this case 4 V. The plot of V_{GS} versus I_D must, of course, be a straight line. This line intersects the V_{GS} axis at the point where $V_{GS} = V_{TH}$ and $I_D = 0$. As I_D increases from zero, V_{GS} becomes smaller and eventually assumes negative values, thus producing the straight line plotted in Fig. 1-33. The intersection of this straight line and the transfer characteristic establishes the quiescent operating value of I_D. You can verify that, for this case, $V_{DS} = 8$ V and $I_D = 3$ mA.

An alternative to the graphical method of determining the operating point for this JFET circuit is given by Prog. 1-3. The transconductance (g_m) of this JFET is 3×10^{-3} mho, a fact that is left for you to verify. Using this value, we can replace the FET of Fig. 1-34 with its small-signal equivalent circuit, as shown in Fig. 1-35. Note that the lowercase symbols for v_i, v_{GS}, and v_o imply that these are *varying* quantities, in contrast to *constant*, or DC, quantities.
We see at once that V_{GS} is equal to v_i. Hence for v_o we get

$$v_o = -g_m v_{GS}(r_d \| 2 \text{ K})$$

We now find A_V to be $A_v = v_o/v_i = -g_m(r_d \| 2 \text{ K})$. If we assume that $r_d = 20$ K, then A_v is numerically equal to

$$A_V \approx -g_m \times 2 \text{ K} = 3 \times 10^{-3} \times 2 \times 10^3 \approx 6$$

This result is fairly typical of FET small-signal amplifiers. That is, the values of A_v are often between 2 and 20, a much lower figure than is found typically for junction transistors. Since FETs provide low voltage amplification and are highly nonlinear, there seems to be only one good reason for using them at all in small-signal applications—their high input resistance.

REFERENCES

1. Jacob Millman and Christos C. Halkias, *Integrated Electronics: Analog and Digital Circuits and Systems* (New York: McGraw-Hill Book Company, 1972).
2. Robert Boylestad and Louis Nashelsky, *Electronic Devices and Circuit Theory, 3rd ed.* (Englewood Cliffs, N.J.: Prentice-Hall, Inc., 1982)

PROBLEMS

1-1.

    (a) The diode has the volt-ampere characteristic shown. Find the voltage v_D and the current i.

    (b) What percent error is made in calculating i if v_D is assumed equal to 0.7 V?

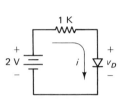 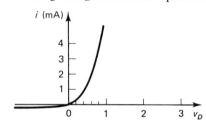

1-2. The two diodes shown have identical characteristics, given by

$$I = I_s \left(e^{V/0.026} - 1 \right)$$

    (a) Find the voltages V_1 and V_2. Do *not* assume that $I = I_s$. Note that while the *magnitude* of the current in one diode must equal that of the other, their algebraic signs do not agree. Why?

    (b) Using the result from (a), find I if $I_s = 10\ \mu\text{A}$.

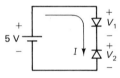

1-3. In very many applications we can assume that semiconductor diodes are *ideal*. An ideal diode is one that, when reverse biased, has infinite resistance and when forward biased, behaves like a short circuit.

    (a) In this circuit, assume that the diodes are ideal and sketch v_o versus v_i, showing the values at breakpoints and slopes.

    (b) If $v_i = 40 \sin \omega t$, sketch v_o, showing values of the angle (ωt) at which breakpoints occur.

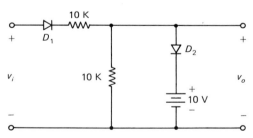

1-4. Diode D is assumed to be ideal and the time constant RC is assumed to be large compared to T.

    (a) To what voltage does C charge? What polarity does this voltage have?

(b) Sketch the waveform of v_o, indicating maximum and minimum voltages.

(c) How would the result of *b* be changed if the diode were reversed?

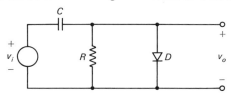

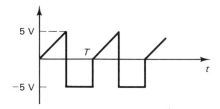

1-5.

    (a) If $v_i = 10 \cos \omega t$, sketch v_o.

        What is the peak value of the reverse voltage seen by each diode?

    (b) Repeat (a) for the circuit shown.

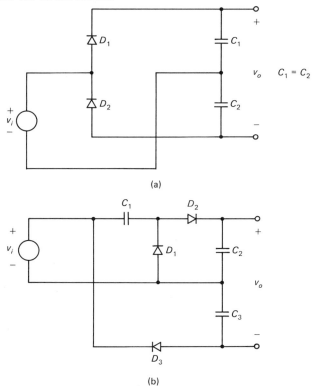

(a)

(b)

1-6. For the diode shown, the reverse saturation current is measured and found to be 5 × 10⁻⁶ A. When $V = 0.2$ V, I equals I_1. When $V = 0.22$ V, I equals I_2. Find the ratio I_2/I_1.

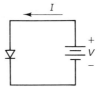

1-7. In the circuit shown, $I = 10$ mA. Find the voltages V_{d_1} and V_{d_2}. $I_S = 10^{-14}$ A.

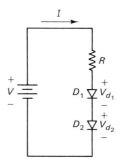

1-8. Given $v_{in}(t)$ as plotted, plot $v_{out}(t)$. Show clearly all breakpoints and voltage levels.

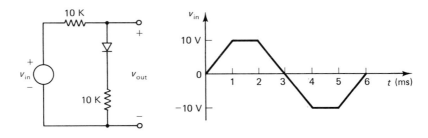

1-9. For the circuit shown, assume that steady state has been reached. Find capacitor voltages V_{c_1} and V_{c_2}. Sketch the diode voltages v_{d_1} and v_{d_2} for one cycle, showing breakpoints and voltage levels.

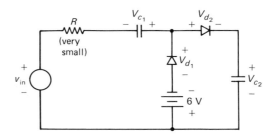

1-10. Idealized collector characteristics are shown for the transistor in the given circuit.
  (a) Calculate the values of V_{CC}, R_B, and R_C to place the quiescent operating point at the center of the load line.
  (b) With the quiescent point located as in (a), what is the largest deviation (swing) in i_B that can be amplified without distortion?

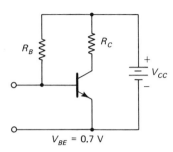

$V_{BE} = 0.7$ V

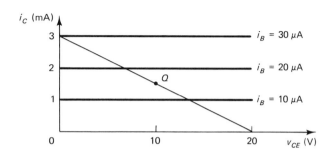

1-11. (a) For the circuit shown below, find the quiescent values of I_B, I_C, and V_{CE}. Assume $h_{FE} = 100$ and $V_{BE} = 0.7$ V.
  (b) Find $A_v = v_o/v_s$. Assume that $h_{oe} = 0$, $h_{fe} = h_{FE}$, and the capacitors are short circuits to time-varying signal voltages.

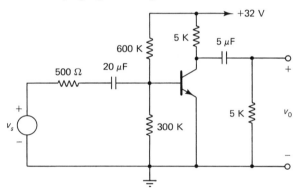

1-12. Find the quiescent operating point for the circuit shown.

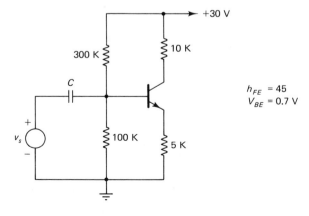

1-13.

   (a) Find the quiescent value of I_B.
   (b) Find the small-signal voltage gain $A_v = v_o/v_s$.

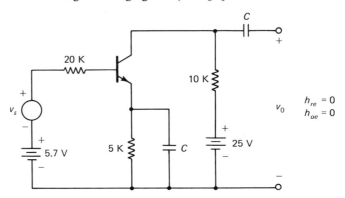

1-14. Find the input and output resistances of the transistor amplifier shown below. Assume $X_C = 0$ for each capacitor.

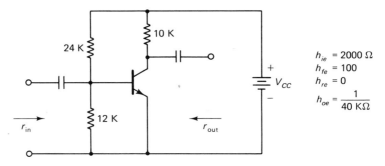

1-15. In the given circuit, $h_{fe} = h_{FE} = 100$ and $V_{BE} = 0.7$ V.
   (a) With $v_s = 0$, find V_{BB} so that $V_{CE} = 20$ V.
   (b) Find $|A_v| = |v_o/v_s|$. Assume $h_{re} = 0$ and $h_{oe} = 0$.
   (c) What is the maximum undistorted peak-to-peak output voltage?

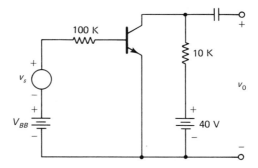

1-16. For the transistor in the given circuit, $h_{FE} = 80$ and $V_{BE} = 0.7$ V. Find V_{CE}.

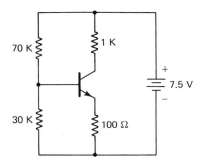

1-17. For the JFET in the amplifier circuit above, $I_{DSS} = 9$ mA and $V_P = -6$ V.
   (a) Find V_{GS}, I_D, and V_{DS} at the quiescent operating point.
   (b) Find A_v.
   (c) If $v_s(t) = 0.1 \sin \omega t$, find $v_o(t)$.

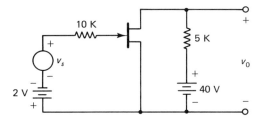

1-18.
   (a) Find V_{DS}, I_D, and V_{GS} at the quiescent operating point.
   (b) Find A_v when the circuit operates near this quiescent point.
   (c) If $v_s(t) = 0.1 \cos \omega t$, find $v_0(t)$.

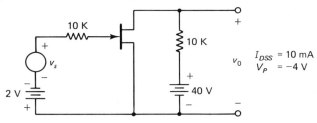

1-19. Find I_D, V_{GS}, and V_{DS} at the quiescent operating point.

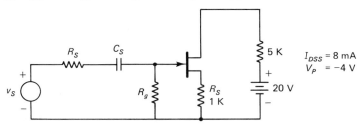

1-20. Assume $X_C = X_{CS} = 0$.
   (a) Find R_D and R_S.
   (b) Find A_v.

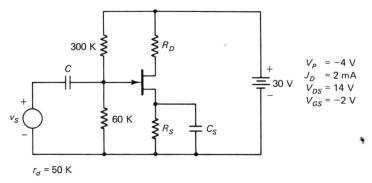

$V_P = -4$ V
$J_D = 2$ mA
$V_{DS} = 14$ V
$V_{GS} = -2$ V

$r_d = 50$ K

1-21. Idealized drain characteristics for the JFET are shown. Assume $r_d = \infty$.
  (a) Find the value of V_{GS} that will set the operating point such that $V_{DS} = 14$ V.
  (b) At the operating point determined in (a), what A_v will this circuit produce?

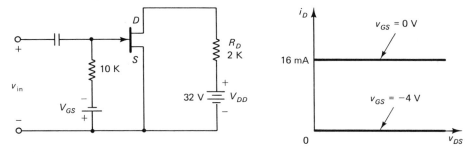

1-22. The three-terminal device shown above is a MOSFET. It turns on only when the gate voltage is more negative than -2 V with respect to the source voltage.
  (a) What type (enhancement or depletion) of MOSFET is it?
  (b) Identify the source terminal in the circuit shown.
  (c) Make a rough sketch of the cross section of such a MOSFET.
  (d) What is the value of V_T?

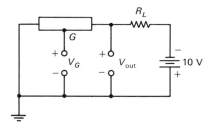

1-23.
  (a) Determine the value of R_b that will make $I_{C_2} = 1$ mA.
  (b) Find h_{ie} for each transistor.
  (c) Draw the equivalent small-signal circuit for the amplifier and state whether h_{oe} can be neglected for either Q_1 or Q_2.
  (d) Find R_{in}.
  (e) Find A_v.

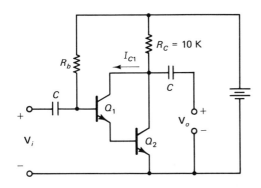

Q_1 and Q_2 are identical with
$h_{FE} = h_{fe} = 50$ and $h_{oe} = 1/10 \ K\Omega$
Assume both $X_C = 0$.
$V_{BE} = 0.7 \ V$.

CHAPTER 2

Transistors and Diodes
as Switching Devices

2.1 INTRODUCTION TO SWITCHING CIRCUITS

Switching circuits have only two conditions or states: on and off. We can get some idea of how this works by considering the simple relay circuits of Fig. 2-1.

In both these circuits, the inputs are taken to be the switch voltages at the left. When one of these switches is manually operated to connect the battery to one of the relay coils, we consider the relay input of that coil to be a logic 1. If the coil is grounded, the input is a logic 0. There is no condition in which the coil inputs are floating. A little study of Fig. 2-1(a) shows that unless both relays A and B are energized, the output will be *zero*. The situation is readily summarized in Table 2-1. In terms of logic levels (0 Volts = 0, V Volts = 1), Table 1-2 can be rewritten as shown in Table 2-2. Such a table is called a *truth table*.

It is easy to determine that with either relay A or relay B energized, the output of the circuit of Fig. 2-1(b) will be V volts or logic 1. You can verify that the truth table for this circuit is as shown in Table 2-3.

A circuit in which all inputs must be 1 or true (as it is sometimes stated) to yield a 1 output is called an AND gate and is represented by the symbol shown in Fig. 2-2. Also shown is the OR gate, whose output is 1 if at least one input is 1.

While Fig. 2-2 shows only gates that have two inputs, there is no reason why multiple-input gates could not be constructed along similar lines.

For the purposes of this discussion, remember that neither input nor output terminals are left in a floating, or disconnected, state. Instead, they must be connected either to ground or to a low-impedance voltage source. There is an important exception to this rule that will be discussed later.

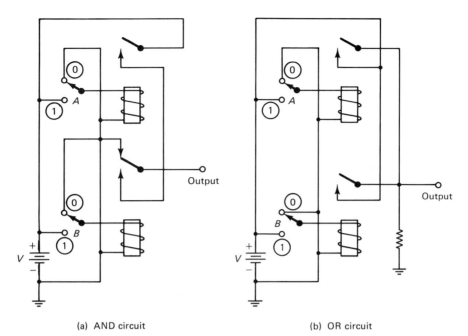

(a) AND circuit   (b) OR circuit

Fig. 2-1   Basic logic circuits.

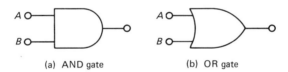

(a) AND gate   (b) OR gate

Fig. 2-2   Basic logic gate symbols.

TABLE 2-1

Voltage inputs		
A	B	Voltage output
0	0	0
0	V	0
V	0	0
V	V	V

TABLE 2-2

Logic inputs		
A	B	Logic output
0	0	0
0	1	0
1	0	0
1	1	1

TABLE 2-3

Logic inputs		Logic output
A	B	
0	0	0
0	1	1
1	0	1
1	1	1

In the early days of logic circuitry, switches and relays were actually used to implement logic functions, and in some applications they may be found even today. But by far the larger number of modern logic circuits use electronic gates.

Before we turn to a detailed consideration of some of these gates, we need to define another logic function—the NOT, or complementing, operation. You can easily devise a relay-switch circuit whose logic output is a 1 when the input is a 0, and vice versa. Such a gate is represented by the symbol shown in Fig. 2-3(a). When an AND or an OR gate incorporates an inverter in its output, it becomes either a NAND (NOT AND) or a NOR (NOT OR) gate. The complementing action is symbolized by the small circle on the output lead, as shown in Fig. 2-3(b) and 2-3(c).

Other logic gates exist and have appropriate symbols to represent them. Since it is not our purpose to go into the problem of implementing complex logic functions, we shall not bother with these gates. Rather, we merely note that such gates—and indeed any logic functions—can be realized by using suitable combinations of NANDs and NORs.

2-2 DIODE-TRANSISTOR LOGIC CIRCUITS (DTL)

As the name implies, diode-transistor logic circuits involve combinations of transistors and diodes to form the various logic gates. The operation of such gates depends largely on the idea of switching the diodes into operating states above or below their cut-in voltage levels. You should recall that very little current flows through a silicon diode until its forward voltage exceeds about 0.6 V. When heavy conduction takes place, the forward voltage may be taken to be about 0.7 V. The lower of these two values, 0.6 V, is called the *cut-in* voltage and is denoted by the symbol V_γ. The higher value of 0.7 V, which we shall call the *sustaining voltage*, is denoted by V_d.

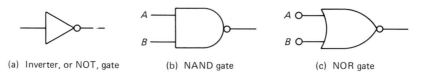

(a) Inverter, or NOT, gate   (b) NAND gate   (c) NOR gate

Fig. 2-3 Inverting to complementing gates.

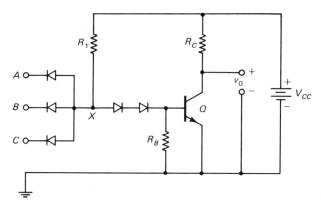

Fig. 2-4 Positive-logic, DTL, NAND gate.

We could utilize the switching properties of diodes exclusively (that is, without using transistors) to design logic gates. A family of such diode logic circuits exists and at one time was widely used; it is used occasionally even today. But the current-amplifying property of the transistor makes its incorporation in logic circuits advantageous when compared to circuits using only diodes because with transistors, small input currents can control large output currents. Hence several outputs can be taken from one gate, and each of these outputs can drive still more gates.

Fig. 2-4 shows a positive-logic NAND gate using a DTL configuration. By positive logic we mean that a positive voltage represents a logic-1 level, whereas a voltage that is nearly zero represents logic-0.

If any one of the three diodes connected to A, B, or C is grounded (so that its input is a logic 0), current will flow from V_{CC}, through R_1, and through the grounded diode. The potential of point X will then be equal to the sustaining voltage V_d. Although the sustaining voltage is greater than the cut-in voltage of one of the diodes to the right of point X, it is less than the sum of two cut-in voltages of the two series-connected diodes. Consequently, transistor Q receives no significant base current and is cut off. Output voltage v_o will be close to the value of V_{CC}, or at logic 1.

Only when all three inputs (A and B and C) are at a potential equal to V_{CC} (logic 1) will current flow through the series-connected diodes. Under this condition, Q receives substantial base current and becomes saturated, lowering v_o to a small positive value (logic 0).

The result of this analysis is summarized in Table 2-4.

It is easy to see that this truth table represents the operation of a three-input NAND gate. Suppose that $V_{CC} = 5$ V, $R_C = 2$ KΩ, $R_B = 4$ KΩ, and $R_1 = 5$ KΩ. Let us assume further that the base-emitter junction of transistor Q behaves much like the diodes in the circuit, so that V_{BE} has a cut-in value of $V_\gamma = 0.6$ V. However, when Q saturates, the voltage V_{CE} is only 0.3 V.

If one input, say A, is connected to another logic gate so that its logic-0 level is derived from another transistor and is therefore equal to 0.3 V, we have an equivalent input circuit, as shown in Fig. 2-5.

Clearly, voltage V_X equals $V_d + 0.3$, or $0.7 + 0.3 = 1.0$ V.

TABLE 2-4

	Inputs		
A	B	C	Output
0	0	0	1
0	0	1	1
0	1	0	1
0	1	1	1
1	0	0	1
1	0	1	1
1	1	0	1
1	1	1	0

Looking to the right of point X, we see the circuit shown in Fig. 2-6, which we use to compute the voltage level necessary to turn on transistor Q.

It is evident that for Q to conduct, V_X must be at least 3 V_γ, or 3 (0.6) = 1.8 V. Since the value of V_X obtained from Fig. 2-5 is less than that required for turn on by 0.7 V, transistor Q is cut off and output voltage v_o = 5 V (logic 1).

We might wonder why *two* series-connected diodes are used in this circuit. Indeed, if only one diode were used, the turn-on value of V_X would be $2V_\gamma$, or 1.2 V, and since, from Fig. 2-5, we calculated that V_X was only 1.1 V, we find Q to be still cut off. However, in this case the margin between V_X (actual value) and V_X (turn-on value) is only 0.1 V.

Then why use the additional diode? The answer lies in the possibility that a positive noise voltage might somehow be present in the input circuit. The situation with only one diode is illustrated in Fig. 2-7. The developed voltage $V_X = V_d + v_n + 0.3 = 0.7 + v_n + 0.3 = 1.0 + v_n$. The required turn-on value of V_X is $2V_\gamma$, or 1.2 V. Hence $1.0 + v_n \geq 1.2$, or $v_n \geq 0.2$V for turn on. That means that only 0.2 V of noise can cause a malfunction of the NAND gate. With two diodes, we have $1.0 + v_n \geq 1.8$, or $v_n \geq 0.8$ V, a very significant improvement in noise immunity.

Next we investigate the behavior of the circuit when all its inputs are at the 5-V, or logic-1, level. Under this condition the input diodes are open, and the base circuit of the transistor appears as shown in Fig. 2-8. Base current will certainly flow in this case; the question to be resolved is whether the base current will be

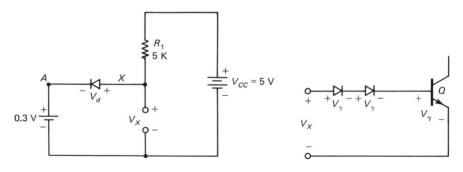

Fig. 2-5 Equivalent input circuit.   **Fig. 2-6** Equivalent base circuit.

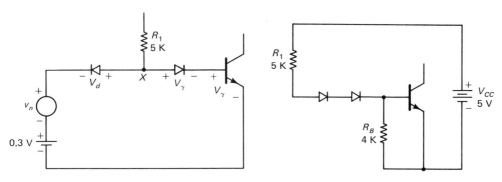

Fig. 2-7 Single-diode circuit including noise voltage V_n.

Fig. 2-8 Base circuit with all inputs at logic-1 level.

sufficient to make the transistor saturate. With base current flowing, we can take V_{BE} to be 0.7 V. That means that a current equal to V_{BE}/R_B, or $0.7/4000 = 0.175$ mA, flows in R_B. The current in R_1 must be equal to $(V_{CC} - 2V_d - V_{BE})/R_1$, or $(5 - 1.4 - 0.7)/5000 = 0.58$ mA. Kirchhoff's current law tells us then that $I_B = 0.58 - 0.175 = 0.405$ mA.

Next we must determine the collector current necessary to produce saturation. If we assume that when the transistor is operating in saturation, its collector-emitter voltage V_{CE} is 0.3 V, we see that I_C has to be at least equal to $(V_{CC} - V_{CE})/R_C = (5 - 0.3)/2000 = 2.35$ mA. That means that the *minimum* value of β has to be

$$\beta_{min} = \frac{2.35}{0.405} = 5.8$$

Values of β almost always fall between 20 and 300, so that this condition is easily satisfied; therefore, I_B is more than adequate to ensure saturation.

Although we have examined the effect of noise voltage on this circuit when one or more of its inputs is at logic 0, we must now determine the effect of noise when all inputs are at logic 1. The problem boils down to that of determining what value of noise voltage is necessary to pull the transistor out of saturation. Once again we refer to Fig. 2-7, which we must imagine to be modified to include two series-connected diodes in the base circuit and an input voltage of 5 V instead of 0.3 V. What we must find to begin with is the value of v_n that will just cause the left-hand diode to begin conducting. The potential V_X is established by the sum $2V_d + V_{BE}$, which represents the voltages due to the conducting diodes and the base-emitter junction. This voltage is

$$V_X = 2(0.7) + 0.7 = 2.1 \text{ V}$$

When the left-hand diode enters its cut-in condition, its voltage is V_γ and the combined voltages to the left of X must just equal V_X. Hence

$$V_{X\text{left}} = V_\gamma + v_n + 5 = V_X$$

Therefore, $v_n = V_X - V_\gamma - 5 = 2.1 - 5 - 0.6$, or

$$v_n \leq -3.5 \text{ V}.$$

We see then that *negative* excursions of noise voltage in excess of 3.5 V are the potential causes of trouble in this case. Note that the 2.1-V value of V_X comprises *two* units of V_d as well as one of V_{BE}, which means that the effect of using two series-connected diodes rather than one is to *reduce* the noise immunity when all inputs are at the logic-1 level. This is exactly contrary to the effect observed when at least one input is at logic 0. However, the logic 0 case is more critical, since small noise voltages (0.1 V to 0.8 V) caused trouble here, as compared to fairly large voltages (-3.5 V) in the logic-1 case. Clearly it is advantageous to use two or more series-connected diodes.

In the foregoing analysis of collector current, no account was taken of the current flowing in an external load. With the transistor conducting (in fact, in saturation), I_C comprises not only the current through R_C but also a component flowing into the collector from the load. In such operation, the NAND gate is said to act as a *sink* for current from the load. (If current were sent from the gate into the load, the gate would be said to act as a *source* for load current.) We shall assume that the loads connected to the gate we are studying consist simply of more gates of the same construction. The situation is illustrated in Fig. 2-9.

If inputs A, B, and C of the right-hand (load) gate were all at the logic-0 level, then the current through R_1 would be shared equally by these three inputs. However, if B and C are at logic 1, then all the current through R_1 enters the collector terminal of Q. This current I_A must be equal to

$$\frac{V_{CC} - V_X}{R_1} = \frac{5 - 1.0}{5000} = 0.8 \text{ mA}$$

We are now obliged to correct the previously calculated value of the collector current of Q by including I_A. But before doing so, let us consider the possibility that the load sustained by Q may consist not of just a single gate but of n gates.

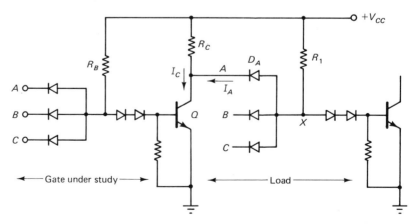

Fig. 2-9 A DTL NAND gate loaded by another gate.

That means that the current I_C must be given by

$$I_C = \frac{V_{CC} - 0.3}{R_C} + nI_A \qquad (2\text{-}1)$$

The question to be answered now is, What is the maximum value of n?

   We have already found that the base current of Q is 0.405 mA. We can make use of the information stated earlier concerning the range of values of β. The worst case occurs when β has its minimum value; then, for a fixed value of I_B, the value of I_C is limited to a corresponding minimum, that is, $I_C + \beta_{min} I_B$. Therefore, from Eq. (2-1),

$$\frac{V_{CC} - 0.3}{R_C} + nI_A = \beta_{min} I_B$$

or

$$n = \frac{\beta_{min} I_B - (V_{CC} - 0.3)/R_C}{I_A}$$

Using the numerical values of the right-hand terms, we get

$$n = \frac{20 \times 0.405 - 4.7/2}{0.8} = \frac{8.1 - 2.35}{0.8} = \frac{5.75}{0.8} = 7.19$$

Since n has to be an integer, we conclude that this NAND gate can drive 7 similar gates, or we say that the gate has a *fanout* of 7.

   The term *fanin* is also used. It refers simply to the number of input terminals of the gate.

   The fanout of this gate can be increased by substituting another transistor for the first of the series-connected diodes. This modification takes advantage of the current amplifying capability of the transistor. The modified circuit is shown in Fig. 2-10.

   If we assume that I_{B1} for the new transistor Q_{D1} has the same value that I_B

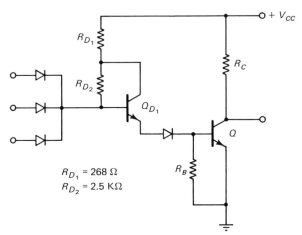

$R_{D_1} = 268\ \Omega$
$R_{D_2} = 2.5\ K\Omega$

Fig. 2-10   Modified NAND gate circuit.

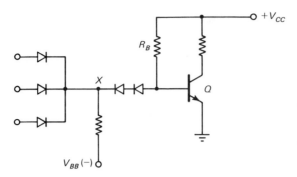

Fig. 2-11 Positive-logic, DTL, NOR gate.

had for Q in the original circuit and that $\beta = 20$, we see that I_B now becomes $20 \times 0.405 - 0.175 = 7.925$ mA. The fanout now becomes

$$n = \frac{20 \times 7.925 - 2.35}{0.8} = \frac{156.2}{0.8} \approx 195$$

which represents an enormous increase.

Shown in Fig. 2-11 is the circuit of a DTL, positive-logic, NOR gate. In this simplified version of the circuit, a negative base-bias voltage is required. You should verify that if any one input is at the logic-1 level, transistor Q is turned on and that only when all inputs are at logic 0 is the transistor cut off. Detailed numerical analysis is left to the problems at the end of the chapter.

2-3 TRANSISTOR-TRANSISTOR LOGIC (TTL)

Closely related to the DTL family of devices is the TTL family. The input diodes and the first of the series-connected diodes are replaced by a single transistor of unusual construction. The second series-connected diode is replaced by a conventional transistor. The unusual transistor has three (or more) emitters instead of the usual one. Its construction is illustrated in Fig. 2-12.

When the substitutions mentioned earlier are made, the DTL NAND gate is converted into the TTL circuit shown in Fig. 2-13.

We assume that this gate is driven by another similar one, so that a logic-0 input is obtained from a saturated transistor like Q_3. Hence the logic-0 level is taken to be 0.3 V. Consequently, if any one of the inputs, A, B, or C are at the

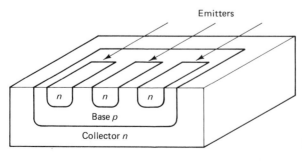

Fig. 2-12 Transistor with three emitters.

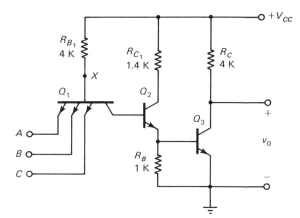

Fig. 2-13 A positive-logic, TTL, NAND gate.

logic-0 level, the corresponding base-emitter junction of Q_1 will be forward biased and the voltage at X will be

$$V_X = 0.7 + 0.3 = 1.0 \text{ V}$$

Since V_γ for the base-emitter junction of Q_2 and Q_3 is at least 0.6 V, the potential of the collector of Q_1 cannot be less than 1.2 V. Hence we see that the collector-base junction of Q_1 is reverse biased, and virtually no collector current flows. As a result, both Q_2 and Q_3 are cut off, and v_o is, therefore, close to the level of V_{CC} (5 V, or logic 1).

Now suppose that all inputs are at the logic-1 level. In this case the potential at X will rise toward V_{CC} until the collector-base junction of Q_1 is forward biased. When that occurs, the voltage at X will become

$$V_X = 0.7 + 0.7 + 0.7 = 2.1 \text{ V}$$

This value will certainly ensure that the emitter junctions of Q_1 are reverse biased. Since the collector-base junction is forward biased, both Q_2 and Q_3 are turned on, and v_o will undoubtedly reach its saturation level of 0.3 V.

Since Q_1 seems to behave merely as a configuration of diodes in either of these two static conditions, we are led to ask why we use a transistor at all. To answer this question let us assume that Q_2 and Q_3 are turned on and V_X is 2.1 V, as discussed earlier. Now assume that one of the inputs is suddenly switched to logic 0. As V_X falls below 1.4 V, the collector-base junction voltage of Q_1 reverses; Q_1 is, although briefly, in the normal, active, operating region for a junction transistor. While this condition holds, the collector current of Q_1 reverses and the stored charge in Q_2 and Q_3 is rapidly pulled out of these transistors. Thus a more rapid switching action than is possible using DTL construction is obtained.

2-4 TOTEM-POLE OUTPUT STAGES

We consider next a modification of the output stage of the TTL circuit shown in Fig. 2-13. The modified circuit appears in Fig. 2-14.

The reason for the modification lies in the fact that the output of the gate

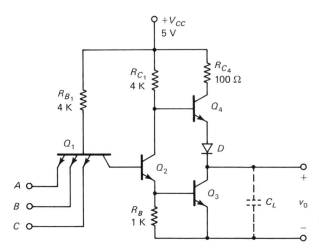

Fig. 2-14 TTL NAND gate with totem-pole output stage.

drives other gates that present, in addition to their input resistances, junction
capacitances that are associated with their emitter-connected input circuits. In
addition to these capacitances, there are also stray wiring capacitances. The capacitance C_L in Fig. 2-14 represents the sum of all these capacitive components.
If the circuit of Fig. 2-13 were used and a sudden change of state were to occur,
causing Q_3 to be switched off, C_L would have to be charged through R_C (4 KΩ)
and a significant time delay would occur before the output voltage v_o achieved its
logic-1 level. Instead of a simple resistor, the circuit of Fig. 2-14 uses another
transistor Q_4, which sits atop Q_3 (hence the name *totem pole*). Here, Q_4 is supposed
to saturate when Q_3 is switched off; when Q_3 is switched on, Q_4 is supposed virtually
to be cut off. We shall now verify that these conditions do occur.

   With all inputs at logic 1, Q_2 and Q_3 are turned on, as we already know. For
Q_2, V_{CE} is 0.3 V, since Q_2 is saturated, and for Q_3, V_{BE} is 0.7 V $= V_d$. These
values mean that the collector voltage of Q_2 with respect to ground must be 0.3
+ 0.7 = 1.0 V. Since V_{CE} for Q_3 is also 0.3 V, we can readily compute the base
voltage (with respect to ground) necessary to cause Q_4 to cut in. This voltage must
be $V_{\gamma(Q4)} + V_{\gamma(D)} + 0.3 = 0.6 + 0.6 + 0.3 = 1.5$ V. Since the actual base voltage
is 1.0 V, whereas 1.5 V is required for cut-in, it is clear that Q_4 is cut off. If diode
D were omitted, the required cut-in voltage would be only 0.6 + 0.3 = 0.9 V, so
that the base voltage of 1.0 V would cause Q_4 to turn on.

   When one or more inputs are at logic 0, Q_2 and Q_3 are cut off. The collector
voltage of Q_2 rises toward V_{CC}. Since the output voltage v_0 occurs across capacitance
C_L, this voltage cannot change instantaneously. Moreover, since the base of Q_4 is
tied to the collector of Q_2, the base voltage of Q_4 rises, while its emitter voltage
changes very little. As a result, Q_4 begins to conduct. Ultimately, the base voltage
of Q_4 must be equal to $V_{BE(Q4)} + V_D + V_{CE(Q3)} = 0.7 + 0.7 + 0.3 = 1.7$ V.
Since Q_2 is cut off, all the current in R_{C1} enters the base of Q_4. This means that
$I_{B(Q4)} = (5 - 1.7)/1400$, or 2.36 mA. If Q_4 is just at the point of saturation,
$V_{CE(Q4)}$ will equal 0.3 V. On that assumption we can calculate the collector current
of Q_4. This yields $I_{C(Q4)} = (5 - 0.3 - 0.7 - 0.3)/100 = 37$ mA. The corresponding

value of β must be $\beta = 37/2.36 = 15.68$. By assumption $\beta_{min} = 20$, so that we are assured that Q_4 does indeed go into saturation.

With Q_4 saturated, the charging resistance associated with C_L must be 100 $+ R_{C(sat, Q_4)} + R_{D(sat)}$. The last two terms represent the saturation resistances of Q_4 and D, respectively. They are usually of the order of only a few ohms. Thus we see that the addition of Q_4 to the circuit yields a great improvement by sharply reducing the time constant associated with C_L.

One might suppose that this time constant could be reduced still further if the 100-Ω collector resistance were deleted from the circuit associated with Q_4. Unfortunately, that resistor must be retained. It turns out that Q_4 turns on faster than Q_3 cuts off. As a consequence there is a short period during which both transistors are conducting, which means that a momentary short circuit of the power supply could occur if the 100-Ω resistor were absent. With this resistor in place, the peak power-supply current is simply the sum of the base and (saturation) collector currents of Q_4, or $37 + 2.36 = 39.36$ mA.

2-5 SCHOTTKY DIODES AND TRANSISTORS

TTL circuits have an inherent limitation that can be troublesome in some applications—their somewhat slow switching speeds. The time required for a gate to switch from one state to its opposite one is referred to as the *propagation delay* of the gate. When several gates are cascaded, as in complex logic circuits, these propagation delays combine and can seriously limit high-speed performance.

A major source of propagation delay in TTL circuits arises from the fact that one or more transistors must be switched from a condition of saturation into cut off. In saturation, a junction transistor actually stores charge in the region of the collector-base junction, and this charge must be removed for switching to occur. Since this process takes time, we might guess that if the transistor were prevented from saturating when in the ON condition, faster switching would be possible.

Two remedies to the problem of transistor saturation are available: Modify the circuit design to prevent saturation, or incorporate a semiconductor clamping device for that purpose in the design of the transistors. We shall consider alternative, nonsaturating, circuit designs later when we take up emitter-coupled logic (ECL). For the moment we shall turn our attention to the semiconductor clamping devices, or *Schottky diodes*.

When metallic aluminum is deposited onto n-type silicon, the aluminum acts as a region of p material, and a metal-semiconductor junction diode is formed. Such a diode is called a *Schottky barrier diode*. When the diode is biased in its forward direction (aluminum side positive), electrons flow from the n material into the aluminum, where they become majority carriers. (Remember that electrons are abundant in a metal). Thus in forward conduction electrons are not stored near the junction, as is the case with semiconductor pn junctions. Therefore, the charge-storage delay time of a Schottky diode when it is switched off is negligible.

Now we recall that an *npn* transistor in saturation has, typically, a collector-emitter voltage of $+0.3$ V and a base-emitter voltage of about $+0.7$ V. This suggests that when it is heavily saturated, the collector-base junction is actually forward biased, although at a level of voltage somewhat less than the cut-in value.

A Schottky diode has a cut-in voltage of about 0.3 V, so that if one were placed between the base and collector of a transistor, as is shown in Fig. 2-15, the diode would cut in and prevent saturation of the transistor. Note the special symbol for the Schottky diode.

A modified transistor symbol, as in Fig. 2-15(b), is used when the Schottky diode is physically incorporated into the transistor structure. The means for accomplishing this are shown in Fig. 2-16.

Since the aluminum behaves as a region of p material, the connection to the p base is simply resistive (or, as it is called, *ohmic*). But the right-hand connection to the n collector produces a Schottky diode. The left-hand connection to the collector is made through an $n+$ region, which causes this connection also to be ohmic. Incorporating the Schottky diode into the structure is no more difficult than making a conventional npn transistor.

2-6 MOSFET LOGIC CIRCUITS

In Chap. 1 we pointed out that all FETs are highly nonlinear devices. Nevertheless, they are sometimes used in linear circuits, where their high input resistance makes it desirable to use them in spite of the design effort required to minimize the effect of their nonlinearity. An example of such an application is in the design of audio power amplifiers using high-power MOSFETs. By far the most common use of FETs is in logic circuitry—in particular, in large-scale integrated (LSI) circuitry like that found in microprocessors and digital memory devices. The FET of choice in these applications is often the MOSFET because of its extremely high input resistance and its low power dissipation.

We consider first an MOS inverter (NOT) circuit, as shown in Fig. 2-17.

This simple circuit has one feature common to most MOSFET circuits: It includes FETs but has no resistors and no capacitance except for the internal capacitance of the devices and that associated with interconnections and wiring. The upper transistor, Q_2, is the load for the lower one, Q_1. This arrangement is an example of DCTL, or *direct-coupled transistor logic*. It resembles the totem-pole arrangement discussed in the last section.

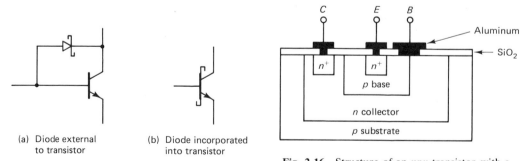

(a) Diode external to transistor

(b) Diode incorporated into transistor

Fig. 2-15 An *npn* transistor with a Schottky diode.

Fig. 2-16 Structure of an *npn* transistor with a Schottky diode between base and collector.

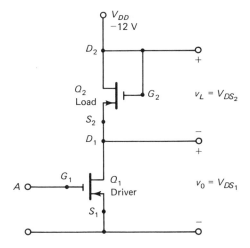

Fig. 2-17 MOS, negative-logic inverter using p-channel MOSFETs.

The variation of drain current I_D with gate-source voltage is given by Eq. (2-2), which is repeated from Chapter 1 with V_T substituted for V_P.

$$I_D = I_{DSS} \left(1 - \frac{V_{GS}}{V_T} \right)^2 \qquad |V_{GS}| \geq |V_T| \qquad (2\text{-}2)$$

The important things to note about Eq. (2-2) are that the variation of I_D with V_{GS} is nonlinear and I_D becomes zero when $V_{GS} = V_T$.

Since, in the circuit of Fig. 2-17, the gate of Q_2 is connected to the drain, $V_{GS2} = V_{DS2}$ for that transistor. The drain characteristic of a MOSFET like Q_2 is shown in Fig. 2-18.

The curve superimposed on this characteristic represents I_{D2} versus V_{DS2} for $V_{GS2} = V_{DS2}$. From Fig. 2-17 we see that the voltage $V_{DS_1} = V_{DD} - V_{DS_2} =$

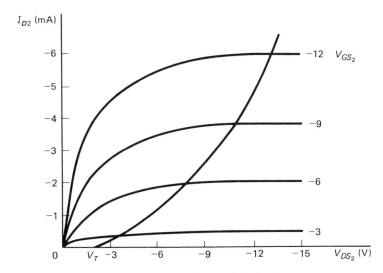

Fig. 2-18 Drain characteristic for Q_2.

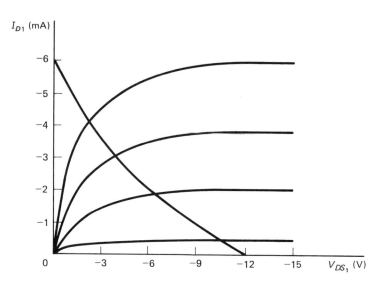

Fig. 2-19 Drain characteristic for Q_1 with load curve of Q_2 superimposed.

$-12 - V_{DS_2}$. To determine the operating range of Q_1, we would ordinarily superimpose a load line on the drain characteristic curves of Q_1. In this case, however, we shall use a load *curve*. This curve is obtained simply by plotting $V_{DS_1} = -12 - V_{DS_2}$ versus I_{D_1}. (Of course, $I_{D_1} = I_{D_2}$ for this circuit). Such a curve is plotted in Fig. 2-19. The intersection of the load curve with the characteristic curves yields the instantaneous operating points.

Since V_{GS_1} is identical with V_A, we see that if the input voltage V_A moves toward -12 V (logic 1), the output voltage V_o, which is identical to V_{DS_1}, moves toward 0 V (logic 0), and vice versa. Thus the circuit of Fig. 2-17 is indeed an inverter. Note that the nonlinear load curve suggests that this inverter might be of limited use in linear applications, although it would be entirely suitable for switching or logic operations.

Only minor modifications are needed to turn the inverter circuit of Fig. 2-17 into a NAND or a NOR gate. Thus, for the circuit shown in Fig. 2-20, we can easily verify that only with both A and B inputs at the logic 1 level will the output V_o be at logic 0.

Similarly, we can show that if either or both inputs are at logic 1 in the circuit of Fig. 2-21, the output will be at logic 0.

In an *n*-channel MOSFET, current is normally made to flow into the drain and out of the source. That means that the source is the more negative of these two terminals. Let us suppose that the potential of the substrate is held at the same level as that of the source. Since the potential between source and substrate is zero, no current flows between them. Furthermore, the drain potential will now be positive with respect to the substrate, so that the drain-substrate junction is reverse biased. Hence there is negligible current flow between these two regions. For a *p*-channel device, we can reverse all polarities and types of material and arrive at the same conclusion: One suitable scheme for biasing the substrate is to connect it to the source terminal.

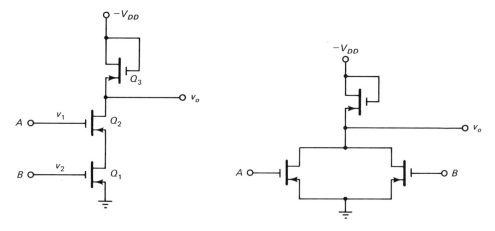

Fig. 2-20 A two-input, negative-logic, NAND gate.

Fig. 2-21 A two-input, negative-logic, NOR gate.

Obviously, if acceptable substrate biasing in an n-channel MOSFET can be achieved by making the potential of the substrate zero with respect to the source, it will also be satisfactory to bias the substrate to a negative voltage with respect to the source.

All this discussion concerning symbols, construction, and substrate biasing of MOSFETs serves to introduce another important family of logic circuits—the *complementary-symmetry MOSFET family*, usually designated CMOS. As its name implies, this family makes use of both p-channel and n-channel devices in the same circuit. These circuits are constructed in integrated form, and it is therefore helpful to consider the physical construction of a simple CMOS inverter before considering complex configurations. The device and its integrated-circuit structure are shown in Fig. 2-22.

When V_i switches to $-V_{DD}$, -10 V, for instance, corrresponding to the logic-1 input condition, Q_1 switches on while Q_2 switches off. This brings V_o to a potential close to the ground level. When V_i switches to 0 V (logic 0), Q_1 turns off, while Q_2 goes on. Thus in either state one transistor is off. This causes the current flow

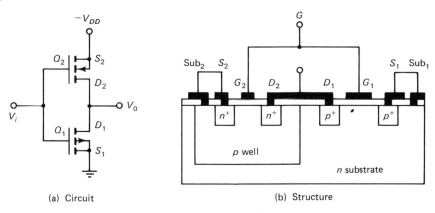

(a) Circuit   (b) Structure

Fig. 2-22 Negative-logic CMOS inverter.

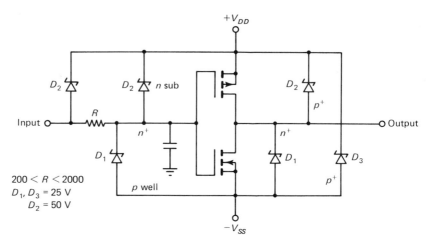

Fig. 2-23 A CMOS inverter with its protective diodes (Courtesy RCA).

through Q_1 and Q_2 to be negligible, and therefore the static power consumed by the inverter is the product of this negligible current and V_{DD}. A typical level of power consumption for this inverter is 50 nW. It is the low power consumption of CMOS logic circuits that makes them attractive for large-scale integration, such as that used in memory and microprocessor chips. Unfortunately this advantage is offset by the rather slow switching speed of this family of devices.

The structure shown in Fig. 2-22 is somewhat idealized. In fact, a typical inverter takes advantage of internal junctions that form protective diodes. These are needed because the extremely thin layer of SiO_2 beneath the gate electrode could be easily damaged by the small gate-to-substrate potentials that might be applied accidentally just in handling the devices. The circuit of an inverter with its protective diodes is shown in Fig. 2-23. These diodes are produced inherently in the manufacturing process.

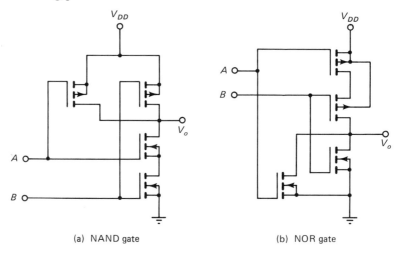

(a) NAND gate   (b) NOR gate

Fig. 2-24 CMOS NAND and NOR gates.

The schematic diagrams of a two-input positive NAND gate and a two-input positive NOR gate are shown in Fig. 2-24.

The upper transistors of the NAND gate are *p*-channel devices, whereas the lower ones are *n*-channel devices. If input *A* is near ground potential, the corresponding upper MOSFET is virtually in saturation. At the same time, the lower MOSFET that is connected to *A* is cut off. This causes V_o to be nearly at the potential V_{DD}. The same argument applies to input *B* and its associated MOSFETs. Only when *both A* and *B* are at a potential close to V_{DD} will both lower transistors saturate. Since these are connected in series, they must both saturate to bring V_o close to ground. Thus we verify that the circuit is indeed that of a NAND gate. You can apply similar arguments to verify the operation of the NOR gate shown in Fig. 2-24(b).

2-6 EMITTER-COUPLED LOGIC (ECL)

Yet another logic family has been devised in the effort to increase switching speed. Emitter-coupled logic (ECL) circuits resemble those linear circuits that use bipolar junction transistors (rather than FETs). The operation of these circuits is arranged to prevent saturation from occurring, thus avoiding the problem of charge storage. The basic configuration of all ECL circuits centers around the differential amplifier—a scheme in which two transistors share a common emitter resistor, as shown in Fig. 2-25. We shall not study this circuit in much detail at this point; such a study is reserved for the next chapter.

Operation of the circuit of Fig. 2-25 depends upon the fact that if R_E is made fairly large, then the current through it is approximately constant. Under that assumption, if Q_1 is made to pass a little more current, Q_2 will carry a little less, so that I_E is held nearly constant. We can evaluate this current I_E by assuming the extreme case in which Q_1 carries no current and Q_2 carries the total. If V_1 is made equal to zero while V_2 is maintained at some fixed level in excess of V_{BE}, then for the base-emitter loop of Q_2, Kirchhoff's voltage law requires that

$$-V_2 + V_{BE} + I_E R_E - V_{EE} = 0$$

I_E must then be equal to $(V_2 + V_{EE} - V_{BE})/R_E$. Since V_{BE} is the base-emitter

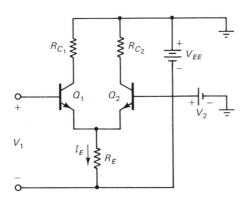

Fig. 2-25 Basic differential amplifier circuit.

Sec. 2-6   Emitter-Coupled Logic (ECL)   **55**

diode sustaining voltage (equal to about 0.7 V for silicon), if V_2 and V_{EE} are fixed, then I_E will be determined by the value of R_E.

As long as V_2 is held constant, the voltage across R_E will be nearly constant, which justifies our assumption that I_E is constant. Clearly then, if V_1 is increased from zero, Q_1 will begin to conduct, causing its emitter to supply a component of current to I_E. In turn, this must result in a decrease in the current passing through Q_2, as stated earlier.

In the next chapter we shall show that the resistance seen at one base of a differential amplifier is approximately $2h_{ie}$. If $\beta = 100$ and I_E is 1 mA (0.5 mA for each transistor), this means that the input resistance must be about 10.4 KΩ. Thus we see that high input resistance is an important feature of ECL circuits. Remember that the higher the input resistance, the more circuits that may be driven by a single gate, that is, the fanout is increased.

Up to this point we have examined the basic emitter-coupled circuit, but we have not yet seen how it performs the function of a logic gate. Figure 2-26 shows a typical ECL gate circuit.

The differential amplifier section of the circuit of Fig. 2-26 comprises transistors, Q_1, Q_2, and Q_3. It may seem surprising that a differential amplifier should require three transistors rather than two, in view of the earlier discussion. However, we must note that transistors Q_1 and Q_2 are connected in parallel and thus, although their inputs are independent, they serve the same purpose as the left-hand transistor Q_1 of Fig. 2-25.

The right-hand section of the differential amplifier circuit of Fig. 2-26 uses two transistors, Q_3 and Q_4, instead of one, as in the case of Q_2 of Fig. 2-25. Actually, Q_3 in Fig. 2-26 performs exactly the same function as Q_2 did in Fig. 2-25. But Q_4, together with diodes D_1 and D_2 and their associated resistors, sets the fixed voltage

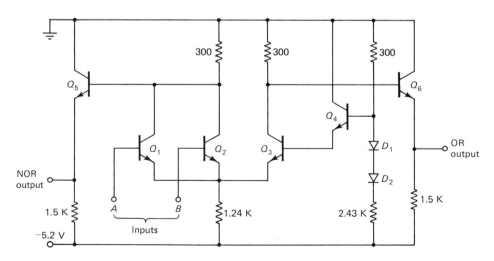

Fig. 2-26 Basic ECL gate circuit.

V_2 that establishes the emitter current I_E. For the moment we shall defer discussion of the function of Q_5 and Q_6.

If we assume that the base current of Q_4 is negligible compared to the current through D_1, D_2, and their associated resistors, we can calculate the base voltage of Q_4. The current through D_1 and D_2 is equal to $(5.2 - 0.7 - 0.7)/2730$, or 1.39 mA. Therefore, the base of Q_4 is $1.4 + 1.39 \times 10^{-3} \times 2.43 \times 10^3 = 4.78$ V above the level of V_{EE}, or $-5.2 + 4.78 = -0.42$ V with respect to ground. The emitter of Q_4 must, therefore, be at a potential of $-0.7 - 0.42 = -1.12$ V with respect to ground, and the emitter of Q_3 must be 0.7 V lower than that, or -1.82 V. We now see that I_E must be about $(5.2 - 1.82)/1.24$ K, or 2.73 mA.

With both inputs A and B at the logic-0 level, let us calculate the output levels at both the left and right output terminals. For ECL circuits the typical logic-0 level is about -1.6 V. Since the emitters of Q_1, Q_2, and Q_3 are all at about -2 V, it is clear that the base-emitter voltages of Q_1 and Q_2 are well below the cut-in level of about 0.6 V. Therefore, neither Q_1 nor Q_2 conducts, and their collectors must rise very nearly to ground potential.

Both Q_5 and Q_6 are emitter followers that enable the ECL gate circuit to deliver more current to a load than it could otherwise do. Hence these transistors help to increase the fanout. In the situation we have just postulated, since the collectors of Q_1 and Q_2 are close to ground potential, we see that the emitter of Q_5 must be at about -0.7 V. This is approximately the logic-1 level for ECL circuits.

If either or both inputs A and B are set to the logic-1 level, the potential of the emitter connection of Q_1, Q_2, and Q_3 is pulled up to about -1.6 V instead of the -1.82 V previously calculated. Since the base of Q_3 is maintained at a potential of -1.12 V, the base-emitter voltage of Q_3 now becomes 0.48 V, which falls below the cut-in value; Q_3 is now cut off. The current I_E, previously assumed to be constant, actually changes and becomes equal to $(5.2 - 1.6)/1.24$ K, or 2.90 mA. With this current flowing through the collector resistor shared by Q_1 and Q_2, the base potential of Q_5 is found to be -0.89 V. Therefore, the output from the emitter of Q_5 is -1.59 V, which is within the range of the ECL, logic-0 level.

The foregoing discussion shows that the output of Q_5 functions as a NOR output. It is left for you to show that the output of Q_6 is an OR output.

As mentioned earlier, ECL circuits never switch into a saturated condition. (In fact, they could be regarded as large-signal, linear circuits.) As a result, there is no charge storage problem, and very high switching speeds can therefore be obtained. However, the swing from the 0 level to the 1 level is only about 0.8 V, which means that ECL circuits have poorer noise immunity than other families of logic circuits. Furthermore, the current levels of ECL circuits are higher than those of other types, which means that for a given IC chip area, more power must be dissipated. Thus ECL circuits do not lend themselves to high-level integration, such as that used in memories and microprocessors.

Despite these disadvantages, the fast switching performance of ECL circuits makes them strong contenders in the design of high-speed, mainframe computers.

REFERENCES

1. Jacob Millman and Christos C. Halkias, *Integrated Electronics: Analog and Digital Circuits and Systems* (New York: McGraw-Hill Book Company, 1972).
2. M. Morris Mano, *Digital Logic and Computer Design* (Englewood Cliffs, N.J.: Prentice-Hall, Inc., 1979).
3. Paul M. Chirlian, *Analysis and Design of Digital Circuits and Computer Systems* (Champaign, Ill.: Matrix Publishing Company, 1976).

PROBLEMS

2-1. This circuit is an inverter (NOT) circuit.
   (a) Assume $I_o = 0$ and calculate V_o when $V_i = 0$ V and when $V_i = 10$ V.
   (b) Find the minimum value of h_{FE} at which operation will be satisfactory.

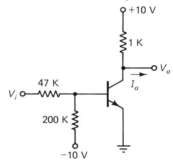

2-2. (a) For the circuit of Prob. 2-1, what current flows into the input terminal when $V_i = 0.3$ V?
   (b) If the collector resistor is reduced to 500 Ω and $h_{FE} = 100$, how many inverter circuits identical to this one can this inverter circuit drive?
   (c) Assuming that the design were modified as in (b) what would happen if the number of *driven* gates were reduced substantially? Is this inverter design practical? Explain.

2-3. Design a two-input NOR gate using the given circuit. Specify R_1 and R_2.

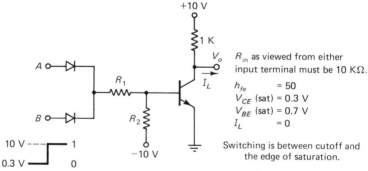

2-4. For the NOT gate of Prob. 2-1, find the noise voltage that will just begin to turn on the transistor when $V_i = 0$. Assume $V_\gamma = 0.6$ V for the base-emitter junction.

2-5. For the DTL gate shown, $h_{FE} = 50$, $V_\gamma = 0.6$ V, $V_d = V_{BE(sat)} = 0.7$ V, and $V_{CE(sat)} = 0.3$ V.
   (a) What logic function (NAND or NOR) does this gate perform?
   (b) Determine R such that Q will just saturate.

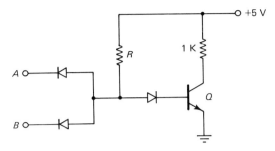

2-6. Find the value of R that will allow the gate of Prob. 2-5 to have a fanout equal to 5.

2-7. To increase its noise immunity, the circuit of Problem 5 is modified as shown. Assume that the fanout is 5. Determine R_C.

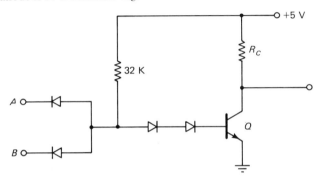

2-8. Suppose that a gate like the one in Prob. 2-5 is producing a logic-1 output (5 V). Assume that with a logic-1 input connected to, for example, terminal A of an identical gate, an input current of 50 μA is observed. How many of these gates can be connected to the output of the first one before its logic-1 level is reduced from 5 V to 3 V?

2-9. (a) Assume that when transistor Q in the circuit shown is at cutoff, $V_{BE} = 0$. What would be the value of V_X in this case? If all inputs A, B, and C are at logic 0 ($+0.3$ V), is this assumption reasonable?

   (b) Assume that all inputs are at logic 1 ($+5$ V). What will be the value of V_X? What now is the value of V_{BE}?

   (c) With V_A, V_B, and V_C at 0.3 V (logic 0), find V_{BE} and V_{CE} assuming D_1 conducts.

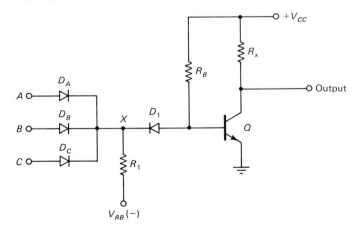

2-10. Find the magnitude of the noise voltage that will just begin to affect the circuit of Prob. 2-9 (a) when the output is at logic 0 and (b) when the output is at logic 1.

2-11. A modification to the circuit of Prob. 2-9 is shown.
   (a) Will the circuit function as a NOR gate?
   (b) Will this circuit have better or poorer noise immunity than that of Prob. 2-9? Explain.

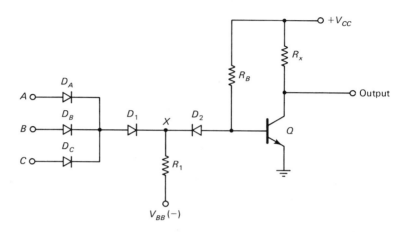

2-12. **(a)** For the circuit shown, calculate the input current to terminal A when B and C are at logic 0 and A is at logic 1 and when A is at logic 0.
   (b) Calculate the fanout for this gate when its output is logic 1 and when its output is logic 0. Assume that the minimum acceptable voltage level for logic 1 is 3.3 V.

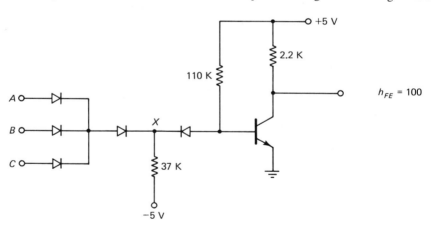

2-13. The circuit of a DTL gate is shown.
   (a) With both A and B inputs at logic 1 ($+5$ V), find the base current of Q_1. Assume $h_{FE} = 20$ for all transistors.
   (b) What is the base current of Q_2?
   (c) Is Q_3 saturated or cut off?
   (d) With either A or B at logic 0 ($+0.3$ V), what base current flows at Q_3?

(e) Find the minimum value of R that will just provide a fanout of 20.

(f) What logic function does this gate perform?

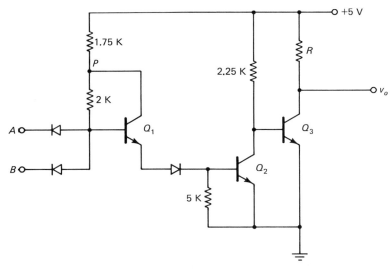

2-14. Determine the noise voltage necessary to disturb the gate of Prob. 2-13 (a) when its output is at logic 1 and (b) when its output is at logic 0.

2-15. (a) If inputs A and B are both at logic 1 ($+5$ V), estimate the emitter current of Q_2 and the base current of Q_3.

   (b) Is Q_4 cut off or saturated? Explain.

   (c) Will the output be at logic 1 or logic 0? Why?

   (d) Repeat (a), (b), and (c) assuming that either A and B or both are at logic 0.

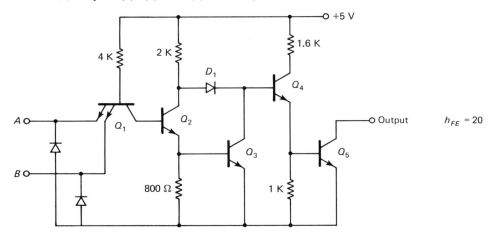

2-16. Calculate the fanout of the gate of Prob. 2-15.

2-17. (a) Find the output *voltage* levels corresponding to the two logic output states.

   (b) Find the voltage levels at points A and B when all inputs are at logic 1 and at logic 0.

   (c) What logic function does this circuit perform?

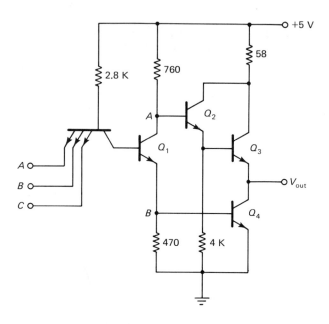

2-18. **(a)** Find the output voltage levels corresponding to logic 1 and logic 0.
    (b) Consider Q_2, Q_3, Q_4, and D_1. When one transistor is on, D_1 keeps the other two off. Which transistor(s) are on and which are off? Explain the role of D_1.
    (c) What logic function does this circuit perform?

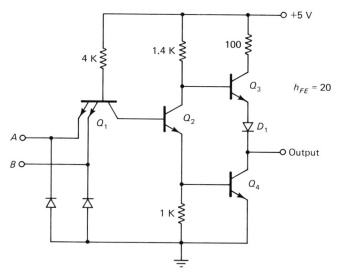

2-19. **(a)** Assume that Q_1 and Q_2 are not conducting while Q_3 conducts. Calculate the voltage at P and the output voltage V_o.
    (b) What are the logic-1 and logic-0 output-voltage levels?

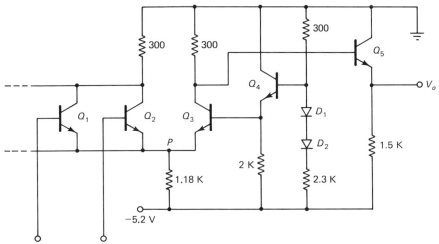

2-20. Using as input-logic levels, the two values found in Prob. 2-19(b), calculate the noise margin for the circuit of Prob. 2-19.

2-21. Increased (more positive) voltage corresponds to logic 1. What voltage inputs at A and B will produce a logic-1 output? What logic function does this CMOS gate perform?

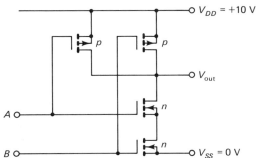

2-22. The circuit shown is a transmission gate that—when placed in cascade with logic gates—permits the realization of tri-state logic. (Tri-state logic, as the name implies, has three states: logic-1, logic-0, and high-impedance, or open, states).

E and $\bar{E}$ are complementary enable terminals: When $E = 1$, $\bar{E} = 0$, and vice versa. Find the voltage levels for E and $\bar{E}$ that will cause the gate to be closed. Find the values of E and $\bar{E}$ that will cause the gate to be open.

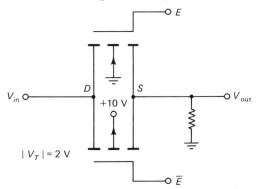

CHAPTER 3

Integrated Circuits
for Analog Applications

3-1 DIRECT COUPLING

As you probably know, the typical audio amplifier built with discrete semiconductor components consists of several stages that are usually coupled together through RC networks. Such circuitry is simple and is certainly well suited to applications in which the lowest frequency encountered is not less than 15 Hz. But for certain applications, an amplifier may have to contend with much lower frequencies. For example, in an electrocardiograph, frequencies as low as that of the human heart-beat may be encountered. For the heart of the normal person at rest that can be a frequency of 72 beats per minute or 1.2 Hz. In industrial applications, such as the measurement of a slowly varying temperature, the frequencies encountered may be very nearly zero (DC). In such a situation, no practical interstage coupling arrangement using capacitors is possible, and we must therefore use circuits that employ direct coupling. Moreover, in the fabrication of integrated circuits, it is not practical to produce large capacitors on a silicon chip along with the various transistors, diodes and resistors. Small capacitors may be formed on a chip, but even the number of these is usually kept to a minimum.

Thus, for low-frequency applications and in integrated-circuit technology, amplifier stages are usually coupled without the use of capacitors—in other words, they are said to be direct coupled.

While it has the advantage of permitting the amplification of signals that vary extremely slowly, direct coupling has the disadvantage of also allowing the amplification of the slowly drifting operating points of the various transistors. This is especially true of the operating point of an input transistor, since such a variation gets more amplification than those occurring in later stages. Such drifts of operating point can result from slow temperature changes of the transistors or even slow changes of component values.

Fortunately there are two factors that offset these undesirable effects. One is the use of differential amplifiers; the other is the use of integrated circuits as opposed to discrete circuits.

Circuit integration reduces the effect of drift because of the miniaturization of circuit elements. This allows close matching of transistors to be achieved. Since all transistors on a single chip (which may have a surface area of, say, a 1.25-mm square) are fabricated together, it is easy to achieve a very close match in transistor parameters. In addition, the proximity of the transistors assures us that temperature changes will be experienced almost identically by all units on the chip. As a result, the differential amplifiers that are fabricated in integrated form on a single chip minimize operating-point drift due to temperature changes. To appreciate the role of differential amplifiers in achieving operating-point stability we must consider these amplifiers in detail.

3-2 GENERAL PRINCIPLES OF DIFFERENTIAL AMPLIFIERS

A differential amplifier is a device that has two inputs and one or more outputs. Its output (usually voltage) is some multiple of the *difference* between the two inputs. The amplifier can be symbolically represented as shown in Fig. 3-1.

For the input voltages v_a and v_B applied as shown, the output voltage v_o is given by

$$v_o = A_{vd}(v_A - v_B) \qquad (3\text{-}1)$$

A_{vd} is the voltage gain of the amplifier.

One can readily see that if v_A equals v_B, the output voltage v_o will be zero. Unfortunately such behavior is an idealization and in actual performance, a small output voltage may be observed.

Before we analyze this performance imperfection quantitatively, let us first consider some applications of differential amplifiers. Aside from their incorporation into integrated circuits, these amplifiers find wide use in biological and industrial

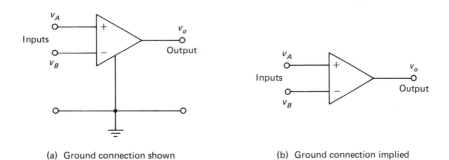

(a) Ground connection shown   (b) Ground connection implied

Fig. 3-1 Differential amplifier symbolizations.

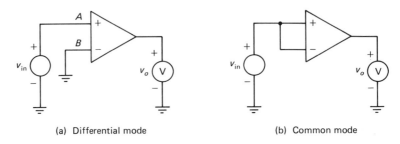

(a) Differential mode   (b) Common mode

Fig. 3-2 Measurements to determine CMRR.

measurements, where the potential difference between two points in a system is the quantity of interest and where the potential that each of these points has with respect to, say, ground, is ignored.

Suppose that in obtaining an electrocardiogram, electrodes are connected to two points on a patient's body and the difference in potential between these points is recorded. As we know from experience in the laboratory, the 60-Hz voltage between any point on the body and ground may be in the order of volts. (Recall what you see when you touch the ungrounded input terminal of an oscilloscope with your finger.) In fact, the 60-Hz voltage relative to ground at each of the two points of measurement may be considerably greater than the differential voltage due to the potentials produced by the heart. It is clear, therefore, that we need the differential amplifier to magnify the differential voltage and to reject, as far as possible, the 60-Hz voltage that is common to both inputs. This property is spoken of as common-mode rejection. A measure of the quality of performance of a differential amplifier is its *common-mode rejection ratio* (CMRR). This ratio, expressed in db, is defined as

$$\text{CMRR (db)} = 20 \log_{10} \frac{A_{vd}}{A_{vcm}} \tag{3-2}$$

where A_{vd} was defined in connection with Eq. (3-1) and A_{vcm} is the gain that the amplifier provides for common-mode signals.

Measuring the CMRR is a simple procedure. Two gain measurements are made as shown in Fig. 3-2. The ratio v_o/v_{in} from Fig. 3-2(a) yields A_{vd} while the same ratio determined from the measurements obtained in Fig. 3-2(b) yields A_{vcm}. At first glance it might appear that since v_{in} is applied to both inputs in Fig. 3-2(b), the output should be zero. That would certainly be true if the amplifier were ideal. We must remember that the *observed* value of v_o is a measure of the degree to which the amplifier fails to be ideal.

3-3 DIFFERENTIAL AMPLIFIER CIRCUITS

The basic differential amplifier circuit using bipolar junction transistors is shown in Fig. 3-3.

If we assume that the bias-network resistors R_1 and R_2 are very large compared

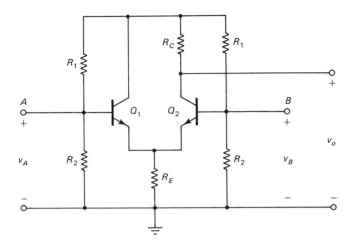

Fig. 3-3 Differential amplifier using junction transistors.

to other resistances in the circuit, we can omit them as we undertake an approximate analysis of the amplifier. In the differential mode of operation, the most extreme (and the simplest) case occurs when one input is zero and the other carries a test signal. Therefore, we take v_B to be zero (which is equivalent to grounding input terminal B) and apply a voltage v_A to terminal A. We wish to calculate the output voltage, v_o.

One way to treat this network is to consider that Q_1 and R_E form an emitter follower that is driving a common-base amplifier Q_2. (Remember that with terminal B grounded, Q_2 does function as a common-base circuit). From the discussion in Chap. 1, we know that the resistance looking into the emitter of a common-base amplifier is $h_{ie}/(1 + h_{fe})$ if we neglect h_{oe}. This resistance in parallel with R_E is the load seen by Q_1, the emitter follower. Again, from Chap. 1, we can find the input resistance of Q_1. This is

$$r_{\text{in}} = h_{ie} + (1 + h_{fe})\left(R_E\|\frac{h_{ie}}{1 + h_{fe}}\right)$$

We are assuming identical values of h_{ie} and h_{fe} for Q_1 and Q_2.

Evaluating r_{in}, we obtain

$$r_{\text{in}} \approx h_{ie} + (1 + h_{fe})\left[\frac{(R_E h_{ie})/(1 + h_{fe})}{R_E + h_{ie}/(1 + h_{fe})}\right]$$

Dividing numerator and denominator by R_E, we get

$$r_{\text{in}} = h_{ie} + (1 + h_{fe})\left[\frac{h_{ie}/(1 + h_{fe})}{1 + h_{ie}/R_E(1 + h_{fe})}\right] \tag{3-3}$$

Typical values for $h_{ie}/(1 + h_{fe})$ are on the order of a few kilohms divided by 100 or 200, or on the order of *tens* of ohms. Also, R_E is on the order of kilohms, so that $h_{ie}/R_E(1 + h_{fe})$ may be equal to a few hundredths of an ohm at the most. As a consequence, we can neglect the term $h_{ie}/R_E(1 + h_{fe})$, so that Eq. (3-3) is approximated quite well by

$$r_{\text{in}} = h_{ie} + h_{ie} = 2h_{ie} \tag{3-4}$$

Knowing the input resistance, we are in a position to calculate the approximate differential-mode voltage gain. For this purpose we use the equivalent circuit of Fig. 3-4. Note the common-base portion that represents Q_2.

We can easily analyze this network on a step-by-step basis. First of all, $i_b = v_A/r_{\text{in}}$, and from the previous calculation, $r_{\text{in}} \approx 2h_{ie}$. Hence $i_b \approx v_A/2h_{ie}$. By the current division rule, i_e is equal to $h_{fe}i_b[R_E/(R_E + h_{ib})]$, or

$$i_e = h_{fe}\left(\frac{v_A}{2h_{ie}}\right)\left(\frac{R_E}{R_E + h_{ib}}\right)$$

Since R_E is ordinarily made large [certainly large with respect to $h_{ib} = h_{ie}/(1 + h_{fe})$], the ratio $R_E/(R_E + h_{ib})$ is very nearly equal to unity. Hence $i_e \approx h_{fe}v_A/2h_{ie}$.

We see that $v_o = -h_{fb}i_eR_C$. The quantity h_{fb} (or α) is very close to -1 so that $v_o \approx +i_eR_C$. Hence $v_o = h_{fe}R_Cv_A/2h_{ie}$.

We therefore find the approximate differential voltage gain to be

$$A_{vd} = \frac{v_o}{v_A} \approx \frac{h_{fe}R_C}{2h_{ie}} \tag{3-5}$$

You should note that this is one-half the value found for the magnitude of the voltage gain of a single-transistor, common-emitter amplifier.

In Fig. 3-3 a collector resistor was assumed to be present only in the circuit of Q_2 and not in that of Q_1. This is an unnecessary restriction, and collector resistors are often found in both locations. With such a modification the circuit of Fig. 3-4 becomes that shown in Fig. 3-5. In this circuit v_o remains at the same value just calculated. However, a new output voltage, v_o', appears and is found to be equal to

$$v_o' = -h_{fe}R_Cv_A/R_{\text{in}} \approx -h_{fe}R_Cv_A/2h_{ie}$$

This value is equal in magnitude but 180° out of phase with the quantity determined for v_o. Thus if the output voltage were taken *between* the two collector terminals, the gain would be very nearly the same as that found for a common emitter

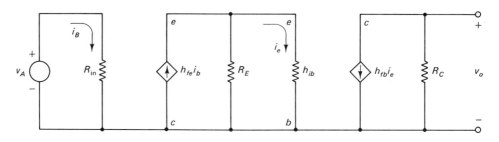

Fig. 3-4 Equivalent circuit for differential amplifier.

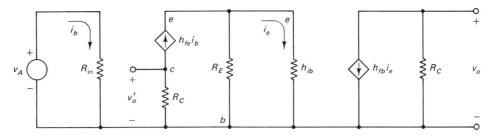

Fig. 3-5 Differential amplifier with equal resistors in both collector circuits.

amplifier. This floating (ungrounded) differential output is often used to drive a second differential amplifier.

We next seek to determine the common-mode gain of the circuit of Fig. 3-3. For this purpose, terminals A and B are assumed to be tied together and driven from a single source, v_A. Since terminal B is no longer grounded, we can no longer consider Q_2 to be in a common-base configuration. Figure 3-6 shows one equivalent circuit that can be used for this analysis.

We shall not carry out the analysis; instead we simply indicate how it can be done. You can work out the details if you are interested. We note that i_b is identical for Q_1 and Q_2 and must be equal to $(v_A - v_E)/h_{ie}$. The voltage v_E must equal $(2i_b + 2h_{fe}i_b)R_E$. Combining these quantities, we solve for i_b and then compute $v_o = -R_C h_{fe}i_b$. The ultimate result is

$$v_o = \frac{-R_C h_{fe} v_A}{h_{ie} + 2R_E(1 + h_{fe})}$$

Dividing by v_A we find the common-mode voltage gain,

$$A_{vcm} = -\frac{R_C h_{fe}}{h_{ie} + 2R_E(1 + h_{fe})} \tag{3-6}$$

The ratio of the magnitudes of A_{vd} to A_{vcm} gives the *numerical* value of CMRR. Of course, it is then a simple matter to find the CMRR in decibels.

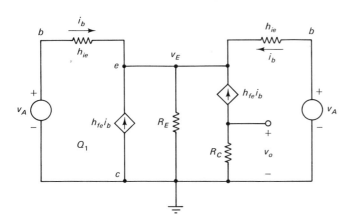

Fig. 3-6 Equivalent circuit to find common-mode gain.

Example 3.1

Suppose we have a differential amplifier using transistors having $h_{ie} = 3\text{K}\Omega$ and $h_{fe} = 100$ in a circuit in which $R_C = 2 \text{ K}\Omega$ and $R_E = 5 \text{ K}\Omega$. From Eq. (3-5), we find

$$A_{vd} = \frac{100 \times 2000}{2 \times 3000} = 33.3$$

From Eq. (3-6),

$$A_{vcm} = \frac{-2000 \times 100}{3000 + 2 \times 5000 \times (101)} = -0.196$$

The CMRR is 33.3/0.196, or 170. Expressed in decibels, the CMRR is 20 log 170 = 44 db. Usually this quantity is expressed as a negative, the idea being that the common-mode gain is *below* the differential-mode gain.

Algebraically the CMRR may be expressed as

$$\text{CMRR} = \frac{h_{fe}R_C/2h_{ie}}{h_{fe}R_C/[h_{ie} + 2R_E(1 + h_{fe})]} = \frac{h_{ie} + 2R_E(1 + h_{fe})}{2h_{ie}}$$

If we divide numerator and denominator by h_{ie} and observe that $2R_E(1 + h_{fe})/h_{ie} \gg 1$, we have the approximate result that

$$\text{CMRR} \approx R_E(1 + h_{fe})/h_{ie} \tag{3-7}$$

Equation (3-7) suggests that for a given transistor, CMRR can be increased by increasing R_E. That idea is the basis for the circuit modifications discussed in the next section.

3-4 MODIFICATIONS TO THE BASIC DIFFERENTIAL AMPLIFIER CIRCUIT

Since the CMRR is increased by making R_E greater, it is logical to suppose that if R_E could be replaced by an ideal constant-current source, the CMRR could be made infinite. Of course this idea rests on the concept that the internal resistance of an ideal constant-current source is itself infinite. In practice, we are obliged to work with nonideal current sources that have finite, though large, internal resistance. Such a current source can be had, of course, by using a transistor.

It is worth noting in passing that if the bottom end of R_E in Fig. 3-3 is returned to a fairly large negative source V_{EE} instead of ground, then the resistor R_E can be made larger and the CMRR increased. We can picture that the resistor R_E and the voltage source V_{EE} could be replaced by an equivalent constant current source and shunt resistor, also equal to R_E. This was the line of thinking used in discussing the ECL circuits of Chap. 2.

A differential amplifier in which R_E is replaced by a transistor acting as a constant-current source is shown in Fig. 3-7. It is instructive to carry out an analysis of this circuit, as is done in Example 3-2.

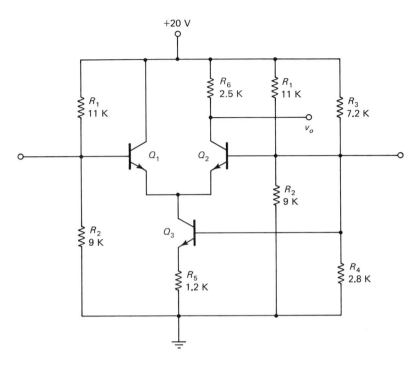

Fig. 3-7 Differential amplifier circuit with current-source transistor.

Example 3-2.

We begin with a DC analysis and attempt first to find the current level at which Q_3, the current-source transistor, is operating. We shall assume that I_B for Q_3 is negligible compared to the current drawn by the voltage divider R_3R_4. In fact, the current drawn by R_3R_4 is 20/10 K = 2 mA, which is very likely to be quite large compared to I_B for Q_3. The base-ground voltage of Q_3 must be 2 mA × 2.8 K = 5.6 V. Assuming V_{BE} = 0.7 V, the emitter of Q_3 is at 5.6 − 0.7, or 4.9 V, above ground. Hence the emitter current of Q_3 must be 4.9/1.2 K = 4.08 mA, which is also close in value to the collector current of Q_3. This value of current is very nearly independent of V_{CE} for Q_3, which is merely another way of saying that Q_3 approximates an ideal current source of 4 mA.

The two voltage dividers R_1R_2 establish the bases of Q_1 and Q_2 at 9 V. For V_{BE} = 0.7 V, their emitters are at 9 − 0.7 = 8.3 V, as is the collector of Q_3. We have thus verified that there is an adequate value of V_{CE} (8.3 − 4.9 = 3.4 V) for Q_3.

If Q_1 and Q_2 are well matched and since their base circuits are biased in an identical manner, their emitter currents will be equal. This means that each transistor passes 2 mA. The collector of Q_2 is thus at a potential of 20 − 2.5 K × 2 mA = 15 V, so that V_{CE} for Q_2 is 15 − 8.2 = 6.8 V.

For Q_1 and Q_2, assuming h_{fe} = 100, h_{ie} is approximately equal to (26/2) × 100 = 1300 ohms. The differential voltage gain is A_{vd} = 100 × 2500/(2 × 1300) = 96.1.

Evaluation of the common-mode gain appears to be a perplexing problem at first, since we seem no longer to have a value of R_E to use. A little reflection soon makes clear that the base of Q_3 operates at a fixed potential and therefore looks like ground so far as AC signals are concerned. Hence Q_3 can be treated as a small-signal, common-base amplifier. The resistance that Q_3 provides between ground and the emitters of Q_1 and Q_2 can be determined by using Eq. (1-5) of Chap. 1, where this type of circuit was analyzed.

Since the DC emitter current of Q_3 is 4 mA, the value of h_{ib} is $26/4 = 6.5 \, \Omega$. If we use the same values of h_{ob} and h_{rb} that were used in Chap. 1—namely, $h_{ob} = 0.49 \times 10^{-6}$ mho and $h_{rb} = 2.9 \times 10^{-4}$—we find

$$g_o = h_{ob} + \frac{h_{rb}}{R_E + h_{ib}}$$

or $g_o = 0.49 \times 10^{-6} + 2.9 \times 10^{-4}/(1200 + 6.5) = 7.304 \times 10^{-7}$ mho. We can therefore take account of R_E in Eq. (3-6) and Eq. (3-7) by using a value of $1/g_o = 1.37 \, \text{M}\Omega$. This produces a value for CMRR equal approximately to $1.37 \times 10^6(1.01)/13.00 \approx 1.06 \times 10^5$, which corresponds to about 100 db.

The DC current I_E, which is the 4-mA constant current drawn from the emitters of Q_1 and Q_2, is determined by the voltage divider network $R_3 R_4$ and by R_5, the emitter resistor of Q_3. Adjusting the values of three resistors to obtain a desired level of I_E is an inconvenience, particularly in an integrated-circuit design. Another way to do this is through the use of a circuit sometimes called a *current mirror*.

Figure 3-8 shows a pair of transistors in a current-mirror circuit. Since the base and collector of Q_1 are connected together, the collector-emitter and the base-emitter voltages of this transistor are the same. Although this means that v_{CE} for this transistor is small, the transistor is nevertheless so biased that it operates in the usual manner. The collector current of Q_1 must be given by

$$I_{C1} \approx (V_{CC} - V_{CE})/R = (V_{CC} - V_{BE_1})/R \tag{3-8}$$

This expression is an approximation, since it neglects the fact that base current for both transistors Q_1 and Q_2 flows through R in addition to I_{C_1}. However, for the usual values of h_{FE}, this approximation will be a very good one.

The fact that the bases and emitters of Q_1 and Q_2 are connected together so that $V_{BE_2} = V_{BE_1}$ is what makes the circuit work. We must recall that the value usually taken for V_{BE}, 0.65, or 0.7, volts, is only an approximation that we use to avoid a messy, nonlinear relationship between I_B and V_{BE}. In the case of Q_1, we have to remind ourselves that V_{BE_1} establishes itself at a value that just corresponds

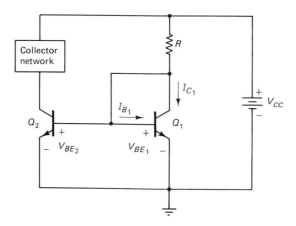

Fig. 3-8 Current-mirror circuit.

to the value of I_{B_1}. But I_{B_1} must be given simply by $I_{B_1} = I_{C_1}/h_{FE}$ while I_{C_1} is given by Eq. (3-8). Thus, we choose V_{CC} and R to fix I_{C_1}. Then I_{B_1} is fixed because of h_{FE}. With I_{B_1} set, V_{BE_1} is also set. Now, since $V_{BE_2} = V_{BE_1}$, I_{B_2} must be equal to I_{B_1} if Q_1 and Q_2 are identical. It must follow that I_{C_2} is equal to, or mirrors, I_{C_1}.

For the circuit of Fig. 3-8 to operate as described, it is necessary that Q_1 and Q_2 be identical. As mentioned earlier, this is an easy requirement to satisfy.

The differential amplifier circuit of Fig. 3-7 is now modified to make use of a current-mirror circuit to replace the simple current-source transistor Q_3. The modified circuit is shown in Fig. 3-9.

Example 3-3

As a variation of our usual circuit-analysis procedure, let us attempt to design this circuit to achieve a desired level of performance. Suppose that $h_{fe} = 150$, $R_C = 5200 \ \Omega$, and we require that the differential-mode voltage gain be 100. From Eq. (3-5), we find that

$$h_{ie} = \frac{h_{fe}R_C}{2Av_d} = \frac{h_{fe} \times 5200}{2 \times 100}$$

From Chap. 1 we know that $h_{ie} = (1 + h_{fe}) \times 0.026/I_E$. Hence

$$(1 + h_{fe})\frac{0.026}{I_E} \approx \frac{h_{fe} \times 5200}{200}$$

and

$$I_E = \frac{200 \times 0.026}{5200} = 10^{-3} = 1 \text{ mA}$$

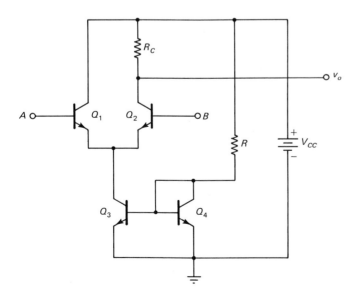

Fig. 3-9 Differential amplifier circuit using current mirror.

Recalling that in this case I_E represents the DC emitter current for *each* transistor (Q_1 and Q_2), we see that Q_3 must pass 2 mA. Hence Q_4, the other element of the current mirror, must likewise pass 2 mA. We can now specify the value of R.

$$R = \frac{15 - 0.7}{2 \times 10^{-3}} = 7.15 \text{ K}\Omega$$

In Example 3-3, use was made of Eq. (3-5), which relates A_{vd} to R_C, h_{fe}, and h_{ie}. If we rearrange this equation, we can write

$$A_{vd} \approx \frac{R_C I_E}{2 \times 0.026}$$

which shows that the level of DC emitter current has a direct and proportional effect on the differential voltage gain. This fact suggests we should be able to vary the gain of the amplifier by varying I_E. One simple way to accomplish this is to substitute a transistor for the series resistor that sets the current of Q_4 in the circuit of Fig. 3-9. Such an arrangement is shown in Fig. 3-10.

The input resistance of the differential amplifier of Fig. 3-7 is approximately the parallel combination of R_1, R_2, and $2h_{ie}$. In some applications—for instance, in operational amplifiers for use in instrumentation—it is desirable to have a much higher input resistance. This can be done by incorporating FETs in the input circuits of Q_1 and Q_2. An FET-input differential amplifier is shown in Fig. 3-11. The resistors R_g, which primarily determine the input resistance, can be made quite large—on the order of 10^5 Ω instead of kilohms, as in the case where $2h_{ie}$ is the principal determinant of the input resistance.

If we assume the following parameters for the FETs and BJTs, we can analyze the performance of this circuit and calculate design values. Let us take I_{DSS} =

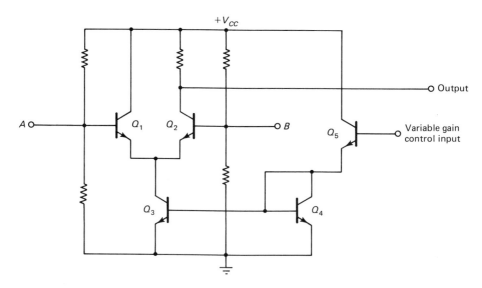

Fig. 3-10   Differential amplifier with variable gain.

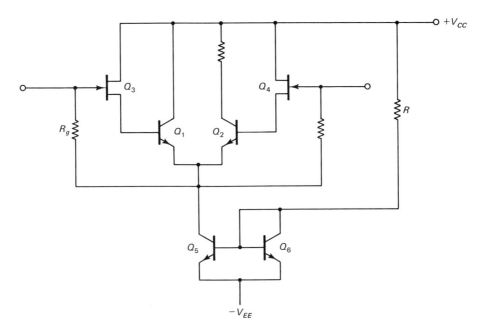

Fig. 3-11 Differential amplifier with FET input.

32 μA, $V_P = -3.5$ V, and $h_{fe} = 50$. Since R_g is returned to the emitters of Q_1 and Q_2, the gate-source voltage $V_{GS} = -0.7$ V. Hence $I_D = 32(1 - 0.7/3.5)^2$, or $32(1 - \frac{1}{5})^2 = 32 \times \frac{16}{25} \approx 20$ μA. This is the base current for Q_1 and Q_2. Then $I_E = h_{fe} \times 20$ μA, or 50×20 μA $= 1$ mA. If $V_{CC} = 10$ V and $V_{EE} = -10$ V, R must be set so that Q_6 and Q_5 each pass 2 mA. Hence $R = (20 - 0.7)/2 \times 10^{-3} = 9.65$ KΩ.

We should note that if the input terminals are operated at potentials that vary about ground (0 V) as a quiescent value, then the emitters of Q_1 and Q_2 and the collector of Q_5 must be at ground (in the quiescent state). With $V_{EE} = -10$ V, V_{CE} of Q_5 is 10 V, as is V_{CE} for Q_1. Also, V_{CE} of Q_2 can be set to 5 V by making $R_C = 5/1$ mA $= 5$ KΩ. Then A_{vd} will be approximately $h_{fe}R_C/2h_{ie}$, or $50 \times 5 \times 10^3/(2h_{ie})$. Since Q_1 and Q_2 operate at I_E equal to 1 mA, $h_{ie} \approx 50(0.026/0.001) = 1300$ Ω. Hence $A_{vd} = 250 \times 10^3/2 \times 1.3 \times 10^3 = 96.1$.

3-5 DIFFERENTIAL TRANSCONDUCTANCE

For integrated-circuit differential amplifiers certain special circuit analysis techniques are advantageous. To appreciate this point, let us consider the circuit shown in Fig. 3-12.

In Fig. 3-12, V_1 and V_2 represent the voltages applied to the bases of Q_1 and Q_2. The corresponding collector currents are I_{C_1} and I_{C_2}. Current-source transistor Q_3 sets the current I_{EE}, which is assumed to be constant.

Recalling that the emitter-base junction of a transistor is forward biased to cause current carriers to be injected into the base, we take it to be reasonable that

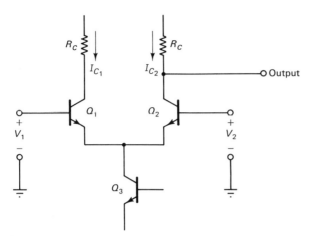

Fig. 3-12 Elements of an integrated-circuit differential amplifier.

emitter current should increase exponentially with increasing forward bias. Expressed mathematically this yields

$$I_E = I_S e^{qV_{BE}/kT}$$

Since $kT/q = 0.026$ V at room temperature, we can write

$$I_E = I_S e^{V_{BE}/0.026}$$

where I_S is the reverse saturation value of I_E. Since I_{EE} is constant, $I_{E_1} + I_{E_2} = I_{EE}$. Furthermore, since the emitters of Q_1 and Q_2 are tied together, $V_{BE_1} = V_{B_1}$ and $V_{BE_2} = V_{B_2}$, where V_{B_1} and V_{B_2} mean, simply, the voltages of the respective bases of Q_1 and Q_2 relative to ground. For any transistor $I_C \approx I_E$. Using the equations we now have available, we can express I_S as

$$I_S = I_{E_2} e^{-V_{B2}/0.026}$$

Then

$$I_{E_1} = I_S e^{V_{B1}/0.026} = (I_{E_2} e^{-V_{B2}/0.026}) e^{V_{B1}/0.026}$$

Since $I_{E_2} = I_{EE} - I_{E_1}$, we have

$$I_{E_1} = (I_{EE} - I_{E_1}) e^{(V_{B1} - V_{B2})/0.026}$$

or

$$I_{C_1} = \frac{I_{EE}}{1 + e^{(V_{B1} - V_{B2})/0.026}}$$

The term $V_{B_1} - V_{B_2}$ that appears in the above equation is simply the differential input voltage. It is convenient to call this V_d. Then the equation becomes

$$I_{C_1} = \frac{I_{EE}}{1 + e^{V_d/0.026}} \qquad (3\text{-}9a)$$

or

$$\frac{I_{C_1}}{I_{EE}} = \frac{1}{1 + e^{V_d/0.026}}$$ (3-9b)

Figure 3-13 is a plot of Eq. (3-9b). As one would expect, when $V_d = 0$, I_{C_1} = I_{C_2} and since these currents must add up to I_{EE}, each must be exactly half of I_{EE}. For large positive values of V_d, $V_{B_1} > V_{B_2}$, I_{C_1} predominates and approaches the value of I_{EE}. When $V_{B_2} > V_{B_1}$, V_d is negative and then I_{C_2} predominates.

To obtain maximum differential-mode output voltage, we should like the change in I_{C_1} with respect to a change in V_d, for example, to be a maximum. To study this requirement more closely, let us differentiate Eq. (3-9a) to obtain

$$\frac{dI_{C_1}}{dV_d} = \frac{-I_{EE} e^{V_d/0.026}}{(1 + e^{V_d/0.026})^2 (0.026)}$$

Rather than differentiate this expression once again to find its maximum, we need only look at Fig. 3-13 to see that dI_{C_1}/dV_d is a maximum when $V_d = 0$. This leads us to define a quantity g_{m_d} such that

$$g_{m_d} \approx \frac{dI_{C_2}}{dV_d}\Bigg|_{V_d=0} = \frac{I_{EE}/0.026}{4}$$ (3-10)

In terms of Eq. (3-10) we see that A_{vd} for the circuit of Fig. 3-12 is given simply by $A_{vd} = g_{m_d} R_C$. Since g_{m_d} depends only on I_{EE}, we see once again the possibility of creating a voltage amplifier having automatic gain control.

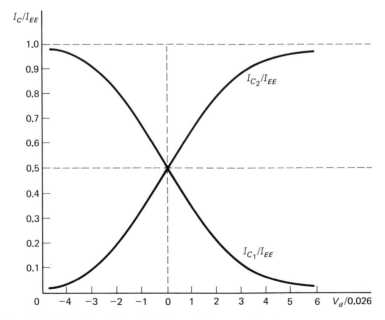

Fig. 3-13 Variation of I_C with V_d (quantities normalized).

We can make some further observations from the graph of Fig. 3-13. We can see that to assure a linear relationship between I_C and V_d, the range of $V_d/0.026$ should be restricted to be between -1 and $+1$. Or, in other words, V_d should lie approximately between $+26$ mV and -26 mV. Furthermore, we see that values of V_d in excess of 0.1 V cause saturation of one transistor and cutoff in the other.

3-6 DC VOLTAGE LEVELS IN INTEGRATED CIRCUITS

Direct coupling, despite its advantages, presents certain problems. The absence of coupling capacitors means that not only signal voltages but also DC voltage levels are interconnected from point to point in an integrated network. To understand the significance of this point, let us consider the circuit shown in Fig. 3-14.

The problem that we wish to consider is that of finding the DC voltage levels at various points in the circuit. We begin be calculating the voltages at the bases of Q_1 and Q_2. Let us assume for the moment that the base currents are negligible; we shall verify this assumption later. The base-circuit voltage dividers provide base voltages to Q_1 and Q_2 of 5 V. The emitter voltage is therefore $5 - 0.7 = 4.3$ V. Therefore, the current in the emitter resistor is $4.3/4.3$ K $= 1$ mA. Each transistor, Q_1 and Q_2, passes $\frac{1}{2}$ mA and the collector voltage of each is therefore $15 - 10^4 \times 0.5 \times 10^{-3} = 10$ V.

If h_{FE} is equal to 50, I_B for either transistor is $(0.5 \text{ mA})/50 = 10$ µA. This current is negligible compared to $15/75$ K $= 200$ µA that flows in the 50 KΩ-25 KΩ voltage divider. Thus we have verified the assumption made earlier.

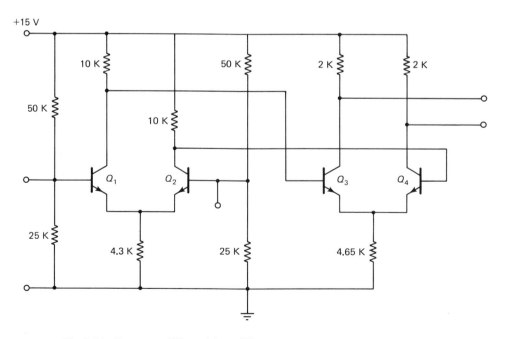

Fig. 3-14 Two-stage differential amplifier.

Because of the direct connection of the collectors of Q_1 and Q_2 to the bases of Q_3 and Q_4, these bases must also have a potential of 10 V with respect to ground. Thus we can carry out a similar analysis for Q_3 and Q_4. The emitters of these transistors must be at a potential of $10 - V_{BE} = 10 - 0.7 = 9.3$ V above ground. Hence the current in the emitter resistor must be 9.3/4.65 K = 2 mA. We again take the current in each transistor to be half that value, or 1 mA. This allows us to calculate that the collectors of Q_3 and Q_4 are at a voltage of $15 - 2 \times 10^3 \times 10^{-3} = 13$ V above ground.

The important conclusion that can be drawn from these calculations is that although the bases of Q_1 and Q_2 are at $+5$ V, those of Q_3 and Q_4 are at $+10$ V. And if another stage followed $Q_3 - Q_4$, its bases would have to be at some still higher potential. All this suggests that in a multistage circuit, we should soon "run out of volts."

Also observe that while the collector voltages of Q_1 and Q_2 can swing up to 15 V and down to about 5 V, the collector voltages of Q_3 and Q_4 range only from 15 V to perhaps 10 V. Here again we see a limitation that is imposed as a result of the direct coupling of stages.

What can be done about this problem? One obvious solution would seem to be to raise the supply voltage. Of course, if there are enough stages, this solution has its own limitations.

There is yet another troublesome question that arises because of direct coupling—the question of how to connect external apparatus to the amplifier. We must remember that direct coupling must be used everywhere if slowly varying signals are to be observed. If, in the example at hand, an external source such as a geophone (a special low-frequency microphone for detecting shock waves in the earth) were to be connected to the bases of Q_1 and Q_2, this instrument would have to be constructed to operate at 5 V above ground. In addition, the output device, which might be a pen recorder, would have to operate at a quiescent potential of 10 V above ground. Although no insurmountable difficulty is involved in operating input and output devices this way, it is often inconvenient.

3-7 LEVEL-SHIFTING CIRCUITS

All the above remarks serve to introduce the topic of *level shifting*. Level shifting involves the techniques that are used to adjust to specified values the quiescent potentials at various points of a direct-coupled circuit.

3-7.1 Zener-diode level shifters

Perhaps the most obvious way of producing a shift in voltage level is by means of a Zener diode. Applied to the interstage coupling of the amplifier of Fig. 3-14, this yields the circuit shown in Fig. 3-15. Only the coupling between Q_1 and Q_3 is shown; that between Q_2 and Q_4 is omitted to avoid cluttering the diagram.

Assuming that Q_1 (and Q_2) operate at the same quiescent point as in the previously analyzed circuit, then we know that the collector voltage of Q_1 is $+10$ V. If we want the base of Q_3 (and of Q_4) to operate at the same level as the base of

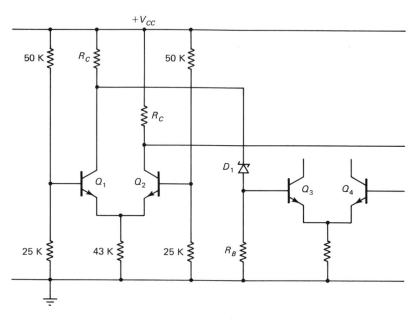

Fig. 3-15 Circuit modified to include Zener diode level shifter.

Q_1 ($+5$ V), then the Zener diode, D_1, must drop a voltage of 5 V. Unfortunately, this diode must pass some current in order to operate past the knee of the Zener region curve, which will ensure that its voltage drop will be nearly constant. Suppose we assume that the diode requires 200 μA for this purpose. Then the collector resistor of Q_1 must pass not only the 0.5 mA that we calculated for the circuit of Fig. 3-14 but also this additional 0.2 mA. For a collector voltage of $+10$ V, this means that R_c must equal $(15 - 10)/0.7$ mA $= 7.14$ KΩ, as compared to 10 KΩ in the original circuit. Clearly, this reduced collector resistance will result in a lower voltage gain.

Since the base current of Q_3 should be about 1 mA/50 $= 20$ μA (assuming $h_{FE} = 50$), some path must be found for the remaining 180 μA of Zener-diode current. Such a path is provided by resistor R_B, whose calculation is left as a problem. This resistor also will tend to reduce the voltage gain of the amplifier.

One way in which a Zener diode may be formed is by the use of the peculiar connection applied to the transistor shown in Fig. 3-16.

Observe that the base (p) is reverse biased with respect to the emitter-collector region (n). Of course, the same configuration could be forward biased, in which case the voltage across the diode would be the familiar V_d equal to 0.7 V, a value that would remain fairly constant if diode current should vary.

3-7.2 Transistor level shifters

Other level-shifting circuits exist and some of these are already familiar to us. In this category is the emitter follower shown in Fig. 3-17.

Here we focus our attention on the DC voltage levels, V_i at the input and V_o

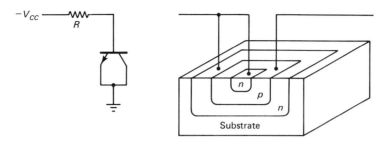

Fig. 3-16 An *npn* transistor connected to form a Zener diode.

at the output. Clearly $V_o = V_i - V_{BE}$, so that V_o is offset or shifted down from V_i by 0.7 V. As in the previously considered circuits, the price we pay for level shifting is a reduction in gain, since A_v for an emitter follower is always less than 1.

A modification of this circuit, which provides a greater shift of level, is shown in Fig. 3-18.

By tapping down on the emitter resistor, the desired output level V_o can be obtained. Once again, however, the price we pay for this adjustment is reduced voltage gain. This undesirable effect can be offset by the circuit modification shown in Fig. 3-19.

When the input voltage V_i increases as a result of an increasing signal component, the emitter-follower transistor attempts to pass more current. Assuming that no current is drawn by the load connected to the output, the emitter current cannot increase, since R is in series with a constant current source I. In other words, since the current through R must be constant, there is no change in the voltage *across R*. Consequently, V_o must exactly follow any changes in V_i.

In reality the situation is not that good, since—instead of an ideal constant current source—we have the transistor Q_3 in the actual circuit. However, the circuit

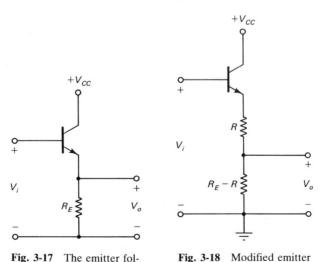

Fig. 3-17 The emitter follower as a level shifter.

Fig. 3-18 Modified emitter follower.

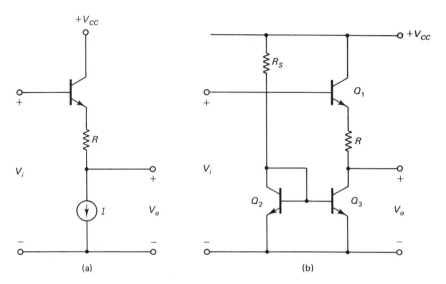

Fig. 3-19 The emitter follower with (a) an added current source and (b) actual circuit arrangement to implement the current source.

does offset much of the loss in gain of the simple, tapped-resistor emitter follower quite effectively.

Yet another modification to this circuit permits positive feedback to be introduced. This is done to increase the overall voltage gain of the amplifier that includes this circuit as one of its stages. The reason that such a procedure works to increase gain is discussed in Chapter 4. Now, we wish to verify that the current I_f flowing in the modified circuit shown in Fig. 3-20 drives V_o in the positive direction.

In this circuit, the base-emitter voltages of Q_2 and Q_3 are not equal, as they were in the circuit of Fig. 3-19. Nevertheless, Q_2 still establishes the level of current passed by current source Q_3. The resistors R_s and R_3 and the voltages V_{CC} and V_{BE} establish the level of current and the base potential of Q_2. Hence the base voltage of Q_3 is also established. That being the case, we need only subtract V_{BE} from this voltage to know what voltage exists across R_1 and R_2. This then establishes the current level for Q_3 and Q_1. When the current I_f is introduced, it flows through R_2, causing the potential of point X to increase. This in turn tends to increase the potential of the emitter of Q_3. But since the potential of the base of Q_3 is fixed, the effect is to decrease the current through Q_3. This decrease reduces the drop across R, driving the potential V_o in the positive direction, which is what we had set out to establish.

Finally, it should be observed that complementary transistors *npn* and *pnp* can be used for level shifting. Applying this concept to the circuit of Fig. 3-14 leads to the circuit shown in Fig. 3-21.

The first stage, comprised of Q_1 and Q_2, is identical to that shown in Fig. 3-14. The second stage, however, uses *pnp* transistors instead of *npn* types. Let us see the effect of this modification. Recall that the collectors of Q_1 and Q_2 are at

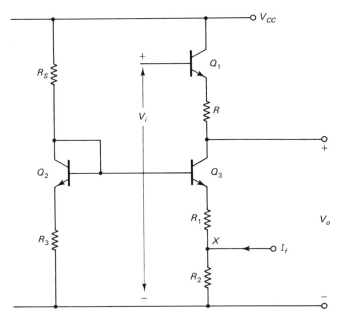

Fig. 3-20 Emitter-follower–current-source circuit modified to allow current feedback.

+10 V. Hence the bases of Q_3 and Q_4 are also at +10 V. The emitters of Q_3 and Q_4 must, therefore, be at a potential of +10.7 V. Hence the drop across the emitter resistor must be 15 − 10.7, or 4.3 V. The current in the emitter resistor must therefore be 4.3/2.15 K = 2 mA, setting the current in Q_3 and Q_4 at 1 mA in each transistor. The collector voltages must therefore be 5 V above ground. Note the improvement in operating point. Also, observe that the output voltage level of 5 V is the same as the level of the inputs to the bases of Q_1 and Q_2.

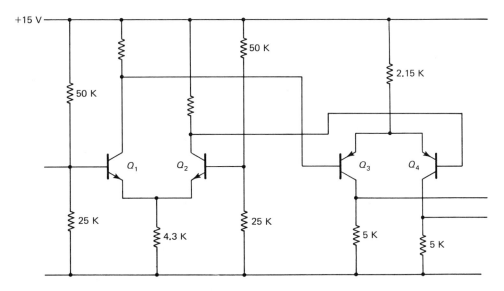

Fig. 3-21 Level shifting by the use of complementary transistors.

3-8 SPECIAL INTEGRATED-CIRCUIT VARIATIONS OF DIFFERENTIAL AMPLIFIER CIRCUITS

The input circuits used in integrated-circuit operational amplifiers almost invariably consist of differential amplifiers. While these amplifiers often include features that we considered, such as current-source transistors in the emitter circuit of the differential pair, there are other circuit arrangements that merit our attention. Indeed, when we first encounter the circuit diagrams of some IC devices, we may be quite perplexed by some of the circuit configurations that we find.

3-8.1 Differential cascode connection

One such circuit is shown in Fig. 3.22. In this circuit Q_1 and Q_2 form the differential pair in the usual fashion. The current source for the emitter circuit is symbolized by I_1 as a convenience to avoid cluttering the diagram by showing the details of the two-transistor current mirror.

The new concept embodied in the circuit shown in Fig. 3-22 is the *cascode connection*. This name is given to the arrangement in which a common-emitter input transistor drives a common-base output transistor. Here Q_1 and Q_3 form such a pair, as do Q_2 and Q_4. It may seem odd to consider Q_3 and Q_4 to be connected in a common-base configuration, but this claim can be verified by referring to Fig. 3-23.

Each transistor in Fig. 3-23 is replaced by its common-emitter equivalent circuit. (*Remember:* Any form of equivalent circuit may be used, provided that the external elements are correctly connected). The emitter terminals, e_3 and e_4, are of course driven by the collectors of Q_1 and Q_2, respectively, although these transistors are not shown. Strictly speaking, current source I_2 should be omitted since

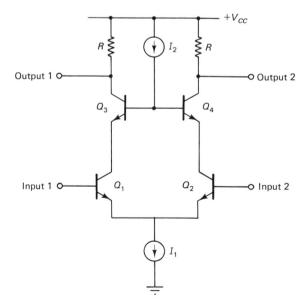

Fig. 3-22 Differential amplifier using cascode connection.

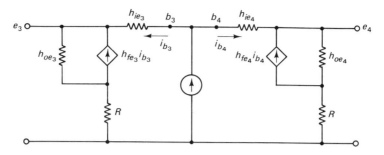

Fig. 3-23 Equivalent circuit for Q_3 and Q_4.

it is a *constant* bias source and as such has no place in a *small-signal* diagram. However, it is included here to emphasize the point that its removal leaves only a very high resistance connecting the b_3-b_4 node to ground. For that reason, $i_{b4} = -i_{b3}$. Therefore, the b_3-b_4 node is at a potential with respect to ground that is exactly halfway between the potentials of e_3 and e_4. Since e_3 and e_4 are exactly equal in magnitude and opposite in phase because of the symmetry of the differential pair Q_1 and Q_2, the potential of the e_3-e_4 node *never varies*. Thus the bases of Q_3 and Q_4 always remain at the same potential; thus these transistors are, in effect, in the common-base configuration.

The collectors of Q_1 and Q_2 see the common-base input resistances of Q_3 and Q_4, respectively. These resistances are the h_{ib} values, which are equal to $h_{ie}/(1 + h_{fe})$. Thus for a transistor that has $I_c = 100$ μA, $h_{ib} \approx (0.026/0.1 \times 10^{-3})$ $= 260$ Ω. This fairly low value of collector-load resistance results in a minimization of the effects of the internal capacitances of Q_1 and Q_2.

Since the collector currents of Q_1 and Q_3 are very nearly equal, as are those of Q_2 and Q_4, the signal voltages developed across resistors R have very nearly the same values they would have were these resistors connected directly to the collectors of Q_1 and Q_2. Thus we get the voltage gain of the standard differential amplifier connection but with improved high-frequency performance. It is for this reason that single-ended cascode circuits are used in "front ends" of television sets.

3-8.2 Differential emitter-follower

The circuit of Fig. 3-24 consists of a common-collector (or emitter-follower) differential pair, Q_1 and Q_2. Each of these transistors drives an emitter load that comprises a common-base-connected transistor with collector load resistor R. The common-base transistors are Q_3 and Q_4; they operate in the same manner as Q_3 and Q_4 of Fig. 3-22, which we have already considered.

If the input terminals of the circuit of Fig. 3-24 are both connected to signal ground—that is, a nonvarying voltage of the proper level—the base-emitter voltages of Q_1 and Q_2 will be equal. As a consequence, the base-emitter voltages of Q_3 and Q_4 will also be equal. Since Q_3 and Q_4 are very closely matched, their quiescent base currents must therefore be equal. The sum of these base currents must equal the current passed by the current source I. In fact, even if these base currents become unbalanced, their sum must still equal I; that is, if the base current

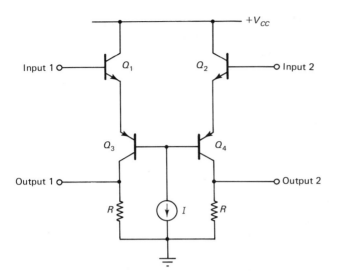

Fig. 3-24 Emitter-follower differential pair driving common-base load pair.

of Q_3 increases by ΔI, the base current of Q_4 must change by $-\Delta I$. The emitter currents of the left-hand transistors must then change by $h_{fe}\Delta I$ and the emitter currents of the right-hand transistors must change by $-h_{fe}\Delta I$. This behavior is exactly like that of conventional differential amplifiers.

The special feature of the circuit of Fig. 3-24 is that the emitter-follower configuration results in high input resistance. As expected, however, this circuit does not produce much voltage gain.

A more exact idea of the performance of the common-base differential pair can be formed by pursuing the following line of reasoning. (This argument closely parallels the one we used earlier in connection with common-emitter differential pairs). Suppose that one emitter input terminal is grounded and a signal voltage is applied to the other, as shown in Fig. 3-25.

In Figure 3-25 the left-hand section represents transistor Q_3 of Fig. 3-24 driving Q_4, which has its emitter grounded. Thus we have a common-base transistor driving a common-emitter transistor.

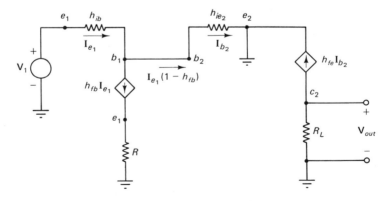

Fig. 3-25 Common-base differential pair equivalent circuit.

Integrated Circuits for Analog Applications   Chap. 3

Note that $I_{b_4} = I_{e_3}(1 - h_{fb})$. The voltage V_1 equals $h_{ib}I_{e_3} + (1 - h_{fb})I_{e_3} h_{ie}$ and the output voltage is equal to $-h_{fe}I_{b_4}R_L = -h_{fe}(1 - h_{fb})I_{e_3}R_L$. Now since $h_{fb} = h_{fe}/(1 + h_{fe})$, the term $1 - h_{fb}$ is equal to $1/(1 + h_{fe})$. We also note that $h_{ie} = (1 + h_{fe})h_{ib}$. Using these facts we can write

$$V_1 = \left[h_{ib} + \frac{(1 + h_{fe})h_{ib}}{1 + h_{fe}} \right]I_{e_3} = 2h_{ib}I_{e_3}$$

We can write V_{out} as $V_{out} = -h_{fe}[I_{e_3}/(1 + h_{fe})]R_L$. Putting these terms together, we obtain

$$A_v = \frac{V_{out}}{V_1} = \frac{-h_{fe}R_L}{(1 + h_{fe})2h_{ib}} \approx -\frac{R_L}{2h_{ib}}$$

We shall have occasion later to need an expression for the output current in terms of V_1. Since, as we have just learned, V_{out} is given by a current times the resistance of R_L, we observe that this current is $-h_{fe}I_{e_3}/(1 + h_{fe})$. This is approximately equal just to $-I_{e_3}$. But in terms of V_1, I_{e_3} is $V_1/2h_{ib}$. Hence the output current I_o is about equal to $-V_1/2h_{ib}$.

We should recall that the relationship between h_{ie} and h_{ib} is given by $h_{ib} = h_{ie}/(1 + h_{fe})$. Making use of this fact, we see that the expression for A_v is approximately $A_v = h_{fe}R_L/2h_{ie}$, whereas the output current expression becomes $I_o \approx -V_1 h_{fe}/2h_{ie}$. These are the same expressions that we obtained for the standard differential amplifier.

3-8.3 Differential amplifier with dynamic loads

In the circuit of Fig. 3-26, the roles of Q_1 and Q_2 are precisely those with which we are familiar from our earlier discussion. The actions of Q_3 and Q_4, however, are not obvious from inspection.

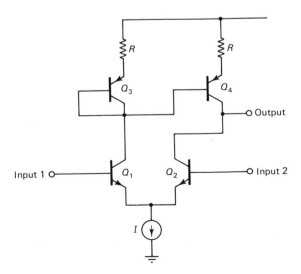

Fig. 3-26 Differential pair with dynamic load.

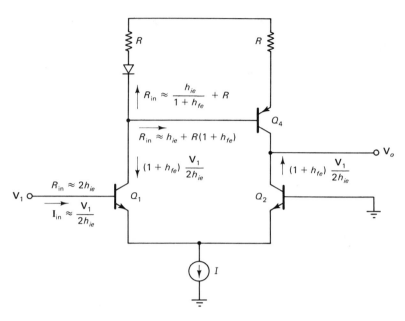

Fig. 3-27 Circuit for approximate analysis.

Perhaps the easiest way to understand how this circuit works is through an approximate analysis. For this purpose, Fig. 3-26 is redrawn with Q_3 represented as an equivalent diode and with certain input resistances and currents shown. The resulting circuit is shown in Fig. 3-27. All transistors are assumed to have equal parameters.

Since we are concerned with the differential-mode voltage gain, Q_2 is shown with its base at signal ground. You should verify that the approximate signal currents and resistances shown on the diagram are correct.

Note that the output voltage $\mathbf{V}_o$ is taken from the node that connects the collectors of Q_2 and Q_4. This means that the Thevenin resistance at this point involves the parallel connection of h_{ob} (for Q_2) and h_{oe} (for Q_4). Of course, h_{oe} must be viewed as tied to the emitter of Q_4, not to ground. Although h_{oe} is much larger than h_{ob}, we can simplify our calculations by neglecting h_{oe} for the moment. Focusing attention solely on Q_4, we can construct the equivalent circuit shown in Fig. 3-28.

Referring to Fig. 3-27, we apply the current divider rule to the collector node of Q_1. This allows us to determine that the base current of Q_4 is given by

$$\mathbf{I}_b = -(1 + h_{fe}) \frac{\mathbf{V}_1}{2h_{ie}} \left[\frac{h_{ie}/(1 + h_{fe}) + R}{h_{ie}/(1 + h_{fe}) + R + h_{ie} + R(1 + h_{fe})} \right].$$

Turning now to the collector node of Fig. 3-28, we see that $\mathbf{I}_x$ must equal $(1 + h_{fe})\mathbf{V}_1/2h_{ie} - h_{fe}\mathbf{I}_b$. Before proceeding further, we simplify the expression

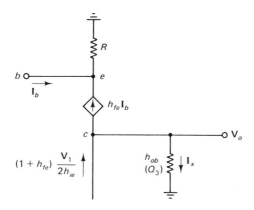

Fig. 3-28 Equivalent circuit for Q_4.

for $\mathbf{I}_b$ to obtain

$$\mathbf{I}_b = -(1 + h_{fe}) \frac{\mathbf{V}_1}{2h_{ie}} \frac{[h_{ie}/(1 + h_{fe}) + R]}{[h_{ie}/(1 + h_{fe}) + R]} \left[\frac{1}{1 + (1 + h_{fe})} \right].$$

Since $(1 + h_{fe}) >> 1$, we obtain the somewhat surprising result that $\mathbf{I}_b \approx -\mathbf{V}_1/(2h_{ie})$. Therefore, $\mathbf{I}_x \approx (1 + h_{fe})\mathbf{V}_1/(2h_{ie}) + h_{fe}\mathbf{V}_1/(2h_{ie})$ or $\mathbf{I}_x \approx h_{fe}\mathbf{V}_1/h_{ie}$. The output voltage must therefore be

$$\mathbf{V}_o \approx (1/h_{ob})h_{fe} \cdot \mathbf{V}_1/h_{ie}.$$

The voltage gain is thus given by

$$A_v \approx \mathbf{V}_o/\mathbf{V}_1 = h_{fe}/(h_{ob} \, h_{ie}). \tag{3-11}$$

If we assume that $h_{fe} = 150$, $h_{oe} = 10^{-5}$ mho, and the DC emitter currents are 100 μA, then h_{ie} must equal $150 \times 0.026/10^{-4}$ or 39 KΩ. The value of h_{ob} must be about $10^{-5}/150$ or 6.7×10^{-8} mho. Thus, the value of A_v must be about $150/(6.7 \times 10^{-8} \times 3.9 \times 10^4)$ or 57,400.

This very high value of A_v is not altogether realistic. For one thing, we have neglected the effect of h_{oe} contributed by Q_4. At first glance, we might suppose that h_{oe} (an admittance) should have been added to h_{ob}. However, the presence of emitter resistor R requires that we use the output resistance of Q_4 in our calculation. Using the methods discussed in Chapter 1, you can show that this output resistance r_o is given by

$$r_o = (1/h_{oe}) \left[1 + h_{fe} \frac{R}{R + h_{ie}} \right] + h_{ie} \| R. \tag{3-12}$$

Taking r_o into account, we have

$$A_v \approx \left(\frac{1}{h_{ob}} \| r_o \right) \frac{h_{fe}}{h_{ie}}.$$

Assuming $R = h_{ie}$, we calculate that $r_o = 7.62$ MΩ. A_v is then found to be 7.62 M$\|(1/6.7 \times 10^{-8}) \times 150/39 \times 10^3$ or 19,400.

In typical applications in integrated circuits, this type of differential amplifier is directly coupled to the base of the transistor of the next stage. Thus, there is an effective external load resistance that further reduces the voltage gain. Nevertheless, with a circuit of this type it is possible to obtain values of A_v approaching 10,000.

3-9 OUTPUT STAGES

The output stage of an integrated-circuit amplifier must provide the following features:

1. Level shifting so that the output voltage swings about the ground, or zero, level.
2. Low output resistance.
3. Power output capability.
4. Current limiting to protect the output transistors against an accidental short circuit of the output terminals.

These features are provided by the circuits of Figs. 3-29 and 3-30.

In the circuit of Fig. 3-29, Q_6 and Q_7 are *npn* and *pnp* transistors, respectively. For this reason, Q_6 conducts more heavily when its base-emitter potential increases,

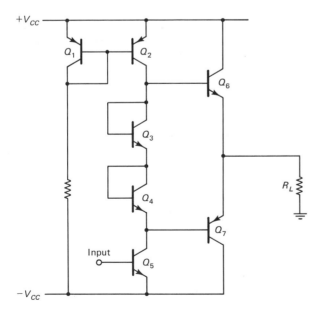

Fig. 3-29 Complementary-symmetry output stage.

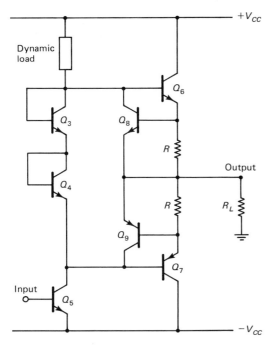

Fig. 3-30 Complementary-symmetry amplifier with current limiting arrangement.

whereas Q_7 conducts more heavily when its base becomes more negative with respect to its emitter. In the zero-signal or quiescent condition, both Q_6 and Q_7 conduct slightly and develop equal collector-emitter voltages so that the voltage across the load, R_L, is very nearly zero. The transistor that drives the bases of Q_6 and Q_7 is Q_5. The collector load for Q_5 is provided by transistor Q_2. Connected in a current-mirror circuit with Q_1, Q_2 acts as a dynamic load. Transistors Q_3 and Q_4 are connected as diodes and provide a voltage offset between the base terminals of Q_5 and Q_6 that closely approximates the sum of the base-emitter voltages of Q_5 and Q_6.

Details of the operation of complementary-symmetry amplifiers of this sort will be discussed in Chapter 10. For the moment we shall merely note that with sinusoidal input, Q_6 and Q_7 conduct on alternate half cycles.

The circuit shown in Fig. 3-30 is a modification of the complementary-symmetry circuit of Fig. 3-29. The modification consists of the addition of two resistors and the transistors Q_8 and Q_9. If the output current were to rise suddenly, as might occur if R_L were accidentally short-circuited, the voltage across one of the resistors marked R would increase. If Q_6 happened to be the output transistor conducting at this moment, then the base-emitter potential of Q_8 would increase, making Q_8 conduct more. Therefore, Q_8 would take base current from Q_6, limiting the output current developed by Q_6.

Both of these circuits are large-signal variants of the basic emitter-follower amplifier. Consequently, while such stages provide current and power amplification, their voltage gains must be no more than 1.

3-10 OPERATIONAL AMPLIFIERS

Perhaps the most common application to integrated-circuit structures of the circuits we have studied is in the operational amplifier. The *operational amplifier*, or op-amp, has many uses as an instrumentation amplifier and as a building block in active filter networks, to name only a few. We shall postpone consideration of applications to a later chapter and concern ourselves here with the internal structure and some of the characteristics of the operational amplifier.

To begin with, we can define an operational amplifier as an amplifier having the following characteristics:

1. The amplifier can amplify DC signals.
2. Input is applied to a pair of differential-input terminals. Signals applied to one input appear at the output without inversion. This input terminal is called the *noninverting* input; the other is called the *inverting* input.
3. Input impedance at either terminal is very high, in the order of 10^5 Ω or higher.
4. Output impedance is low, typically about 100 Ω or lower.
5. Voltage gain is very high. Figures such as 10^4 and 10^5 are common.

3-10.1 The 741 Operational Amplifier

One of the most widely used operational amplifiers is the 741. This amplifier is made by several manufacturers and is listed with various alphabetical and numerical prefixes. The 741 uses both *npn* and *pnp* types. The *pnp* units have low values of h_{fe} (for example, about 4) compared to those of the *npn* transistors. However, the use of both types allows greater flexibility in the circuit design, as we shall see.

Assuming the supply voltages to be $+15$ V and -15 V, we begin the DC analysis of the circuit of Fig. 3-31 by finding the current passed by Q_{11} and Q_{12}. This current, which we denote by $I_{E_{11}}$, is $(30 - 2 \times 0.7)/39$ K $= 733$ μA. To find the current through Q_9 and Q_{10} is unfortunately not as simple. Two factors enter into the determination of this current: the base-emitter voltage of Q_{11} and the ratio of the areas of the base-emitter junctions of Q_{10} and Q_{11}.

The base-emitter voltage of Q_{11}, denoted by $V_{BE_{11}}$, is found from the equation

$$I_{E_{11}} = I_o(Q_{11})(e^{qV_{BE11}/kT} - 1)$$

where $I_o(Q_{11})$ is the saturation current of this *pn* junction. Since the bias voltage $V_{BE_{11}}$ is positive and is more than a few tenths of a volt, we can use the approximate relationship

$$I_{E_{11}} \approx I_o(Q_{11})e^{qV_{BE11}/kT}$$

Solving this equation for $I_o(Q_{11})$, we obtain

$$I_o(Q_{11}) = I_{E_{11}}e^{-qV_{BE11}/kT} \tag{3-13}$$

The saturation currents $I_o(Q_{11})$ and $I_o(Q_{10})$ are unequal because the areas of the base-emitter junctions of transistors Q_{11} and Q_{10} are unequal. In fact, these

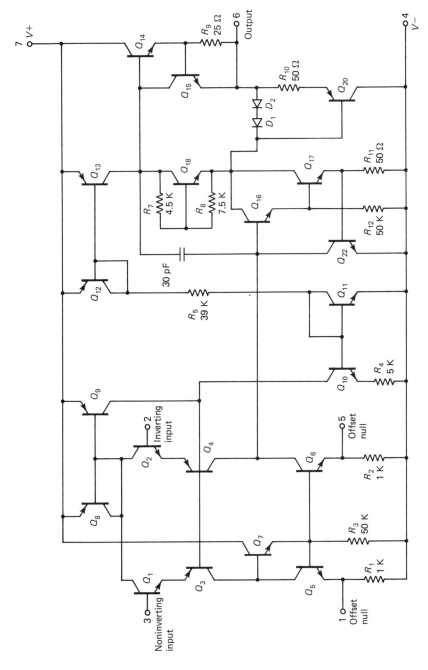

Fig. 3-31 Schematic diagram of 741 operational amplifier.

93

areas—and hence the saturation currents—are in the ratio

$$\frac{I_o(Q_{10})}{I_o(Q_{11})} = 4$$

The current $I_{E_{10}}$ is given by

$$I_{E_{10}} = I_o(Q_{10})e^{qV_{BE10}/kT}$$

Because of the circuit connection, $V_{BE_{10}}$ is given by $V_{BE_{10}} = V_{BE_{11}} - I_{E_{10}}R_4$. Hence

$$I_{E_{10}} = I_o(Q_{10})e^{q(V_{BE11} - I_{E10}R_4)/kT} \tag{3-14}$$

Substituting Eq. (3-13) into Eq. (3-14) and making use of the ratio of saturation currents, we get

$$I_{E_{10}} = 4I_{E_{11}}e^{-qI_{E10}R_4/kT} \tag{3-15}$$

At room temperature $q/kT \approx 40$. We have already found $I_{E_{11}}$ to be equal to 733 μA, and the resistor R_4 has the value 5 KΩ. Substituting these values into Eq. (3-15) yields

$$I_{E_{10}} = 2932e^{-2 \times 10^5 I_{E10}}$$

This equation is transcendental and therefore cannot be solved by a direct analytical procedure, although it yields readily to a trial-and-error approach. A more direct approach is to make use of Prog. 3-1.

```
10   PRINT "CALCULATION OF CURRENT IN 741 MODIFIED CURRENT-MIRROR CIRCUIT B
     Y ITERATION"
20   PRINT "THIS CALCULATION ASSUMES THAT JUNCTION AREAS OF Q10 AND Q11 ARE
     "
25   PRINT "IN THE RATIO OF 4 TO 1."
30   INPUT "ENTER FIRST ESTIMATE OF EMITTER CURRENT OF Q10 IN MICROAMPERES.
     ";X1
40   INPUT "ENTER ACCEPTABLE PERCENTAGE DIFFERENCE BETWEEN SUCCESSIVE CALCU
     LATED VALUES OF IE10.";PCT
50   X2 = X1 + (2920 *  EXP ( - .2 * X1) - X1) / (584 *  EXP ( - .2 * X1) -
     1)
60    IF 100 * ( ABS (X2 - X1) / X2) < = PCT THEN 120
70   X1 = X2
80    GOTO 50
120   PRINT "EMITTER CURRENT OF Q10 EQUALS ";X2
130   END
```

Program 3-1.

The current $I_{E_{10}}$ is thus found to be about 24 μA. About 6 μA of this current comes from the base of Q_3 and Q_4, whereas the remaining 18 μA is provided by the collector of Q_9. Transistors Q_8 and Q_9 are connected in a current-mirror configuration. Because of the dissimilarity of junction areas of these two transistors, the current passed by Q_8 is about 30 μA, so that each leg of the input differential amplifier—Q_1, Q_3, and Q_5 or Q_2, Q_4, and Q_6—passes 15 μA.

Since h_{FE} for transistors Q_1 and Q_2 is about 150, the input current to each base is therefore (15 μA)/150 = 100 nA.

Transistors Q_1, Q_2, Q_3, and Q_4 form an emitter-follower–common-base differential amplifier like the one shown in Fig. 3-24 and discussed in Sec. 3-7.2. Transistors Q_5 and Q_6 act as dynamic loads for Q_3 and Q_4. The circuit is similar to that shown in Fig. 3-26, except that the coupling of the collector of Q_5 to the base of Q_6 is via the emitter follower Q_7. The behavior of this circuit, so far as voltage gain is concerned, is essentially the same as that of the circuit of Fig. 3-26. To compute A_v for this stage we shall make use of Eq. (3-11), modified to include the loading effect of h_{oe} for Q_6 and the input resistance of Q_{16}. The effect of h_{oe} is accounted for by using Eq. (3-12). To find the input resistance of Q_{16} we carry out the following analysis.

The current-mirror circuit involving Q_{11}, Q_{12}, and Q_{13} establishes the current in Q_{13} at 730 μA. This current (except for the base currents of Q_{14} and Q_{20}) is essentially the current that passes through Q_{18} and Q_{17}. Therefore, h_{ie} for Q_{17} must be about 150.26/0.71 = 5.5 KΩ. Seen from the base of Q_{17}, the emitter resistor R_{11} looks like a resistance of 150 × 50 = 7.5 KΩ. Thus the input resistance of Q_{17} is about 13 KΩ.

This input resistance appears as a load in parallel with the emitter-resistor R_{12}. The resistance of this combination is 40 K‖13 K = 10.3 KΩ. The emitter current of Q_{16} is about 710 μA/150, or about 5 μA. Hence, $h_{ie_{16}}$ is about 750 KΩ. The input resistance of Q_{16} is, therefore, approximately 750 K + 160 × 10.3 K = 2.25 MΩ.

The DC emitter current of Q_6 is, as we have seen, about 15 μA. Hence h_{ie} for this transistor must be about $150(0.026/15 \times 10^{-6})$ or 26×10^4 Ω. Taking h_{oe} for these transistors to be 10^{-5} mho, we have, for h_{ob}, $h_{ob} \approx 10^{-5}/150 = 6.67 \times 10^{-8}$ mho. From Eq. (3-12) we get

$$ r_o = \frac{1}{10^{-5}} \left[1 + 150 \left(\frac{1}{1 + 260} \right) \right] + 26 \times 10^4 \| 10^3 $$

or $r_o \approx 10^5(1.57) + 10^3 \approx 158$ KΩ.

Making use of Eq. (3-11) and including the effects of Q_6 and Q_{16} we obtain

$$ A_v \approx \frac{h_{fe}}{h_{ie}} \left[\frac{1}{6.67 \times 10^{-8}} \| 158 \text{ K} \| 2.25 \text{ M} \right] $$

$$ A_v \approx \frac{150}{26 \times 10^4} \times 1.46 \times 10^5 = 84. $$

Transistors Q_{16} and Q_{17} form a Darlington stage whose load consists of the input resistances of Q_{14} and Q_{20} together with the collector resistance of Q_{13}, which acts as a dynamic load. Transistor Q_{18} provides a fixed offset voltage between the bases of Q_{14} and Q_{20} and also couples the collectors of Q_{16}–Q_{17} to the collector of Q_{13}.

The base-emitter voltage of Q_{18} is about 0.7 V, so that the current through R_8 must be around 100 μA. The collector-emitter current of Q_{18} is, therefore, about 730 − 100 = 630 μA. This means that the base current of Q_{18} is around 5 μA and is thus a negligible component of the current in R_7 and R_8. Hence the voltage across R_8 must be about 0.45 V and the collector-emitter voltage of Q_{18} is 0.7 + 0.45 = 1.15 V.

For full turn-on of transistors Q_{14} and Q_{20}, the voltage between their bases would have to be at least 2 × 0.7 = 1.4 V. Thus Q_{18} maintains a voltage between these bases that is somewhat less than the full turn-on value. This causes Q_{14} and Q_{20} to conduct a little at the no-signal condition. This means that only one of these transistors conducts significantly at any time while a signal is applied. Each operates in effect as an emitter follower.

For a load resistance at the output equal to, for instance, 1000 Ω and with a value of h_{fe} for the output transistors assumed equal to 100, we can estimate that the input resistance to Q_{14} or Q_{20} is about $1000 \times 100 = 10^5 \, \Omega$. The voltage gain for the Darlington stage may now be estimated from the equation

$$A_v \approx -\frac{(h_{fe})^2 R_L}{R_{in}} = -\frac{(150)^2 \times 10^5}{2.25 \times 10^6} = 1000$$

The overall gain of this operational amplifier is now seen to be about $84 \times 1000 = 84,000$. This figure falls well within the manufacturer's specification of 50,000 (minimum) and 200,000 (typical).

Two transistors have been ignored until now, Q_{15} and Q_{22}. These transistors provide protection against short circuits that may occur at the output. If the output is connected through a low resistance to the V^- line, excess current will be passed by Q_{14}. As soon as the current exceeds about 25 mA, the voltage drop across R_9 will exceed 0.6 V, causing Q_{15} to begin to conduct. When this occurs, Q_{15} will take base current from Q_{14}, thus turning this transistor off.

If the output is connected through too low a resistance to the V^+ line, excessive current flows through Q_{20}. When this current reaches 25 mA, the voltage across R_{10} becomes 1.25 V. This voltage is in series with the 0.7-V emitter-base voltage of Q_{20}. The total voltage is more than sufficient to turn on diodes D_1 and D_2. When this occurs, current is directed toward the base of Q_{20}, tending to turn that transistor off. More important, however, is the fact that a large part of this current is passed through Q_{17} and R_{11}. When the current through R_{11} reaches about 15 mA, the base-emitter voltage of Q_{22} reaches the turn-on value, so that Q_{22} takes base current from Q_{16}. This results in an amplified shut-down action for Q_{20}. Under normal operating conditions, both Q_{15} and Q_{22} are cut off.

3-10.2 The MC 1556 Operational Amplifier

As a second example of operational amplifier circuitry, we consider the MC 1556, which embodies certain advances beyond the design of the 741. Design objectives for the MC 1556 were to increase the slew rate and to decrease the input bias current. The topic of slew rate is discussed in a later chapter and will not be dealt with here. However, we shall consider the reduction of input bias current, since this reduction results from the use of *super-beta* transistors, which we have not encountered before.

Super beta transistors are produced by a double diffusion process that results in an extremely narrow base region. In general, the narrower the base region of a junction transistor, the higher are the values of h_{FE} and h_{fe}. However, the use of narrow bases requires a severe restriction on the collector-base voltage. This restriction is needed because, with a narrow base, the depletion region due to the reverse bias at the collector-base junction can extend all the way across the base to the emitter region even when the bias is very small. Furthermore, because of the narrow base, the electric-field intensity in the depletion region can become very large so that avalanche breakdown can occur, destroying the transistor. For this reason, it is necessary to operate super beta transistors with collector-base voltages that are almost zero.

The advantage of the super beta transistor is that, because of the high value of h_{FE}, reasonable values of I_C can be obtained, whereas very small values of I_B are required, since $I_B = I_C/h_{FE}$. In the circuit of Fig. 3-32, Q_1 and Q_2 are the super beta transistors. They form the input pair for the differential amplifier that comprises Q_1, Q_2, Q_3, Q_4, Q_6, Q_7, Q_8, and Q_9.

Transistors Q_1 and Q_2 are shunted by Q_3 and Q_4, respectively, together with their associated 1.5-KΩ emitter-resistor. Calculation of the collector currents of Q_{10}, Q_6, and Q_7 are left to Prob. 3-22. Calculation of the input-bias current is left

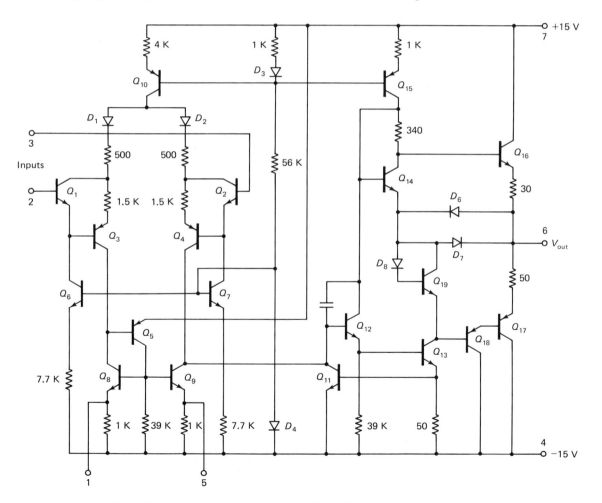

Fig. 3-32 The MC 1556 operational amplifier. (Courtesy Motorola Semiconductor Products, Inc.)

Sec. 3-10   Operational Amplifiers

to Prob. 3-23. Note that the collector-emitter voltage of Q_1 equals the base-emitter voltage of Q_3 plus the small voltage across the 1.5-KΩ resistor.

To estimate the voltage gain of the amplifier, we begin with the output and work back to the input. Output transistors Q_{16} and Q_{17} function much the same as their counterparts in the 741. Hence we treat these transistors as emitter-followers and estimate their gain as 1.

Going back from Q_{16} and Q_{17}, we are at first perplexed about which transistors to consider next. We can resolve the puzzle by noting first that Q_{14} normally operates in saturation; that is, V_{CE} for this device is nearly constant and equal to about 530 mV. Consequently, for signal purposes, Q_{14} acts as a short circuit. Similarly, D_5 and Q_{19} form a composite diode, in effect, and also appear to time-varying signals as a short circuit. This means that the collector of Q_{13} sees the 340-Ω resistor and the collector of Q_{15}, the base of Q_{16}, and the base of Q_{18} as its loads.

Recall that Q_{16} is active when the output voltage swings positive with respect to ground, whereas, Q_{17} is active on negative swings of output voltage. Therefore, the collector of Q_{13} *always* sees the 340-Ω resistor plus the collector of Q_{15} but, depending upon the polarity of the output voltage, it sees *either* the base of Q_{16} *or* the base of the cascade-connected transistor Q_{18}. (Clearly, Q_{18} is an emitter-follower driving the output transistor Q_{17}.) The resistance looking into the collector of Q_{15} is very large. Hence, it is reasonable to neglect it in comparison to the base resistance of Q_{16} or of Q_{18}.

Thus we are left with Q_{13}, Q_{16}, Q_{17}, and Q_{18}. The last three, as we have seen, are essentially emitter-followers and, therefore, it appears that only Q_{13} provides gain greater than 1. In turn, Q_{13} is driven by the emitter-follower Q_{12}, whose base is driven by the first stage. Transistor Q_{11} is nearly cut off in normal operation. Its function is to provide short-circuit protection, as we shall see later.

To find the second-stage gain, we need only to find the gain provided by Q_{13}. To do so, we need to know the values of resistance looking into the bases of Q_{16} and Q_{18}. The value of h_{fe} for Q_{16} is about 75. The composite value of h_{fe} for Q_{17} and Q_{18} taken together is also 75. Therefore, for a load resistor of 2 KΩ, the resistance at the bases of Q_{16} or Q_{18} is approximately $75 \times 2000 = 150$ KΩ. Since the DC operating level of collector current for Q_{13} is 500 μA (see Prob. 3-19), h_{ie} for this transistor is equal to $h_{fe} \times 26/0.5 = 52h_{fe}$. Therefore, A_v for this stage equals.

$$h_{fe} \times 150 \text{ K}/(h_{fe}R_E + 52h_{fe}) \approx (150 \text{ K})/(50 + 52) = 1470$$

To find the gain of the first stage, we need to know what load it drives. This load must certainly be equal to the resistance looking into the base of Q_{12}. To find this resistance, we start with the base of Q_{13}. For $h_{fe} = 100$, the resistance looking into the base of Q_{13} must be approximately $h_{fe}(50 + 26/0.5) = 100 \times 102 \approx 10$ KΩ. In parallel with the 39-KΩ emitter-resistor of Q_{12}, this yields (10 K)$\|$ (39 K) ≈ 8 KΩ. Assuming that h_{fe} for Q_{12} equals 75, the input resistance to the base of Q_{12} must be about 75×8 K $= 600$ KΩ. This is the load seen by the first stage.

The first stage consists of Q_1 and Q_2 acting as differential emitter-followers driving the common-emitter differential pair Q_3 and Q_4. Transistors Q_5 and Q_7 are simply current sources for Q_1 and Q_2. Transistors Q_8 and Q_9 are dynamic loads for Q_3 and Q_4. From Chap. 1, the conductance looking into the collector of either Q_8 or Q_9 is given by $g_o = h_{ob} + h_{rb}/(R_E + h_{ib})$. Since Q_9 is passing a quiescent emitter current of about 50 μA, h_{ib} equals $26/0.05 = 520$ Ω. Assuming $h_{ob} = 0.5 \times 10^{-6}$ and $h_{rb} = 10^{-4}$, then $g_o = 0.5 \times 10^{-6} + 10^{-4}/(1000 + 520) = 5.66 \times 10^{-7}$ mho. Hence the resistance looking into the collector of Q_9 is $1/5.66 \times 10^{-7} = 1.77$ MΩ. This resistance is in parallel with the 600 KΩ seen at the base of Q_{12}. Hence the total load resistance seen by the collector of Q_4 is (1.77 M)∥(600 K) = 448 KΩ.

Finally, we compute the gain of the Q_3–Q_4 differential amplifier, first noting that the dynamic resistance of diode D_2 is about 400 Ω. Assuming h_{fe} for Q_4 is 100 and knowing that the quiescent collector current is 50 μA, we have $h_{ie} \approx 100 \times 26/0.05 = 52$ KΩ. The gain of the Q_3–Q_4 differential pair is therefore equal to $h_{fe} \times (448 \text{ K})/2[h_{fe}(500 + 400) + 52 \text{ K}]$, or $A_v = (448 \text{ K})/[2(900 + 520)] = (448 \text{ K})/2840 = 158$. The overall gain must equal this figure times the gain of the second stage, or $158 \times 1470 = 232,260$, or 107 db, which compares well with the nominal value of 106 db.

If the output is short-circuited accidentally, Q_{17} would not be affected, but Q_{16} might very well be destroyed. However, the voltage drop across the 30-Ω resistor in the emitter of Q_{16} will cause diode D_7 to conduct. This increases the current passed by Q_{14} (which is drawn from the constant-current source Q_{15}), thus taking current from the base of Q_{16}. This tends to turn off Q_{16}, thereby protecting it.

Similarly, diode D_6 conducts if an effort is made to push excessive current into the output terminal. This occurs when an excessive voltage drop appears across the 50-Ω resistor in the emitter of Q_{17}. At the same time, when Q_{13} is called upon under this condition to pass more current, the increased drop across its 50-Ω emitter-resistor turns on Q_{11}, which then takes the base current from Q_{12}, turning off Q_{12}, Q_{13}, Q_{18}, and Q_{17}.

REFERENCES

1. Jacob Millman, *Microelectronics: Digital and Analog Circuits and Systems* (New York: McGraw-Hill Book Company, 1979).
2. L. J. Giacoletto, *Differential Amplifiers* (New York: Wiley-Interscience, a division of John Wiley & Sons, Inc., 1970).
3. Adel S. Sedra and Kenneth C. Smith, *Microelectronic Circuits* (New York: Holt, Rinehart and Winston, 1982).
4. R. J. Widlar, "Some circuit design techniques for linear integrated circuits," *IEEE Transactions on Circuit Theory*, CT-12(1965):586–90.
5. R. J. Widlar, "Design techniques for monolithic operational amplifiers," *IEEE Journal of Solid-State Circuits*, SC-9(1969):184–91.
6. Douglas J. Hamilton and William G. Howard, *Basic Integrated Circuit Engineering* (New York: McGraw-Hill Book Company, 1975).

PROBLEMS

3-1.

  (a) With $v_s = 0$ (short-circuited input) find the DC values of I_B, I_E, and V_{CE}.
  (b) Find the differential-mode gain.
  (c) Find the CMRR.

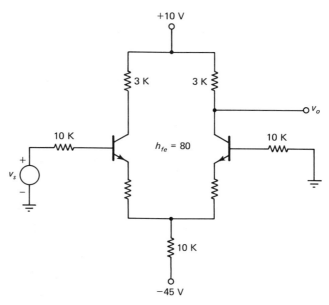

3-2.

  (a) Find the quiescent current in R_E.
  (b) Find the differential-mode gain, common-mode gain, and the CMRR.

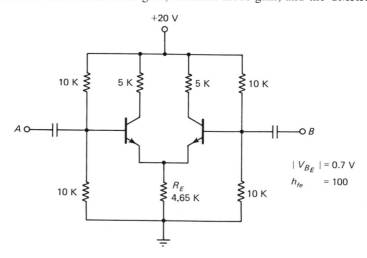

3-3.

  (a) Find quiescent values of I_C for Q_1, Q_2, Q_3, and Q_4.
  (b) Find A_{vd}. Assume $V_{BE} = 0.7$ V.

(c) Find A_{vcm} if $h_{FE} = h_{fe} = 100$.

(d) Find CMRR if $1/h_{ob} = 2 \times 10^6$. Neglect base currents.

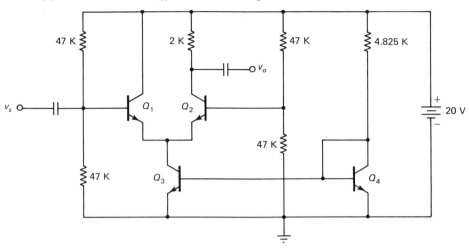

3-4.

    (a) Find the DC operating level of v_o.

    (b) Find A_{vd}, A_{vcm}, and CMRR.

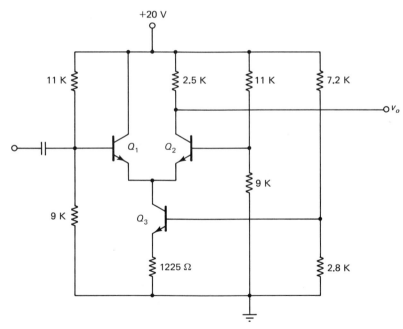

3-5. You are given that $I_C = 1$ mA for each transistor when A and B are both grounded.

    (a) Find the required value of R_E.

    (b) Find the DC voltage to ground at the collectors.

    (c) Find the input resistance from A to ground.

    (d) Find the differential-mode gain from A to C.

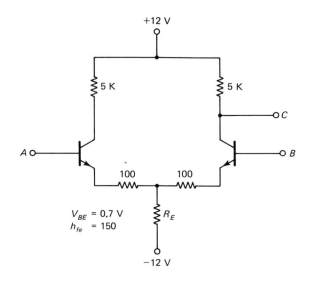

+12 V

5 K   5 K

A   C

100   100   B

V_{BE} = 0.7 V R_E
h_{fe} = 150

−12 V

3-6. Find I_C and V_{CE} for Q_2 when $V_A = V_B = 0$. $h_{fe} = 100$, and $h_{oe} = 10^{-5}$ mho.

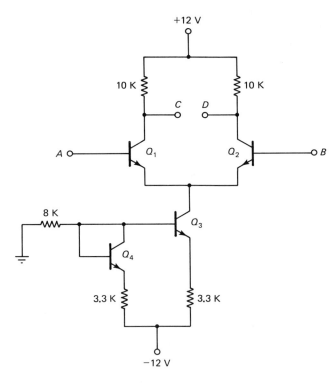

+12 V

10 K   10 K

C   D

A Q_1 Q_2 B

8 K Q_3

Q_4

3.3 K   3.3 K

−12 V

3-7. For the circuit of Prob. 3-6, find the values of V_C, V_D, and $V_D - V_C$ when V_A = 1.000 V and V_B = 1.001 V.

3-8.

(a) Find R so that the DC level of the output voltage will be 0 V.

(b) Find the collector voltages with respect to ground of Q_1, Q_2, Q_3, and Q_4.

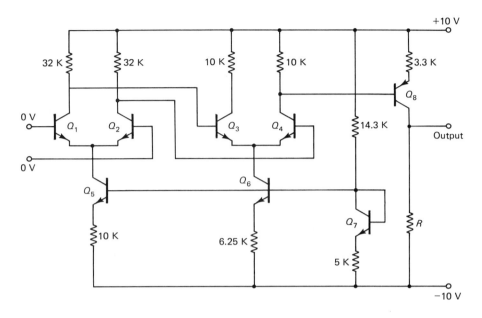

3-9.

Find R_c so that the output voltage level will be 0 V.

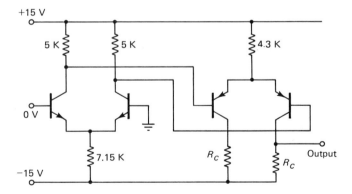

3-10.

The output voltage level is 0 V. Q_3 and Q_4 are operated as Zener diodes. The Zener voltage is 10 V. Find the required values of R to make the output level equal to 0 V. Neglect base currents.

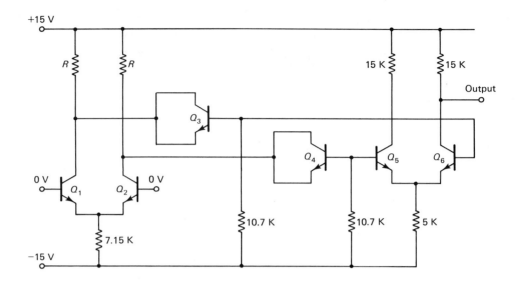

3-11. The bases of Q_1 and Q_2 are at the same potential. Find V_{out}.

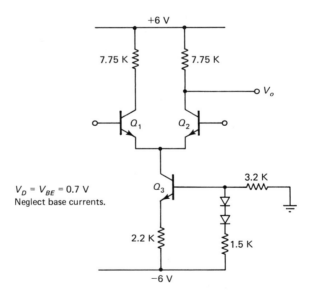

$V_D = V_{BE} = 0.7$ V
Neglect base currents.

3-12. Find A_{vd} for the circuit shown. For Q_3 and Q_4, $h_{fe} = 50$.
For Q_1 and Q_2, $h_{fe} = 100$.

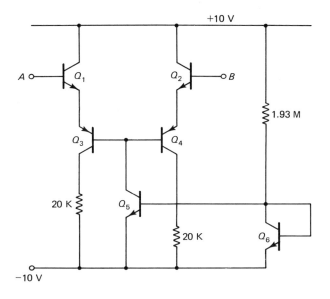

3-13. Find A_{vd} for the circuit shown.

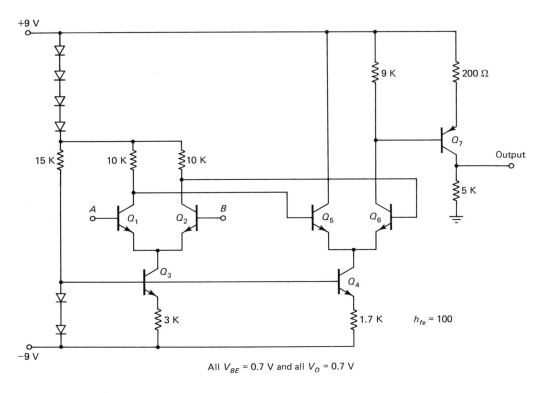

All $V_{BE} = 0.7$ V and all $V_D = 0.7$ V

3-14. The schematic circuit diagram of an integrated-circuit audio amplifier is shown. The 15-K and 16-Ω resistors and the 500-μF capacitors are connected externally. Assume that in the quiescent state, V_{in} = 2.6 V (DC) and the voltage at pin 6 is 10 V (DC). Assuming that the base currents of Q_5 and Q_7 are negligible, estimate the value of the internal resistor R.

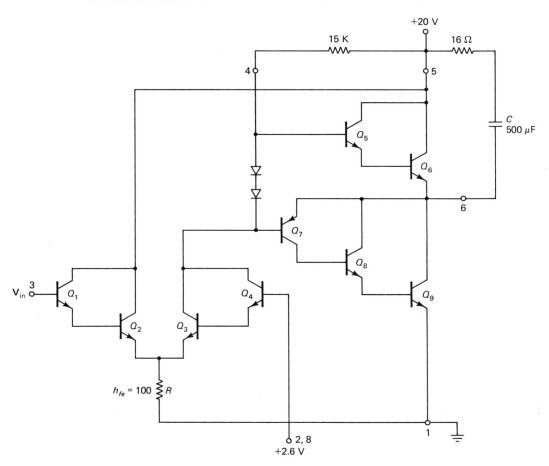

3-15. For the circuit of Prob. 3-14, consider that the output stage comprised of Q_5, Q_6, Q_7, Q_8, and Q_9 looks like a single transistor that has an effective h_{fe} equal to $(20)^2$ and is connected as an emitter-follower. Assume that the other transistors have h_{fe} = 50, and estimate A_v between pin 3 and pin 6. Assume X_c = 0.

3-16. For the operational amplifier circuit shown, assume that the input terminals are at 0 V and calculate the quiescent voltage at the base of Q_{10}. Assume Q_1, Q_2, Q_3, Q_4, Q_7, Q_8, and Q_9 are identical *npn* transistors, while Q_5 and Q_6 are identical *pnp* types.

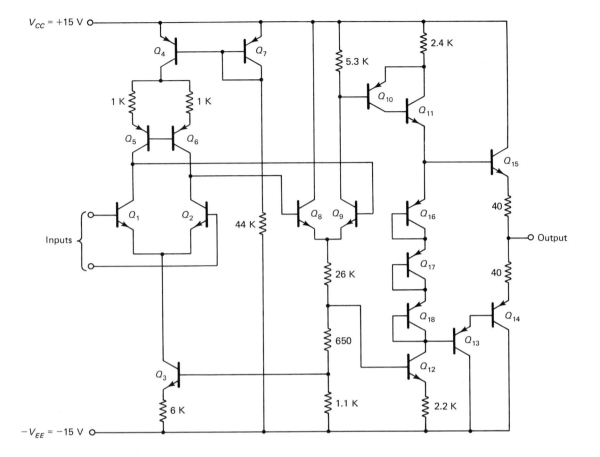

3-17. For the circuit of Prob. 3-16, assuming that the emitter-base junctions of Q_{13}, Q_{14}, Q_{15}, Q_{16}, Q_{17}, and Q_{18} are all identical, find the quiescent current flowing through Q_{14} and Q_{15}.

3-18. In the circuit of Prob. 3-16, with one input held at 0 V and an input signal applied to the other, calculate the gain of this amplifier. Assume $h_{fe} = 150$ for Q_1 and Q_2, $h_{fe} = 100$ for Q_8 and Q_9, and $h_{fe} = 50$ for Q_{10}. Assume also that the load seen by the collector of Q_{10} is equal to 200 KΩ and that the circuit beyond Q_{10} has a gain of one.

3-19. A direct-coupled, DC amplifier has an input signal of 100 mV and a voltage gain of 32 db. If the input signal drifts by $+20$ ppm/h, what is the shift in the output voltage after 10 h?

3-20. (a) Find the value of R that will set the value of I_{E2} to 200 μA. Assume $V_{BE} = 0.7$ V.
(b) Assuming $h_{fe} - 200$ and $h_{oe} = 5 \times 10^{-5}$ mho for all transistors, find $A_v = v_{out}/v_{in}$.

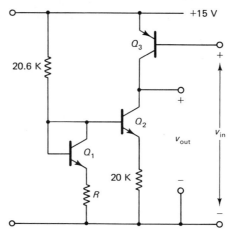

3-21. (a) With $v_1 = 0$ (short circuit) find V_o (the quiescent value of v_o) and V_{DS}, neglecting R_L.

(b) Find A_v.

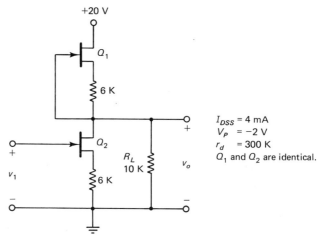

$I_{DSS} = 4$ mA
$V_P = -2$ V
$r_d = 300$ K
Q_1 and Q_2 are identical.

3-22. The reverse saturation current of the base-emitter junctions of Q_6 and Q_7 is 6.137 $\times 10^{-16}$ A.

(a) Find the emitter currents of Q_6 and Q_7. (Do not assume $V_{EE} = 0.7$ V for Q_6 and Q_7. However, that value may be used for all other transistors.)

(b) Neglecting base currents, find collector (and emitter) currents of Q_1, Q_2, Q_3, Q_4, Q_8, and Q_9.

(c) Find V_{CE} for Q_1 and Q_2.

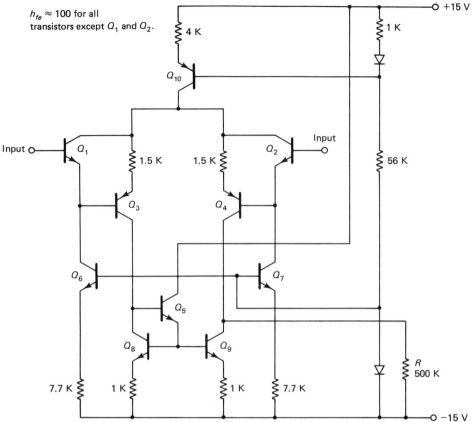

$h_{fe} \approx 100$ for all transistors except Q_1 and Q_2.

3-23. The transistors Q_1 and Q_2 of Prob. 3-22 are super beta transistors and have h_{FE} = 1500. Transistors Q_6 and Q_7 are current sources that establish the collector currents for Q_1 and Q_2 at 12 μA. The emitter-follower differential pair, Q_1 and Q_2, drive a common-emitter differential pair, Q_3 and Q_4. Also, Q_8 and Q_9 are dynamic loads for the collectors of Q_3 and Q_4.

    (a) Assuming $1/h_{ob}$ for Q_8 and Q_9 equals 2 MΩ and $h_{rb} = 3 \times 10^{-4}$, estimate A_v for this amplifier.

    (b) What is the quiescent input current to the bases of Q_1 and Q_2?

3-24.

    (a) Estimate the resistance seen by the collector of Q_2.

    (b) Calculate A_v for the circuit.

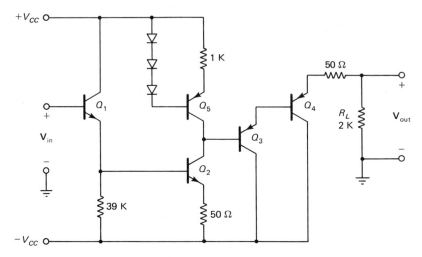

3-25.

    (a) Find the collector current of transistors Q_9, Q_5, and Q_6.

    (b) What is the approximate emitter current of Q_1 and Q_4?

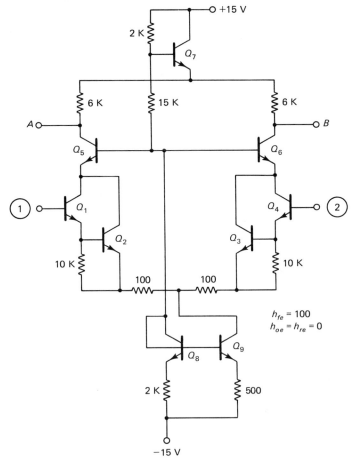

                                 Integrated Circuits for Analog Applications   Chap. 3

3-26. For the circuit of Prob. 3-25 estimate the voltage gain if the output is taken *between* the two terminals A and B.

3-27. Assume that the output voltage is 0 V. Also assume that all base currents are negligible (high h_{FE}) and that all $V_{BE} = 0.7$ V.
(a) Find the collector current of Q_5.
(b) Find the voltages at the collectors of Q_1 and Q_2.
(c) Determine the required value of R, the collector resistors of Q_1 and Q_2.

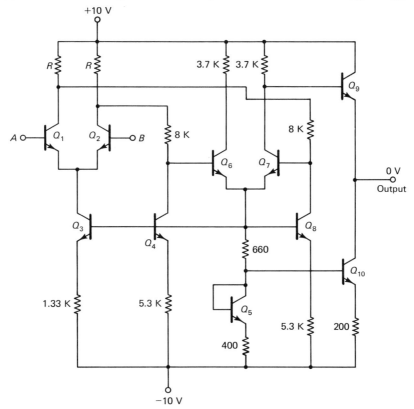

3-28. Estimate A_v between the differential inputs to Q_6 and Q_7 and the output terminal for the circuit of Prob. 3-27. Assume $h_{fe} = 200$ for all transistors.

3-29. Using the results of Prob. 3-27(c), estimate A_v between differential input terminals A and B and the bases of Q_6 and Q_7. Using this result and the result of Prob. 3-28, find A_v overall.

3-30. For the circuit of Prob. 3-27, assume that the bases of Q_6 and Q_7 are connected together so that this stage is driven solely by a common-mode signal. Find the maximum and minimum allowable base-voltage levels. (Limits occur when Q_6 and Q_7 either become saturated or cut off.) Assume that V_{CE} (saturated) $= 0.3$ V. Note that in this mode of operation, the current through Q_5 will not be constant.

3-31. For the circuit of Prob. 3-27 assume that the bases of Q_1 and Q_2 are connected together so that this stage is driven solely by a common-mode signal. In the same manner as in Prob. 3-30, find the maximum and minimum allowable base voltage levels, if possible. If you claim that the determination of one or both conditions is impossible, state why.

3-32.

    (a) Find the dynamic load resistance seen looking into the base-collector terminal of Q_3.

    (b) Making use of the equivalent circuit of Fig. 3-27, find A_v.

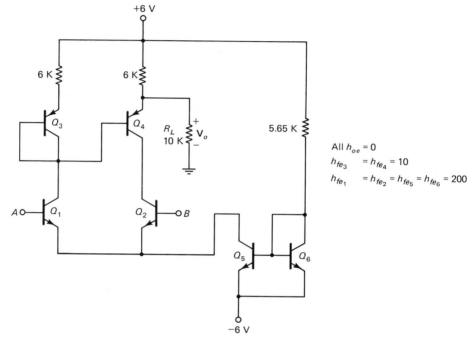

All $h_{oe} = 0$

$h_{fe_3}\;\;\;= h_{fe_4} = 10$

$h_{fe_1}\;\;\;= h_{fe_2} = h_{fe_5} = h_{fe_6} = 200$

3-33. For the MOSFETs in the given circuit, $i_D = K(V_{GS} - V_T)^2$, $g_m = 2\tilde{K}(V_{GS} - V_T)$, $V_T = 3$ V, and $K = 0.3$ mA/V^2.

Find the quiescent value of V_{GS} and V_o.

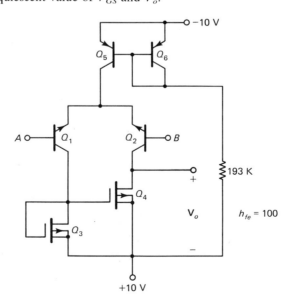

$h_{fe} = 100$

3-34. For the circuit of Prob. 3-33, $r_d = 10$ KΩ and h_{oe} may be neglected. Calculate A_{vd}.

3-35. Estimate the emitter current of Q_6. Assume that the emitter-base junction current of Q_6 is given by $I \approx 10e^{(V_{BE6}/0.026)}$ μA.

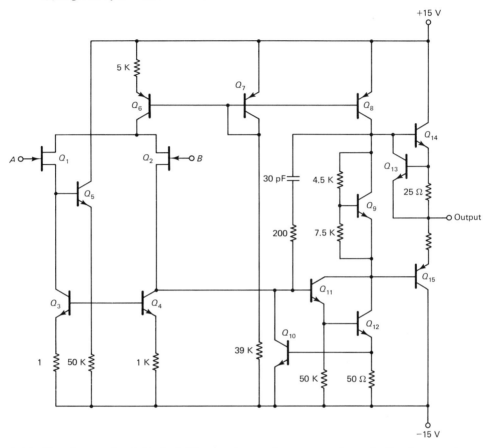

3-36. Use the result of Prob. 3-33 to estimate the current through Q_5. Assume $I_{DSS} = 254$ μA, $V_P = -4$ V. Assume also that base currents may be neglected. Assume input terminals at A and B are grounded.

CHAPTER 4

Feedback Principles

4-1 BASIC CONCEPTS

The term *feedback* is now widely used even in everyday vocabulary. Its original use was probably to describe electronic amplifiers in which some part of the output signal is returned to the input circuit. However, the practice of utilizing an output signal to control the input long predates the beginning of electronics. We have only to think of the governors used on steam engines to recall that fact. Indeed, feedback mechanisms exist in great number in nature itself; the iris of the eye, for example, responds to increased light input by closing. This is an excellent example of negative feedback, since when an increased stimulus produces a nerve signal that is beyond a comfortable level, the system produces a muscular response that reduces the stimulus.

A feedback system, such as an amplifier, may be represented by the block diagram of Fig. 4-1.

Block A represents an amplifier having voltage gain A, while block β represents the feedback network. This diagram has to be studied with certain assumptions in mind: first, that transmission in each block is only in one direction; and second, that the feedback network, β, does not load the output of amplifier A. In practice, if the β structure consists of linear elements like resistors, capacitors, and inductors, there is no reason why a signal at its left-hand terminal pair (or port) should not produce some effect at its right-hand port. Similarly, some current must be drawn from the output of A by the β network connected across it. Nevertheless, to simplify the analysis at this point in the discussion, we shall neglect these two nonideal conditions. It is often possible, in practical cases, to justify doing so.

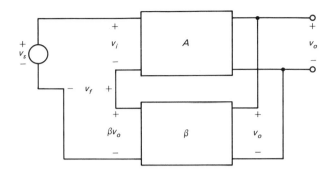

Fig. 4-1 Feedback system.

At the input of A, Kirchhoff's voltage law yields $-v_s + v_i + v_f = 0$, from which we get $v_i = v_s - v_f$.

Since $v_o = Av_i$ and $v_f = \beta v_o$, we can combine these equations to obtain

$$v_o = A[v_s - \beta v_o]$$

The ratio of v_o to v_s represents the voltage gain *with feedback*, designated by A_{v_f}, and is found to be

$$A_{v_f} = \frac{v_o}{v_s} = \frac{A}{1 + A\beta} \tag{4-1}$$

In this example, the feedback is said to be negative, since v_f *is subtracted from* v_s to yield v_i. Notice that the effect of negative feedback is to produce a resultant voltage gain, A_{v_f}, that is smaller than the gain without feedback, A, by a factor of $1/(1 + A\beta)$. This effect can be stated as a general principle: If the gain is reduced by the application of feedback, then the feedback is negative. The opposite case, positive feedback, is possible and has important uses in constructing oscillators, for example.

You should be aware that other textbooks may use different conventions in dealing with feedback, which can alter the form of Eq. (4-1). In a typical voltage amplifier, there is usually an odd number of stages. As a result, v_o is in phase opposition to v_i so that A is a negative number. Some writers prefer to let A take its own sign (usually negative), adjusting the β network connections to provide the negative feedback. This leads to an expression for A_{v_f} involving $1/(1 - A\beta)$. In this book, however, *A is taken as a positive number and phase reversal is accounted for by the connection of the β network*. In any event, one rule is immutable: *Negative feedback reduces voltage gain*.

A variety of other connections to produce feedback are possible: The signal fed back to the input could be a current rather than a voltage. In such a case, current would be fed back by the β network and injected into an input node. We could then apply Kirchhoff's current law to study the summation of input signals to A. Or the output signal could be taken as the load current, so that the right side of the β network would be connected in series with R_L.

As long as we are dealing with block diagrams, it is not too difficult to handle any of these cases. But when practical networks are considered, the mere identification of the type of feedback network can be tricky. For that reason we shall not bother with such an approach, but instead we shall develop an alternate, general method of attacking these problems.

4-1.1 Feedback Applications

Before developing a general method, however, let us see why we should be interested in feedback at all. As we can see from Eq. (4-1), the effect of negative feedback is to reduce the voltage gain, a result that might seem at first glance not at all beneficial. But suppose that A is subject to some variation. (This might occur in various ways in a practical case. For example, in a production run of transistor amplifiers, h_{fe} may vary by as much as 3 to 1, thereby affecting A.) What is the resulting variation in A_{vf}? This is easily found by considering the derivative of A_{vf} with respect to A—that is, dA_{vf}/dA.

You can easily verify that

$$\frac{dA_{vf}}{dA} = \frac{(1 + A\beta) - A\beta}{(1 + A\beta)^2} = \frac{1}{(1 + A\beta)^2}$$

Using the approximation that $\Delta A_{vf}/\Delta A \approx dA_{vf}/dA$ and factoring the right-hand denominator, we get

$$\frac{\Delta A_{vf}}{\Delta A} \approx \frac{1}{(1 + A\beta)(1 + A\beta)}$$

If we multiply numerator and denominator by A, we get

$$\frac{\Delta A_{vf}}{\Delta A} \approx \frac{1}{A} \cdot \frac{A}{1 + A\beta} \cdot \frac{1}{1 + A\beta}$$

Since $A/(1 + A\beta) = A_{vf}$, we can write

$$\frac{\Delta A_{vf}}{\Delta A} = \frac{1}{A} A_{vf} \frac{1}{1 + A\beta}$$

which can be rearranged to yield

$$\frac{\Delta A_{vf}}{A_{vf}} = \frac{\Delta A}{A} \cdot \frac{1}{1 + A\beta} \tag{4-2}$$

Here $\Delta A_{vf}/A_{vf}$ is the *fractional change in A_{vf}*, whereas $\Delta A/A$ is *the fractional change in A*.

Example 4-1

Let us apply some numbers to the results we have obtained thus far in order to see the significance of these results. Suppose $A = 100$, $\beta = 0.1$, and A is subject to a 50% change. Let us find A_{vf} and the fractional change in A_{vf}.

From Eq. (4-1) we get

$$A_{vf} = \frac{100}{1 + 100(0.1)} = \frac{100}{11} = 9.1$$

Using Eq. (4-2) with $\Delta A/A = 0.5$, we get $\Delta A_{vf}/A_{vf} = (0.5)(\frac{1}{11}) = 0.0454$. Expressed as a percent, this is 4.54%. Thus we see that a 50% change in A produces only a 4.5% change in A_{vf}, so that although we have sacrificed absolute gain, we have achieved a great improvement in gain variation.

Next we consider the effect of feedback on bandwidth. Let us start with the case of a low-pass amplifier. Such an amplifier is assumed to pass all frequencies from 0 Hz to the *cutoff frequency* f_c with a gain equal to A_o. At the cutoff frequency, the gain is 3 db below its value at low frequencies. This situation can be represented by the frequency-response curve shown in Fig. 4-2.

Mathematically, we can express the behavior of A by writing

$$A = \frac{A_o}{1 + jf/f_c}$$

When $f = f_c$, the denominator of this expression becomes $1 + j1$, so that the *magnitude* of A becomes $A_o/\sqrt{2}$, which is exactly the correct value for the 3-db point.

When feedback is applied, we use Eq. (4-1) to obtain

$$A_{vf} = \frac{\dfrac{A_o}{1 + jf/f_c}}{1 + \dfrac{A_o}{1 + jf/f_c} \cdot \beta}$$

We can multiply numerator and denominator by $1 + jf/f_c$ to obtain

$$A_{vf} = \frac{A_o}{1 + jf/f_c + A_o\beta}$$

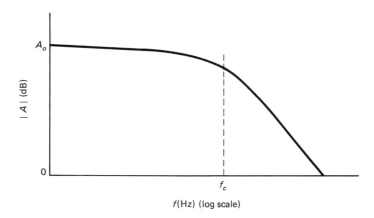

Fig. 4-2 Frequency response of a low-pass amplifier.

This equation can be rearranged to yield

$$A_{vf} = \frac{A_o}{(1 + A_o\beta)\left[1 + j\dfrac{f}{f_c(1 + A_o\beta)}\right]}$$

The factor $A_o/(1 + A_o\beta)$ is simply the gain with feedback at low frequencies. Let us call this factor A_{vfo}. If we define a new frequency, f_h, to be equal to $f_c(1 + A_o\beta)$, we can rewrite the last equation in the form

$$A_{vf} = \frac{A_{vfo}}{1 + jf/f_h} \tag{4-3}$$

Equation (4-3) shows that the effect of feedback is to make the low-frequency gain become $A_{vfo} = A_o/(1 + A_o\beta)$, which agrees with Eq. (4-1), and to cause the cutoff frequency to become equal to $f_h = f_c(1 + A_o\beta)$. In other words, the effect of negative feedback is to reduce the gain and increase the bandwidth, both by the factor $(1 + A_o\beta)$.

To see the effect on a high-pass amplifier, consider Fig. 4-3. The response shown in Fig. 4-3 can be represented mathematically by the expression

$$A = \frac{A_o}{1 + jf_c/f}$$

Considering the effect of negative feedback and proceeding in the same way as in the case of the low-pass amplifier, we arrive at the equation

$$A_{vf} = A_{vfo} \cdot \frac{1}{1 + j\dfrac{f_c}{f(1 + A_o\beta)}}$$

Here A_{vfo} has precisely the same significance as before. But this time we define a frequency $f_l = f_c/(1 + A_o\beta)$. The equation above then becomes

$$A_{vf} = \frac{A_{vfo}}{1 + jf_l/f} \tag{4-4}$$

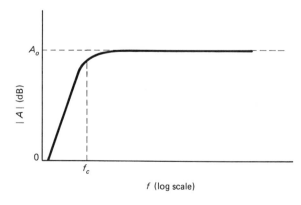

Fig. 4-3 Frequency response of a high-pass amplifier.

Feedback Principles   Chap. 4

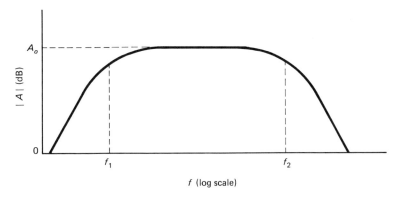

Fig. 4-4 Frequency response of a band-pass amplifier.

We can now conclude that the effect of negative feedback in this case is to reduce the gain by a factor of $1 + A_o\beta$, as before, and to lower the cutoff frequency by the same factor. Of course, for a high-pass amplifier, lowering the cutoff frequency widens the bandwidth so that for either amplifier we can say that gain is reduced and bandwidth increased.

For a band-pass amplifier which has both upper and lower cutoff frequencies, the response curve is as shown in Fig. 4-4. The behavior of this amplifier for frequencies in the vicinity of f_1 is described by Eq. (4-4), whereas for frequencies in the vicinity of f_2, Eq. (4-3) applies. The overall effect is perhaps best understood by considering a numerical example.

Example 4-2

Suppose $A_o = 100$, $\beta = 0.1$, $f_1 = 200$ Hz and $f_2 = 20$ KHz. Let us find f_l and f_h.

$$f_l = \frac{200}{1 + 100 \times 0.1} = \frac{200}{11} = 18.2 \text{ Hz}$$

$$f_h = 20 \times 10^3 \times (1 + 100.1) = 20 \times 11 \times 10^3 = 220 \text{ KHz}$$

The bandwidth in this case equals $f_h - f_l$, or $220 \times 10^3 - 18.2 \approx 220$ KHz.

Notice that regardless of our possible interest in the lowering of f_1 to f_l by means of negative feedback, we can certainly neglect this effect in considering bandwidth. In other words, the effect of feedback upon bandwidth is almost exclusively determined by what happens to the upper cutoff frequency.

As we have seen, in negative feedback some portion of the output signal is subtracted from the input signal, the difference signal then being amplified. This suggests that if distortion is introduced by the amplifier, there must be a signal component at the output that is not present at the input. Therefore, feeding back this component in reversed phase to the input ought to make it possible to cancel out, at least partially, the unwanted component.

Assume that the input to the feedback amplifier is short circuited and suppose that an unwanted signal v_n is generated internally somewhere in the amplifier. Since we cannot assign the location of this source to the input, we cannot claim that the gain seen by v_n is A. Therefore, let us suppose that this peculiar gain has a value equal to α. At the output, the unwanted signal component, which we shall call v_{no}, must be equal to the algebraic sum of the internally generated and amplified signal, αv_n, and the normally amplified portion of the feedback signal. This feedback signal must equal βv_{no}. Hence we can write

$$v_{no} = \alpha v_n - A\beta v_{no}$$

Rearranging the equation, we obtain

$$v_{no} = \frac{\alpha v_n}{1 + A\beta}$$

Without feedback, the unwanted signal at the output would simply be $v_{no} = \alpha v_n$. Thus we see that the effect of negative feedback is to reduce this output component by the now familiar factor of $1 + A\beta$.

Some caution must be exercised in applying this result. We can easily see why this is so if we consider again Eq. (4-1). This equation can be rearranged to read

$$v_o = \frac{Av_s}{1 + A\beta}$$

What this equation is telling us is that the amplified signal voltage Av_s is reduced by the effect of feedback and by the same factor $1 + A\beta$. These results suggest that no improvement in signal-to-noise or signal-to-distortion ratios appears to occur.

Fortunately things are not quite as bad as they might seem. If we increase the input signal v_s, for instance, by such an amount that the output signal is brought to the level it would have had without feedback, then the internally generated signal v_n remains nearly unchanged.

Example 4-3

We can get a clearer picture of what happens in this case by considering an example. Suppose that we have an amplifier with gain of 200 that produces 5% second-harmonic distortion at the output. This means that if a 50-mV, 1000-Hz signal is applied to the input without feedback, a 10-V, 1000-Hz signal plus a 0.5-V, 2000-Hz signal would appear at the output. Now suppose negative feedback using a feedback factor of $\beta = 0.045$ is applied. The 2000-Hz output component is now reduced by a factor $(1 + 200 \times 0.045)$, or 10, to yield $0.5/10 = 0.05$ V, assuming that the output component of 1000 Hz is maintained at the same 10-V level. Of course, to maintain this level, the input signal has to be increased to 10 times 50 mV, or 0.5 V. Distortion at the output has now been reduced to 0.5%.

A similar improvement in signal-to-noise ratio can sometimes be achieved using negative feedback. But once again a word of caution must be offered. As we have seen, both the desired and the undesired signals are reduced by the same factor when negative feedback is applied. From this discussion it might appear that all we have to do to offset this difficulty is to increase the input signal. Such a procedure will work well enough for an audio power amplifier, for example, where there is presumably plenty of input signal amplitude available. But what happens in the case of a high-gain amplifier such as an audio preamplifier? An amplifier of this type usually operates with the typically weak input signal at its maximum possible voltage level so that no further increase is possible. In such a case, while feedback may be of use in assuring uniform performance for a production run of amplifiers or in adjusting the bandwidth, it offers no help in combating noise.

4-2 ANALYSIS OF FEEDBACK AMPLIFIERS

As stated earlier, practical feedback amplifiers do not consist of boxes labeled A and β. It is therefore necessary to consider the techniques that are needed in analyzing real amplifiers.

We begin by representing the feedback amplifier by the diagram shown in Fig. 4-5. Notice that this diagram is supposed to represent the entire amplifier complete with feedback network. That is, Fig. 4-5 represents the same system as is shown in Fig. 4-1. In Fig. 4-5, a controlled source $\mathbf{x}_b$ is embedded somewhere inside the amplifier and feedback structure. All that we know about this source is that the quantity $\mathbf{x}_b$ (which in this case is shown as a voltage, although a current would work just as well) is determined by another internal quantity $\mathbf{x}_a$. The controlling parameter is k so that $\mathbf{x}_b = k\mathbf{x}_a$. We also know the relationships between $\mathbf{x}_a$, $\mathbf{x}_b$, $\mathbf{v}_s$, and $\mathbf{v}_o$. These relationships are given by the following equations:

$$\mathbf{v}_o = a\mathbf{v}_s + b\mathbf{x}_b \tag{4-5a}$$

$$\mathbf{x}_a = c\mathbf{v}_s + d\mathbf{x}_b \tag{4-5b}$$

Equation (4-5b) expresses the fact that feedback is present, since it indicates that the input to the controlled source is derived from the output of that source as well as from the overall amplifier input.

The constants a, b, c, and d are called *transmission parameters*. They are used here only to relate the input and output variables, $\mathbf{v}_s$ and $\mathbf{v}_o$, to the variables that describe the controlled source, $\mathbf{x}_a$ and $\mathbf{x}_b$.

With a little algebraic manipulation, it is easy to show that

$$\frac{\mathbf{v}_o}{\mathbf{v}_s} = \frac{(a - bc/d)(-kd) + a}{1 - kd} \tag{4-6}$$

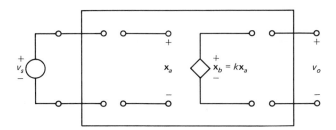

Fig. 4-5 Representation of a feedback amplifier.

What is the meaning of this odd-looking equation? To answer this question let us begin by first letting the source voltage, $\mathbf{v}_s$, become zero (short circuit) and replacing the controlled source, $\mathbf{x}_b$, by an independent one whose value is simply k. Then from Eq. (4-5b), we find $\mathbf{x}_a = kd$.

At this point we define a quantity called the *return ratio* as follows: the return ratio $\mathbf{T}$ is found by replacing dependent source $k\mathbf{x}_a$ by an independent source of value k, setting all other independent sources to zero, and computing the value of $\mathbf{x}_a$ that results. Then we define $\mathbf{T}$ by writing

$$\mathbf{T} \equiv -\mathbf{x}_a$$

Using this definition we see that for the condition under discussion, $\mathbf{T} = -kd$.

Another feedback quantity, the *return difference*, is defined as $1 + \mathbf{T}$, which in this case equals $1 - kd$.

We can now rewrite Eq. (4-6) in the form

$$A_{vf} = \frac{\mathbf{v}_o}{\mathbf{v}_s} = \frac{(a - bc/d)\mathbf{T}}{1 + \mathbf{T}} + \frac{a}{1 + \mathbf{T}} \tag{4-7}$$

Comparing Eq. (4-7) with Eq. (4-1), we see that if we let $\mathbf{T} = A\beta$ and $(a - bc/d) = 1/\beta$, then the first term of Eq. (4-7) corresponds to Eq. (4-1). If we redraw Fig. 4-1 to represent block A as a controlled source, we get the diagram shown in Fig. 4-6. In this figure, A now plays the role that was assigned to k in Fig. 4-5.

Following the definition of return ratio, we now replace dependent source $A\mathbf{v}_i$ with an independent source equal to A and calculate the resulting $\mathbf{v}_i$ under the condition that $\mathbf{v}_s$ is short circuited. This procedure yields $\mathbf{v}_i = -A\beta$. From the definition of return ratio, $\mathbf{T} = -\mathbf{v}_i = A\beta$. Thus we see that the correspondence between Eq. (4-1) and Eq. (4-7) has a reasonable physical basis.

A similar result can be gotten in some cases from the laboratory procedure shown in Fig. 4-7. Here, we short-circuit the input, break the feedback loop, insert a signal, and measure the signal returned around the loop.

The voltage $\mathbf{v}_i$ is produced by the oscillator. Clearly, the returned voltage

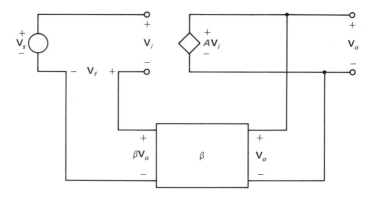

Fig. 4-6 Circuit arrangement for calculating return ratio.

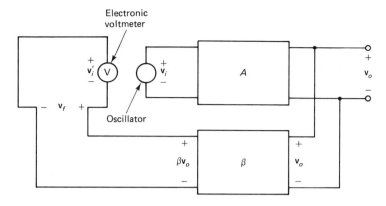

Fig. 4-7   Laboratory procedure for measuring loop gain.

$\mathbf{v}_i'$ will equal $-\beta A \mathbf{v}_i$. We can see that the loop gain will be $\mathbf{v}_i'/\mathbf{v}_i = -\beta A$, which is the negative of the return ratio $\mathbf{T}$. The apparently simple procedure suggested in Fig. 4-7 has a number of hidden pitfalls. First of all, in practical cases block A has some input resistance, which is much less than the infinite resistance implied above. Therefore, in order not to disturb the circuit from its normal operation, the "dangling" leads to the electronic voltmeter must be terminated by a resistance equal to the input resistance of A. Another problem arises when the gain of block A is large. In such a case, with the feedback loop opened, a small signal $\mathbf{v}_i$ may produce a very large signal $\mathbf{v}_o$. This may drive A into saturation or cutoff, so that operation is largely nonlinear. When that occurs, neither the assumptions we have made nor the equations resulting from them are valid.

The analytical procedure for determining $\mathbf{T}$ may be applied at any point inside the feedback loop, but the task is easier if we make use of the natural division of controlling parameter and controlled source that occurs whenever there is a BJT or FET.

Returning to our discussion of Eq. (4-7), we may be led to think at first that the job of untangling the factor $(a - bc/d)$ is hopeless. Fortunately, we do not have to bother to work with the factor. To see what we must do, consider Eq. (4-7) again, this time letting $\mathbf{T}$ approach infinity.

$$\lim_{\mathbf{T}\to\infty} A_{vf} = \lim_{\mathbf{T}\to\infty} \frac{(a - bc/d)\mathbf{T}}{1 + \mathbf{T}} + \lim_{\mathbf{T}\to\infty} \frac{a}{1 + \mathbf{T}}$$

This limit yields

$$\lim_{\mathbf{T}\to\infty} A_{vf} = \frac{a - bc}{d}$$

Let us define this limiting value of A_{vf} as the asymptotic gain K. It is simply the value of A_{vf} for the condition that $\mathbf{T}$ approaches infinity. But, we must ask, How can $\mathbf{T}$ be made to approach infinity? Since for this system $\mathbf{T} = -kd$ and d is a fixed parameter, we can make $\mathbf{T}$ approach infinity by making k grow without limit. While this looks like a simple mathematical technique, it needs further explanation.

After all, what does it mean to apply an independent source whose magnitude is allowed to approach infinity?

To answer this question we go back to Eq. (4-5b) and substitute into it $\mathbf{x}_b = k\mathbf{x}_a$. This yields

$$\mathbf{x}_a = c\mathbf{v}_s + kd\mathbf{x}_a$$

from which we get

$$\mathbf{x}_a = \frac{c\mathbf{v}_s}{1 - kd} = \frac{c\mathbf{v}_s}{1 + \mathbf{T}} \tag{4-8}$$

We now see from Eq. (4-8) that for any finite value of $\mathbf{v}_s$, letting k or $\mathbf{T}$ approach infinity yields

$$\lim_{k \to \infty} \frac{c\mathbf{v}_s}{1 - kd} = \lim_{\mathbf{T} \to \infty} \frac{c\mathbf{v}_s}{1 + \mathbf{T}} = 0$$

Or, in other words, *letting k become infinite requires that $\mathbf{x}_a$ must become zero at the same time.*

There remains the question of the significance of the right-hand term of Eq. (4-7), $a/(1 + \mathbf{T})$. A look at Eq. (4-5a) tells us that the parameter a directly relates a component of the output $\mathbf{v}_o$ to the input $\mathbf{v}_s$. Evidently, the term $a/(1 + \mathbf{T})$ has something to do with the direct coupling of some of the input signal to the output. In Eq. (4-7), if we allow $\mathbf{T}$ to approach zero (which can be done by letting k approach zero), then we find

$$\lim_{k \to 0} A_{vf} = \lim_{k \to 0} \frac{K(-kd)}{1 - kd} + \frac{a}{1 - kd} = a$$

or

$$A_{vf}\big|_{\mathbf{T}=0} = a$$

Let us denote this value of A_{vf} as K_0, the direct transmission gain. Then Eq. (4-7) takes the form of the asymptotic gain formula:

$$A_{vf} = \frac{K\mathbf{T}}{1 + \mathbf{T}} + \frac{K_0}{1 + \mathbf{T}} \tag{4-9}$$

We can now state a step-by-step procedure for finding A_{vf}.

1. Set independent, *external* sources to zero. (Replace current sources by open circuits and voltage sources by short circuits.)
2. Select a convenient controlled source *within* the network and replace the controlled element of the device by an *independent* source of magnitude k.
3. Calculate the resulting value of the controlling current or voltage. This is the quantity that we denoted by the symbol $\mathbf{x}_a$ in the preceding discussion.
4. Find the return ratio by using $\mathbf{T} = -\mathbf{x}_a$.
5. Assume that $\mathbf{v}_o$ has a finite value and, at the same time, assume that the controlling parameter $\mathbf{x}_a$ becomes zero. Calculate the value of $\mathbf{v}_s$ that the circuit requires and then find $K = \mathbf{v}_o/\mathbf{v}_s$.

6. Let $k \to 0$ and find $K_0 = v_o/v_s$.
7. Using Eq. (4-9), evaluate A_{vf}.

To see how the method works, let us consider again the circuit arrangement shown in Fig. 4-6. There is only one controlled source in the system, which we now take to be independent and of value A volts. Taking v_s to be short-circuited, we calculate v_i. This is clearly given by

$$v_i = -A\beta$$

Then $T = -v_i = A\beta$, which agrees with the result previously obtained.

Now, assuming a value for v_o, we let $v_i \to 0$. (Mathematically, this is equivalent to letting $A \to \infty$.) This has no effect on v_o but requires v_s to equal v_f. Hence

$$v_s = v_f = \beta v_o.$$

Therefore, $K = v_o/v_s = 1/\beta$.

When $A \to 0$, $v_o \to 0$, so that we have K_0, the direct transmission term, equal to zero. Substituting these results into Eq. (4-9), we obtain the familiar result

$$A_{vf} = \frac{(1/\beta)A\beta}{1 + A\beta} = \frac{A}{1 + A\beta}$$

Example 4-4

Let us now consider a more practical example using the circuit shown in Fig. 4-8. As before, we want to find the gain with feedback A_{vf}.

We begin by redrawing the circuit using the small-signal model for the transistor. Assuming that the frequency of v_s is in the midband range, we take the capacitors to be short circuits. The equivalent circuit is shown in Fig. 4-9.

To find T, we short-circuit v_s, replace $h_{fe}i_b$ with an independent source of h_{fe} amperes and calculate i_b.

Making two successive applications of the current-division rule, we first find the

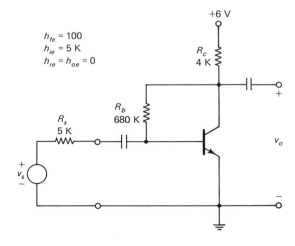

Fig. 4-8 Feedback circuit to be analyzed.

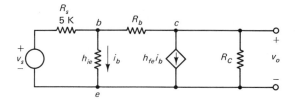

Fig. 4-9 Equivalent circuit.

current into R_b and then find $\mathbf{i}_b$. The current into R_b is given by

$$\mathbf{i}_{R_b} = -h_{fe} \frac{R_c}{R_c + R_b + (R_s \| h_{ie})}$$

The current $\mathbf{i}_b$ is equal to

$$\mathbf{i}_b = \mathbf{i}_{R_b} \frac{R_s}{h_{ie} + R_s} = -h_{fe} \frac{R_c}{R_c + R_b + R_s \| h_{ie}} \cdot \frac{R_s}{h_{ie} + R_s}$$

Since $\mathbf{T} = -\mathbf{i}_b$, we find

$$\mathbf{T} = h_{fe} \frac{R_c R_s}{[R_c + R_b + (R_s \| h_{ie})](h_{ie} + R_s)}$$

With $\mathbf{v}_s$ restored to the circuit, we now find K, the asymptotic gain. For this purpose we allow $\mathbf{i}_b$ to equal zero while the voltage $\mathbf{v}_o$ continues to be present. At the base node, under these conditions, Kirchhoff's current law applies as usual, but there are only two nonzero currents entering the node. With $\mathbf{i}_b = 0$, no current flows through h_{ie} so that the potential of node b is zero. In other words, node b is a *virtual ground*. We shall encounter the virtual-ground concept again later.

We now have the situation symbolized in Fig. 4-10, from which it is clear that $\mathbf{v}_s/R_s + \mathbf{v}_o/R_b = 0$. Therefore, $\mathbf{v}_o/\mathbf{v}_s = -R_b/R_s$. The ratio $\mathbf{v}_o/\mathbf{v}_s$ yields K, which is equal to

$$K = -\frac{R_b}{R_s}$$

To find the direct transmission gain, we let $h_{fe} \to 0$; using Fig. 4-9, we find

$$K_0 = \frac{\mathbf{v}_o}{\mathbf{v}_s} = \frac{R_c}{R_c + R_b} \cdot \frac{h_{ie} \| (R_b + R_c)}{R_s + h_{ie} \| (R_s + R_c)}$$

Putting together the results we have obtained, we get the following:

$$\mathbf{T} = 100 \cdot \frac{4\ \text{K} \cdot 5\ \text{K}}{(4\ \text{K} + 680\ \text{K} + 5\ \text{K} \| 5\ \text{K})(5\ \text{K} + 5\ \text{K})} = 0.291$$

$$K = -\frac{680\ \text{K}}{5\ \text{K}} = -136$$

$$K_0 = \frac{4\ \text{K}}{4\ \text{K} + 680\ \text{K}} \cdot \frac{5\ \text{K} \| (680\ \text{K} + 4\ \text{K})}{5\ \text{K} + 5\ \text{K} \| (680\ \text{K} + 4\ \text{K})}$$

Since $5\ \text{K} \| (680\ \text{K} + 4\ \text{K}) \approx 5\ \text{K}$, K_0 can be approximated by

$$K_0 \approx \frac{20}{(684)(10\ \text{K})} \approx \frac{2}{684} = 0.00292$$

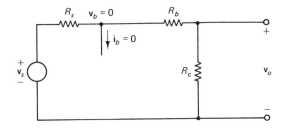

Fig. 4-10 Circuit for calculating asymptotic gain.

Then A_{vf} is equal to

$$A_{vf} = \frac{-136 \times 0.291}{1 + 0.291} + \frac{0.00292}{1 + 0.291} = -30.66 + 0.00226$$

Notice that the second term is less than 10^{-4} times the first. This second term represents direct-signal transmission through the network without regard to the amplifying behavior of the transistor. In this case, which is typical, this term is so small as to be negligible. Thus we arrive at the final value of A_{vf}, which is 30.66.

It is interesting to compare this value with the value that would be obtained if no feedback were present. Since the feedback occurs by virtue of the connection of R_b between the collector and the base, it can be removed simply by connecting the top of R_b to the V_{cc} line. In this way the transistor is properly biased, but no signal is fed back from collector to base.

In this case, the input signal current to the transistor base is about equal to $\mathbf{v}_s/(R_s + h_{ie})$. The magnitude of the output voltage is then $-h_{fe}[\mathbf{v}_s/(R_s + h_{ie})]R_c$. Hence we arrive at

$$A_v = \frac{\mathbf{v}_o}{\mathbf{v}_s} = \frac{-h_{fe}R_c}{R_s + h_{ie}} = \frac{-100 \times 4 \text{ K}}{10 \text{ K}} = -40$$

This number compares closely with the numerator of the first term in the expression for A_{vf} computed earlier, which was $-136 \times 0.291 = -39.58$.

As a second example let us consider a transistor amplifier in which an unbypassed emitter resistor is used. Such a circuit is shown in Fig. 4-11.

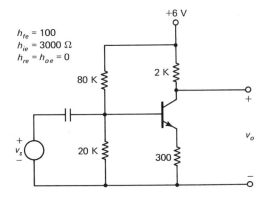

Fig. 4-11 Circuit to be analyzed.

Example 4-5

For simplicity, we shall make the reasonable approximation that the bias resistors can be neglected. This yields the equivalent circuit shown in Fig. 4-12.

With $\mathbf{v}_s$ short-circuited and with $h_{fe}\mathbf{i}_b$ replaced by h_{fe}, we use the current-division rule to find $\mathbf{i}_b$:

$$\mathbf{i}_b = -\frac{R_E}{h_{fe} + R_E}h_{fe}$$

Then

$$\mathbf{T} = \frac{R_E h_{fe}}{h_{ie} + R_E}.$$

When we attempt to determine K by letting $h_{fe} \to \infty$ while $\mathbf{i}_b \to 0$, we are at first perplexed about the current source $h_{fe}\mathbf{i}_b$. Does the product $h_{fe}\mathbf{i}_b$ equal zero? Infinity? Or something finite? The answer is found by picturing that this infinite-gain amplifier acts to produce a finite output, $\mathbf{v}_o$, with zero input (considering $\mathbf{i}_b$ to be this input). Therefore, the presence of $\mathbf{v}_o$ dictates the current in R_c. But since we are assuming no external load current, this current, $\mathbf{v}_o/R_c$, must flow from terminal 2′ to terminal 2 and must be the current produced by the generator $h_{fe}\mathbf{i}_b$. Note that the direction of this current must be opposite to the reference direction shown for $h_{fe}\mathbf{i}_b$. Since $\mathbf{i}_b$ must be zero in this condition, then $\mathbf{v}_s$ must equal the voltage across R_E and must be given by

$$\mathbf{v}_s = -\frac{\mathbf{v}_o R_E}{R_c}$$

From this result we conclude that

$$K = \frac{\mathbf{v}_o}{\mathbf{v}_s} = -\frac{R_c}{R_E}$$

Neglecting the direct transmission term, we obtain

$$A_{vf} = \frac{\dfrac{R_c}{R_E}\dfrac{R_E h_{fe}}{(h_{ie} + R_E)}}{1 + \dfrac{R_E h_{fe}}{h_{ie} + R_E}} = -\frac{h_{fe}R_c}{h_{ie} + (1 + h_{fe})R_E}$$

This is exactly the same result that is found using standard circuit-analysis procedures.

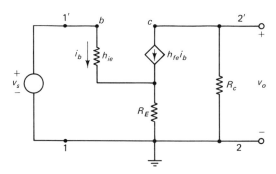

Fig. 4-12 Small-signal equivalent circuit.

Upon evaluating A_{vf} numerically, we obtain

$$A_{vf} = \frac{-100 \times 2000}{3000 + 101 \times 300} \approx \frac{-2000}{330} \approx -6.06$$

Example 4-6

As a final, more complex example, let us consider the circuit shown in Fig. 4-13. Again, we wish to find A_{vf}, so we proceed as before. The small-signal equivalent circuit is shown in Fig. 4-14. Short-circuiting $\mathbf{v}_s$ and replacing $h_{fe_1}\mathbf{i}_{b_1}$ with an independent source of h_{fe_1} amperes, we next find $\mathbf{i}_{b_1}$.

We start with the emitter of Q_2, where the unknown current $\mathbf{i}_{b_2}$ produces a current into R_4 that is equal to $(1 + h_{fe_2})\mathbf{i}_{b_2}$. Using the current-division rule, we find that the current into R_6 is this current times $R_5/(R_5 + R_6 + R_1\|h_{ie})$. Again by the current-division rule, $\mathbf{i}_{b_1}$ must be this last result times $R_1/(h_{ie_1} + R_1)$. Thus, in terms of $\mathbf{i}_{b_2}$, we find $\mathbf{i}_{b_1}$ to be equal to

$$\mathbf{i}_{b_1} = (1 + h_{fe_2})\mathbf{i}_{b_2} \frac{R_5}{R_5 + R_6 + R_1\|h_{ie_1}} \frac{R_1}{h_{ie_1} + R_1}$$

Unfortunately, this result expresses $\mathbf{i}_{b_1}$ in terms of $\mathbf{i}_{b_2}$, which we have now to find. For this purpose we note that the resistance between the emitter of Q_2 and ground is equal to $R_4 + R_5\|(R_6 + R_1\|h_{ie_1})$. If we convert the current source h_{fe_1} and resistor to an equivalent voltage source, we have the circuit shown in Fig. 4-15, from which we can find $\mathbf{i}_{b_2}$.

By Kirchhoff's voltage law,

$$+h_{fe_1}R_2 + \mathbf{i}_{b_2}R_2 + (1 + h_{fe_2})\mathbf{i}_{b_2}[R_4 + R_5\|(R_6 + R_1\|h_{ie_1})] = 0$$

Hence

$$\mathbf{i}_{b_2} = \frac{-h_{fe_1}R_2}{R_2 + (1 + h_{fe_2})(R_4 + R_5\|[R_6 + R_1\|h_{ie_1}])}$$

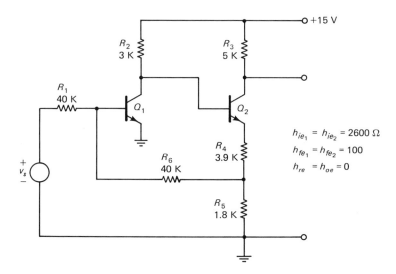

Fig. 4-13 Feedback amplifier to be analyzed.

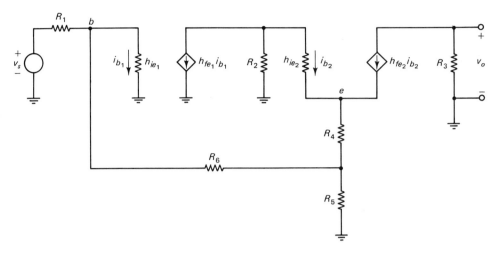

Fig. 4-14 Small-signal equivalent circuit.

Combining these results, we obtain

$$\mathbf{T} = \frac{h_{fe1}(1 + h_{fe2})R_2 R_1 R_5}{\{[R_2 + (1 + h_{fe2})(R_4 + R_5 \| [R_6 + R_1 \| h_{ie1}]\}(R_5 + R_6 + R_1 \| h_{ie1})(R_1 + h_{ie1})}$$

This rather formidable-looking expression is not as bad as it seems. Remember that in solving this sort of problem step by step, we can compute the values of various series and parallel combinations as we go along. Easy approximations will often be readily apparent, so that a numerical solution may be more easily found than the complicated expression above might seem to suggest.

The numerical value of T turns out to be approximately

$$\mathbf{T} \approx \frac{10 \text{ K} \times 3 \text{ K} \times 40 \text{ K} \times 1.8 \text{ K}}{[3 \text{ K} + 100(3.9 \text{ K} + 1.8 \text{ K})][44.2 \text{ K}][42.6 \text{ K}]} = \frac{2.160 \times 10^6}{565.7 \times 44.2 \times 42.6}$$

$$\approx \frac{2.160 \times 10^6}{1.065 \times 10^6} \approx 2.028$$

To find the asymptotic gain K, we return to Fig. 4-14; this time we assume that $h_{fe1} \to \infty$, at the same time that $\mathbf{i}_{b1} \to 0$. As we saw in previous examples, the current source $h_{fe1}\mathbf{i}_{b1}$ produces some finite current such that, with amplification by Q_2, an

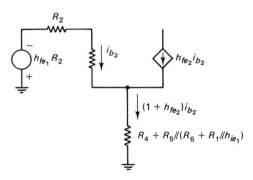

Fig. 4-15 Circuit for computing $\mathbf{i}_{b2}$.

Feedback Principles   Chap. 4

output voltage $\mathbf{v}_o$ will occur across R_3. This permits us to calculate $h_{fe2}\mathbf{i}_{b_2}$, which must be, simply, $-\mathbf{v}_o/R_3$. We can easily see that $\mathbf{i}_{b_2}$ must be $-\mathbf{v}_o/(R_3 h_{fe2})$. The current into R_4 must then be equal to

$$(1 + h_{fe2})\mathbf{i}_{b_2} = -\frac{\mathbf{v}_o}{R_3}\left(1 + \frac{1}{h_{fe2}}\right)$$

Recalling that the requirement that $\mathbf{i}_{b_1}$ be zero makes node b a virtual ground, we calculate the current into R_6 by the current-division rule. This yields

$$\mathbf{i}_{R_6} = \frac{\mathbf{v}_o}{R_3}\left(1 + \frac{1}{h_{fe2}}\right)\frac{R_5}{R_5 + R_6}$$

Kirchhoff's current law applied to node b, with due regard to the fact that $\mathbf{i}_b = 0$, yields

$$\frac{\mathbf{v}_s}{R_1} = \mathbf{i}_{R_6}$$

from which we conclude that

$$\mathbf{v}_s = -\mathbf{v}_o\frac{R_1}{R_3}\left(1 + \frac{1}{h_{fe2}}\right)\frac{R_5}{R_5 + R_6}$$

Therefore,

$$K = \frac{\mathbf{v}_o}{\mathbf{v}_s} = -\left(\frac{h_{fe2}}{1 + h_{fe2}}\right)\frac{R_3}{R_1}\cdot\frac{R_5 + R_6}{R_5}$$

The numerical value of K is equal to

$$K = -\frac{100}{101}\cdot\frac{5\text{ K}}{40\text{ K}}\frac{1.8\text{ K} + 40\text{ K}}{40\text{ K}} \approx -0.13$$

Finally, we evaluate A_{vf} (neglecting K_0).

$$A_{vf} = \frac{K\mathbf{T}}{1 + \mathbf{T}} = \frac{-0.13 \times 2.028}{3.028} = -0.087$$

(Incidentally, you should observe that this amplifier, while suitable for practicing this method of analysis, is not a very effective signal booster. Gain figures, both with and without feedback, are very low.)

4.3 INPUT AND OUTPUT IMPEDANCE

The last subject that must be discussed in this section is the problem of calculating input and output impedances. For this purpose we need an additional tool—Blackman's impedance formula.

As before we consider the feedback amplifier, using the representation shown in Fig. 4-5. This time, however, interest is focused upon a single terminal pair (or port) to which we apply an external current source. The arrangement is shown in Fig. 4-16.

Although Fig. 4-16 seems to imply that the circuit is set up exclusively for the calculation of input impedance, this is not the case. Notice that the internal

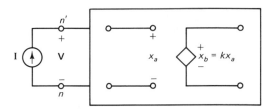

Fig. 4-16 Circuit for deriving Blackman's impedance formula.

circuitry that connects the terminals n-n' to the inside variables $\mathbf{x}_a$ and $\mathbf{x}_b$ is not shown. Therefore, the n-n' terminals may be at either the input or the output.

Both the voltage $\mathbf{V}$ and the variable $\mathbf{x}_a$ may be expressed as linear functions of the other two variables I and $\mathbf{x}_b$. Thus we can write

$$\mathbf{V} = \alpha \mathbf{I} + \beta \mathbf{x}_b \qquad (4\text{-}10a)$$

and

$$\mathbf{x}_a = \gamma \mathbf{I} + \delta \mathbf{x}_b \qquad (4\text{-}10b)$$

where α, β, γ, and δ are unspecified constants. The impedance at terminals n-n' must be $\mathbf{V}/\mathbf{I}$. Hence we have, after several steps of algebraic manipulation,

$$\mathbf{Z}_{nn'} = \frac{\mathbf{V}}{\mathbf{I}} = \alpha \frac{1 - k(\alpha\delta - \beta\gamma)/\alpha}{1 - k\delta} \qquad (4\text{-}11)$$

From Eq. (4-10a) we see that

$$\alpha = \left.\frac{\mathbf{V}}{\mathbf{I}}\right|_{x_b = 0}$$

This is simply the impedance at terminals n-n' when controlled source $\mathbf{x}_b = k\mathbf{x}_a$ is set equal to zero. Let us designate this impedance by the symbol $\mathbf{Z}_{nn'}^0$.

If we set $\mathbf{V}$ to zero and replace controlled source $\mathbf{x}_b = k\mathbf{x}_a$ by an independent source of value $\mathbf{k}$, we obtain—by substituting in Eq. (4-10a) and (4-10b)—

$$0 = \alpha \mathbf{I} + \beta \mathbf{k}$$

$$\mathbf{x}_a = \gamma \mathbf{I} + \delta \mathbf{k}$$

from which it is easy to obtain

$$\mathbf{x}_a = \frac{\mathbf{k}(\alpha\delta - \beta\gamma)}{\alpha}$$

Notice that making $\mathbf{V} = 0$ is equivalent to short-circuiting the terminals n-n'. Then, replacing $\mathbf{x}_b$ by $\mathbf{k}$ and solving for $\mathbf{x}_a$ is exactly the process that we carried out in the preceding examples to find $\mathbf{T}$. In this case, however, to emphasize the fact that the terminals n-n' are short-circuited, we write $\mathbf{T}$ as $\mathbf{T}_{sc}$ and obtain

$$\mathbf{T}_{sc} = -\mathbf{x}_a = \frac{-\mathbf{k}(\alpha\delta - \beta\gamma)}{\alpha}$$

Next we make $\mathbf{I} = 0$ and again set $\mathbf{x}_b$ to $\mathbf{k}$. This simply corresponds to open-circuiting the terminals n-n' and letting $\mathbf{x}_b$ equal $\mathbf{k}$. Substituting again in Eq.

Feedback Principles   Chap. 4

(5-10b), we obtain

$$\mathbf{x}_a = \mathbf{k}\delta$$

Once again, we see that this result enables us to find another special form of the return ratio, which we shall call $\mathbf{T}_{oc}$. Clearly $\mathbf{T}_{oc} = -\mathbf{k}\delta$.

Putting together these results using Eq. (4-11) we obtain Blackman's equation

$$\mathbf{Z}_{nn'} = \mathbf{Z}_{nn'}^0 \frac{1 + \mathbf{T}_{sc}}{1 + \mathbf{T}_{oc}} \tag{4-12}$$

As we did in the case of finding A_{vf}, we now give a procedure for evaluating $\mathbf{Z}_{nn'}$:

1. Choose a port at which the impedance $\mathbf{Z}_{nn'}$ is to be found and disable an internal controlled source by setting it to zero. As before, the choice of which source to use is arbitrary.
2. Calculate the impedance seen at port nn'. This impedance is $\mathbf{Z}_{nn'}^0$.
3. Find the return difference $\mathbf{T}$ in the usual manner with respect to this controlled source. Designate this value of $\mathbf{T}$ by the symbol $\mathbf{T}_{sc}$.
4. Open circuit port n-n' and again find $\mathbf{T}$. This value of $\mathbf{T}$ is $\mathbf{T}_{oc}$.
5. Substitute these results in Blackman's equation, Eq. (5-12), to find $\mathbf{Z}_{nn'}$.

Let us illustrate the application of these ideas by calculating input and output impedances for one of the circuits that we considered above, the one shown in Fig. 4-11.

Example 4-7

Looking at the equivalent circuit, which is shown in Fig. 4-12, we see that with $h_{fe}i_b$ removed, the input impedance $\mathbf{Z}_{11'}^0$ is simply $h_{ie} + R_E$. The return ratio $\mathbf{T}_{sc}$ is the same as $\mathbf{T}$, which we have already calculated. Hence we have

$$\mathbf{T}_{sc} = \frac{R_E h_{fe}}{h_{ie} + R_E}$$

To find $\mathbf{T}_{oc}$ we assume that the input is open. In that case, since there is no path for the flow of i_b, we can see by inspection that $\mathbf{T}_{oc} = 0$. Putting the results we now have into Blackman's equation, Eq. (4-12), we find

$$\mathbf{Z}_{11'} = (h_{ie} + R_E) \cdot \frac{1 + R_E h_{fe}/(h_{ie} + R_E)}{1 + 0}$$

or

$$\mathbf{Z}_{11'} = h_{ie} + R_E + R_E h_{fe} = h_{ie} + R_E(1 + h_{fe}).$$

This agrees with the result that one can find by conventional circuit-analysis methods. The numerical value of $\mathbf{Z}_{11'}$ for this example is

$$\mathbf{Z}_{11'} = 2600 + 101(300) = 32.9 \text{ K}\Omega$$

To find the output impedance of this circuit, we must think carefully about the conditions that are imposed upon it. These conditions depend upon whether the network is driven at the input port by a current source or by a voltage source. When we disable the independent sources, we must leave open the input port if the driving source is a current or short-circuit it if the source is a voltage. In addition to these considerations, we must also recall that in the derivation of Blackman's equation, calculation of the quantity $\mathbf{T}_{sc}$ involved the assumption that the port nn', at which the impedance was sought, was short-circuited.

Applied to the network of Fig. 4-11, these ideas lead us to maintain a short circuit at port 11'. We first calculate $\mathbf{T}_{sc}$ with port 22' short-circuited and then calculate $\mathbf{T}_{oc}$ with port 22' open-circuited. The conditions under which $\mathbf{T}_{oc}$ is to be calculated are identical to those under which $\mathbf{T}_{sc}$ was calculated in finding the input impedance. Hence, for the output impedance calculation, $\mathbf{T}_{oc} = R_E h_{fe}/(h_{ie} + R_E)$.

To find $\mathbf{T}_{sc}$ we short-circuit port 22' and recalculate $\mathbf{T}$. It should be clear that replacing $h_{fe}\mathbf{i}_b$ by a source of h_{fe} amperes and calculating $\mathbf{i}_b$ will yield exactly the same value of $\mathbf{T}$. Hence, in this example, $\mathbf{T}_{sc} = \mathbf{T}_{oc}$.

The impedance $\mathbf{Z}_{22'}^0$ is found by setting $h_{fe}\mathbf{i}_b$ to zero. When this is done, $\mathbf{Z}_{22'}^0$ is simply equal to R_c. Upon substituting in Eq. 4-12, we obtain

$$\mathbf{Z}_{22'} = R_c \frac{1 + R_E h_{fe}/(h_{ie} + R_E)}{1 + R_E h_{fe}/(h_{ie} + R_E)} = R_c$$

Numerically, $\mathbf{Z}_{22'} = 2\ \mathrm{K\Omega}$.

4-4 VOLTAGE REGULATORS

An important application of feedback principles is found in the design of voltage regulators. Every laboratory contains at least a few constant-voltage, DC power supplies and, often, constant-current supplies as well. We shall confine our discussion to the constant-voltage types, since the principles of both types are similar.

The object of a voltage regulator is to maintain a constant voltage output to a load despite variations in either the load current or the line voltage. The basic elements of a regulator are shown in the block diagram of Fig. 4-17.

One way to look at the regulator is to consider the entire circuit as a Thevenin-equivalent voltage source and to find its output resistance. Given changes in load resistance result in changes in output voltage that can be readily calculated. We shall apply this point of view to the analysis of the simple voltage regulator circuit shown in Fig. 4-18.

We begin with the DC analysis. Each Zener diode (Z_1 and Z_2) has in series with it an ordinary diode (D_1 and D_2) that has a temperature coefficient of opposite sign to that of the Zener. In this way, reference voltages are developed that are very nearly independent of temperature. Both Z_1-D_1 and Z_2-D_2 develop reference

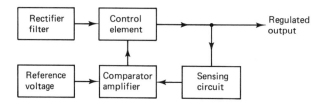

Fig. 4-17 Basic elements of a regulator.

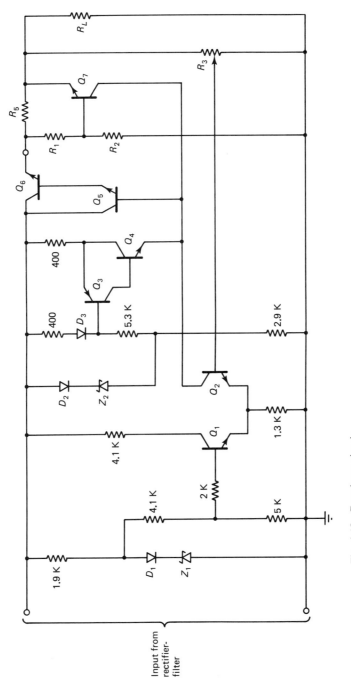

Fig. 4-18 Regulator circuit.

voltages equal to 6 V. This means that the current flowing through D_3 equals 6 V/5.7 K = 1.053 mA.

The diode D_3 has characteristics that match those of the emitter-base junction of Q_3. Since both D_3 and the emitter-base junction of Q_3 are in series with 400-Ω resistors, these two parallel branches function effectively as the parallel elements of a current mirror. Therefore, the current through Q_3-Q_4, a transistor pair that acts as a single, composite *pnp* transistor, is also 1.053 mA. In other words, Q_3-Q_4 acts as a constant current source of 1.053 mA that feeds current to the collector of Q_2, the base of Q_5, and the collector of Q_7.

Because of the 5 K–4.1 K voltage divider, the voltage at the base of Q_1 is very nearly equal to 3.3 V. Therefore, the combined emitter currents of Q_1 and Q_2 are equal to (3.3 − 0.7)/1.3 K = 2 mA. This means that when the sampling potentiometer R_3 is set to provide 3.3 V to the base of Q_2, the collector current of Q_2 will be 1 mA. This is nearly equal to the current provided by the current source, so that very little current is left to drive the base of Q_5, the input element of the Darlington-connected, series-control transistors. The basic elements of this system are thus as shown in the simplified schematic diagram of Fig. 4-19.

Notice that Q_7 and its associated resistor have been omitted in this diagram. The role of these elements will be considered later.

If the load resistor R_L should happen to be reduced so that the output voltage V_o tends to fall, the sample voltage V_s is similarly reduced, causing I_c to decrease. Since less current is being taken by I_c, more is available for the base of Q_5, and thus I_B increases. This increases the current in the series control transistor, thus increasing I_o and restoring V_o to its proper level.

To get a quantitative idea of how this process works, we must study the circuit on a small-signal basis. We begin by considering the way the regulator operates with respect to load variations only. That is, we assume that V_L, the input voltage from the rectifier-filter, is constant. Therefore, for small signal purposes, we can consider V_L to be short-circuited. The differential amplifier and the series-control transistor can be replaced with small-signal equivalent circuits. We are thus led to

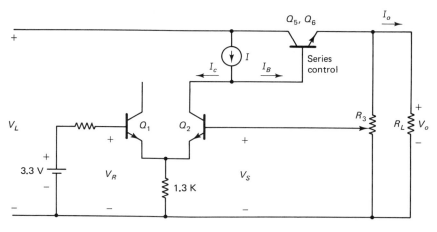

Fig. 4-19 Simplified schematic diagram.

the small-signal representation of the overall regulation circuit that is shown in Fig. 4-20.

In our analysis of this circuit, the parameters h_{ied} and h_{fed} refer to the transistors of the differential amplifier, while h_{ieD} and h_{feD} refer to the composite characteristics that represent the Darlington-connected, series transistors. These composite parameters are related approximately to the parameters of the individual transistors according to the equations

$$h_{ieD} \approx h_{ie1} + (1 + h_{fe1})h_{ie2}$$

$$h_{feD} \approx h_{fe1} + (1 + h_{fe1})h_{fe2}$$

$$h_{oeD} \approx h_{oe2} + (1 + h_{fe2})h_{oe1}$$

To judge the effectiveness of the regulator in maintaining constant voltage in the face of load changes, we need to know the output resistance of the regulator. If we consider the output of the regulator from a Thevenin point of view, then we see that the smaller the output (Thevenin) resistance seen by R_L, the more nearly the regulator approaches the condition of being a constant-voltage source. To find the output resistance, we shall make use of Blackman's equation.

Notice that the differential amplifier is represented as a voltage-controlled *current* source. With R_L removed we disable the Darlington source and find $Z_{11'}^0$. This is done by imagining that an external, phasor test voltage $\mathbf{V}_{\text{test}}$ is applied to port 11′ while we calculate the phasor current $\mathbf{I}_{\text{test}}$. Then $Z_{11'}^0$ will be simply $\mathbf{V}_{\text{test}}/\mathbf{I}_{\text{test}}$. Fig. 4-21 illustrates the procedure.

Let us assume that the potentiometer R_s is set so that the resistance between the tap and ground is kR_s, where k is a fraction less than 1. Then the voltage $\mathbf{V}$ must be given by

$$\mathbf{V}_1 = \frac{(2_{hied}\|kR_s)\mathbf{V}_{\text{test}}}{(1 - k)R_s + 2h_{ied}\|kR_s}$$

Making use of the DC conditions determined earlier and taking h_{fed} to be 100, we find that h_{ied} is equal to 2600 Ω. If we assume that R_s is a 10-KΩ potentiometer and that its slider is set to the midrange position so that k is 0.5, we can calculate individual pieces of this expression. This is advantageous, since these same pieces will turn up repeatedly as our calculations proceed.

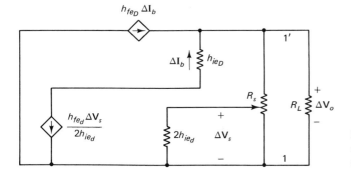

Fig. 4-20 Small-signal equivalent circuit of regulator for studying the effects of load changes.

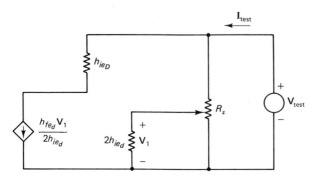

Fig. 4-21 Circuit arrangement to find $\mathbf{Z}_{11}^{0}{}'$.

Thus we find that $2h_{ied}\|kR_s = 5200\|5000 = 2550 \ \Omega$, and $(1 - k)R_s = 5000 \ \Omega$. This leads to the relationship

$$\mathbf{V}_1 = \frac{2550}{5000 + 2550} \mathbf{V}_{\text{test}} = 0.338 \ \mathbf{V}_{\text{test}}.$$

The current $\mathbf{I}_{\text{test}}$ then equals

$$\frac{\mathbf{V}_{\text{test}}}{(1 - k)R_s + 2h_{ied}\|kR_s)} + \frac{h_{fed}\mathbf{V}_1}{2h_{ied}}$$

or

$$\mathbf{I}_{\text{test}} = \mathbf{V}_{\text{test}} \left[\frac{1}{7550} + \frac{100}{2 \times 2600} \times 0.338 \right]$$

This result suggests that $\mathbf{Z}_{11}^{0}{}'$ is equal to 151 Ω.

With the 11' port short-circuited, we find $\mathbf{T}_{sc}$ with respect to the Darlington source. The modified circuit is shown in Fig. 4-22.

Since voltage $\mathbf{V}_1$ is zero because of the configuration of the circuit, the calculation of $\mathbf{T}_{sc}$ becomes trivial, and $\mathbf{T}_{sc}$ is seen at once to be zero.

With the short circuit removed, the circuit takes the form shown in Fig. 4-23. The Darlington source is taken as an independent one of h_{feD} amperes in order to calculate $\mathbf{T}_{oc}$.

We observe that the current $\mathbf{I}_T$ is the sum of $\mathbf{I}_b$ and h_{feD}. Hence the voltage $\mathbf{V}_1$

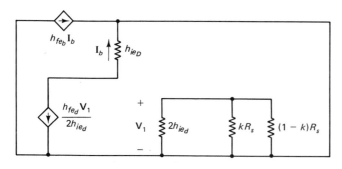

Fig. 4-22 Circuit modified to find $\mathbf{T}_{sc}$.

Feedback Principles   Chap. 4

must equal

$$\mathbf{V}_1 = (2h_{ied}\|kR_s)\mathbf{I}_T = 2500 \cdot (\mathbf{I}_b + h_{feD})$$

This leads to $\mathbf{I}_b = -(h_{fed}/2h_{ied}) \times 2500 \times (\mathbf{I}_b + h_{feD})$
or

$$\mathbf{I}_b = -\frac{100}{5200} \times 2500 \, (\mathbf{I}_b + h_{feD}) = -49(\mathbf{I}_b + h_{feD})$$

Then

$$\mathbf{I}_b = \frac{-49h_{feD}}{1 + 49}$$

Thus we see that $\mathbf{T}_{oc}$, which equals $-\mathbf{I}_b$, must be equal to

$$\mathbf{T}_{oc} = \frac{49h_{feD}}{50}$$

If we assume that h_{fe1} and h_{fe2} both equal 50, then the Darlington, composite h_{feD} must equal $50 + (1 + 50)50 = 50 + 2550 = 2600$. Thus $\mathbf{T}_{oc}$ must be equal to $49 \times 2600/50$, or 2550.

Combining these results, we find that the output resistance is given by

$$R_o = \mathbf{Z}_{11'} = 151 \, \frac{1 + 0}{1 + 2550} \approx 0.0595 \, \Omega$$

To appreciate the meaning of R_o, suppose that the regulator delivers 100 mA into a 1000-Ω load. The Thevenin equivalent of the regulator circuit would be as shown in Fig. 4-24. The load voltage is equal to $1000 \times 0.1 = 100$ V. The voltage drop in R_o is equal to 0.00595 V. Hence, open-circuiting the load would cause the terminal voltage to rise by less than 6 mV.

The quantity $\mathbf{T}_{oc}$ is the return ratio for the regulator circuit. The value found for $\mathbf{T}_{oc}$, 2550, is not exceptionally large. More elaborate regulator circuits can easily surpass this one in providing higher values of loop gain and, hence, better regulation.

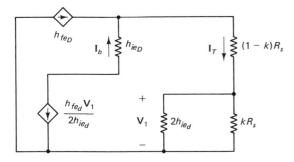

Fig. 4-23 Circuit arranged to find $\mathbf{T}_{oc}$.

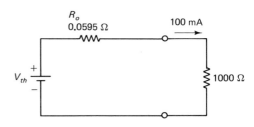

Fig. 4-24 Thevenin-equivalent circuit and load.

To investigate regulation with respect to line-voltage variations, we must modify the circuit in two ways. First, we must consider that a variable input voltage exists because of changes in voltage coming from the rectifier filter. Second, we must now include the effective output conductance h_{oeD} of the Darlington connection. These modifications are incorporated in the equivalent circuit shown in Fig. 4-25.

What we want to find is the phasor output voltage V_o produced by V, which represents the variations of voltage from the rectifier filter. This is essentially the problem of finding A_{vf}, for which we need the asymptotic gain K and the return ratio T.

In gain calculations we saw that T has to be calculated with the input short-circuited. But with V short-circuited, we find that we have the same equivalent circuit that we used previously to find T_{oc} except for the presence of h_{oeD} and R_L. Typical values for h_{oe1} and h_{oe2} are 2.5×10^{-4} mho. Using these values together with the value of 3000 previously assumed for h_{feD}, we obtain h_{oeD} equal to 1.275×10^{-2} mho. This corresponds to resistance of 79.5 Ω, a quantity that does not appear to be negligible. We see from Fig. 4-25 that when V is short-circuited, the left-hand terminal of h_{oeD} is at ground potential. Therefore, we can swing it over to the right and observe that h_{oeD} will then be in parallel with R_L.

The calculation of T for the modified circuit is now somewhat more complicated, but it is not different in principle from the calculations we have done previously. Therefore, we shall simply skip the tedious details and state the result:

$$
T = \cfrac{\dfrac{h_{feD}h_{fed}}{2h_{ied}}\left[\dfrac{kR_s\|2h_{ied}}{(1-k)R_s + kR_s\|2h_{ied}}\right]\left[\dfrac{1}{\dfrac{1}{R_L} + \dfrac{1}{h_{oeD}} + \dfrac{1}{(1-k)R_s + kR_s\|2h_{ied}}}\right]}{1 + \dfrac{h_{fed}}{2h_{ied}}\left[\dfrac{kR_s\|2h_{ied}}{(1-k)R_s + kR_s\|2h_{ied}}\right]\left[\dfrac{1}{\dfrac{1}{R_L} + \dfrac{1}{h_{oeD}} + \dfrac{1}{(1-k)R_s + kR_s\|2h_{ied}}}\right]}
$$

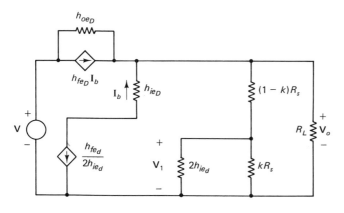

Fig. 4-25 Equivalent circuit of regulator modified to analyze the effect of line-voltage variations.

Despite the messy look of this equation, it improves when numerical values are substituted, since most of the terms are values previously calculated. We then have

$$T = \cfrac{\dfrac{2600 \times 100}{2600} \left(\dfrac{2550}{7550}\right) \left(\dfrac{1}{1/1000 \ + \ 1/79.5 \ + \ 1/7550}\right)}{1 \ + \ \dfrac{100}{2600} \left(\dfrac{2500}{7550}\right) \left(\dfrac{1}{1/1000 \ + \ 1/79.5 \ + \ 1/7550}\right)}$$

$$= \frac{(100)(0.338)(72.7)}{1 \ + \ 0.013(72.7)} \approx 1260$$

(Clearly, h_{oeD} is the most important term among the parallel elements.)

We find the asymptotic gain K by letting the Darlington source become infinite and, at the same time, requiring I_b to become zero. We see immediately that this will force V_1 and, in turn, V_o to become zero. Hence the asymptotic gain turns out to be zero.

This is one of the unusual cases in which the direct transmission term, rather than the asymptotic gain, turns out to be the important term. To find the direct transmission term, we use the circuit of Fig. 4-25, this time letting h_{feD} become zero. The input voltage V will now be in series with the conductance h_{oeD}. Hence we can replace this source by a current source of Vh_{oeD} shunted by the conductance h_{oeD}. This exchange of source results in the circuit shown in Fig. 4-26.

A single node-voltage equation can now be written as follows:

$$-Vh_{oeD} + \frac{h_{fed}V_1}{2h_{ied}} + V_o\left[h_{oeD} + \frac{1}{(1 \ - \ k)R_s \ + \ 2h_{ied}\|kR_s} + \frac{1}{R_L}\right] = 0$$

The term in brackets has been evaluated before and is equal to $1/72.7$. Also, V_1 is related to V_o by the factor calculated previously, so that we have

$$V_1 = 0.338 \ V_o$$

Thus we have

$$-\frac{V}{79.5} + \frac{100}{2600} \times 0.338V_o + \frac{V}{72.7} = 0$$

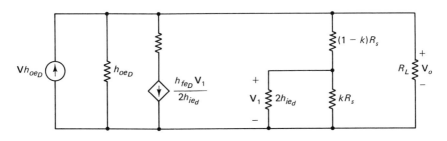

Fig. 4-26 Equivalent circuit to find direct transmission.

Hence K_o turns out to be

$$K_o = \frac{\mathbf{V}_o}{\mathbf{V}} = \frac{1}{79.5 \times 0.0268} = 0.469$$

Finally, we calculate the gain with feedback and find it to be

$$A_{v_f} = \frac{K\mathbf{T}}{1 + \mathbf{T}} + \frac{K_o}{1 + \mathbf{T}} = 0 + \frac{0.469}{1260} = 0.000372$$

To appreciate the significance of this result, assume that the input voltage to the regulator is 150 V and the output is 100 V. Suppose that the input changes by 5%, or 7.5 V. Then the change in the output would be $0.000372 \times 7.5 = 0.0028$ V, which is less than 3 mV.

4-4.1 The foldback current limiter

To complete our study of voltage regulators we now consider the function of transistor Q_7 in the circuit of Fig. 4-18. We have already seen that when the resistance of the load is reduced, the output voltage tends to fall, and the regulator compensates for this by increasing the current delivered to the load. This process can only go so far; at some level of increased load current, the collector dissipation of the Darlington transistors, Q_5 and Q_6, will become excessive and these transistors will burn out. It is the purpose of Q_7 to protect the circuit against the possibility of this kind of failure.

When the load current, which passes through R_s, becomes high enough, the voltage across R_s exceeds the bias voltage developed across R_1 to bring the base-emitter voltage of Q_7 to a level of 0.7 V, which turns on Q_7. Since the source of base current for Q_5-Q_6 is the constant current source Q_3-Q_4, when Q_7 conducts it does so by taking current from the base of Q_5-Q_6. This causes Q_5-Q_6 to reduce the current delivered to the load. This will make the load voltage fall, so that the regulator no longer functions as a regulator. However, both the Darlington-connected transistors Q_5-Q_6 and the load will have been protected.

The base-emitter voltage of Q_7 is given by

$$V_{BE} = I_L R_S - \frac{(V_o + I_L R_S)R_1}{R_1 + R_2} \tag{4-13}$$

(Notice that the potential across the R_1-R_2 voltage divider is $V_o + I_L R_S$.) The value of R_s is set to provide protection against a "dead short" across the load terminals. Therefore, if the short-circuit current is to be limited to a value equal to I_{sc}, the resistance of R_s must be given by

$$R_s = \frac{V_o + I_L R_s}{I_{sc}}$$

Since the term $I_L R_s$ will certainly be small compared to V_o, we can neglect it both in the expression for R_s and in the bracketed part of the expression for V_{BE}. This simplifies matters quite a bit and enables us to write $R_s \approx V_o/I_{sc}$ and

$$V_{BE} \approx I_L R_s - V_o \frac{R_1}{R_1 + R_2} \tag{4-14}$$

Turn-on occurs for $V_{BE} = 0.7$ V. Therefore, for a limiting value of I_L equal to I_{Lmax}, we have

$$\frac{I_{Lmax}}{I_{sc}} \approx \frac{0.7}{V_o} + \frac{R_1}{R_1 + R_2} \qquad (4\text{-}15)$$

Equation (4-15) is useful as a design equation, enabling us to pick values for R_1 and R_2 when I_{Lmax} (the limit on load current) and V_o (the output voltage of the regulator) have been given. However, the behavior of the current limiter when an overload occurs must be understood by using Eq. (4-14). For V_{BE} equal to 0.7 V, this equation suggests that once the current limit has been reached, there are other values of I_L and V_o that will satisfy the equation. Remember that once the limiter switches on, I_L and V_o will both be reduced. This process can go on, so long as V_{BE} continues to be equal to 0.7 V, until finally V_o is reduced to zero. At that point the output voltage at the emitter of the Darlington-connected, series transistor will be just enough so that at the corresponding low value of I_L, 0.7 V will be available for V_{BE}, but the output voltage will be zero. From Eq. (4-14) we see that this zero-voltage current level will be approximately equal to $0.7/R_s$.

We can now plot the curve of regulator voltage versus load current to show the effect of the current limiter Q_7. This is done in Fig. 4-27. From Fig. 4-27 it is clear that the output voltage stays nearly constant at its prescribed value of V_o until the load current reaches the predetermined maximum level. Then the voltage collapses to zero and the current falls to a safe, low value. We must then disable the limiter (for example, by open-circuiting the load) in order to make the regulator operate again in its normal mode. From the shape of the curve, it should be clear why this type of limiter is known as a *foldback current limiter*.

4-5 THE MILLER EFFECT

One manifestation of feedback that is sometimes undesirable is called the *Miller effect*. The Miller effect is the phenomenon by which small parasitic capacitances can look much larger to the associated circuit than they are. To understand this effect, consider the circuit shown in Fig. 4-28. What we wish to find is the input

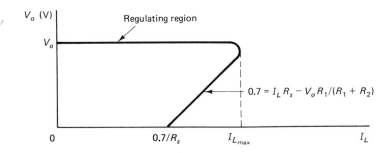

Fig. 4-27 Operation of foldback current limiter.

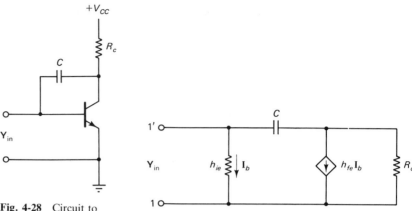

Fig. 4-28 Circuit to demonstrate the Miller effect.

Fig. 4-29 Equivalent network.

admittance $\mathbf{Y}_{in}$. For that purpose we represent the circuit by the equivalent network shown in Fig. 4-29.

To begin with, let us solve the problem on a simplified basis. Imagine that a voltage source $\mathbf{V}$ is applied to the 1-1′ port, as shown in Fig. 4-30. We can calculate the current $\mathbf{I}$ and then find $\mathbf{Y}_{in}$ simply by using the fact that $\mathbf{Y}_{in} = \mathbf{I}/\mathbf{V}$.

The current $\mathbf{I}$ consists of two components. One of these is the current $\mathbf{I}_b$, which equals $\mathbf{V}/h_{ie}$, and the other is $\mathbf{I}_c$, the current through the capacitor. This latter current must be equal to the voltage drop across the capacitor times its admittance, or

$$\mathbf{I}_c = (\mathbf{V} - \mathbf{V}_c)j\omega C$$

The voltage $\mathbf{V}_c$ is the voltage at the collector, which we have now to calculate.

At this point, for the sake of simplicity, we make the assumption that we can calculate $\mathbf{V}_c$ while ignoring the presence of the capacitor. On that basis, we find that $\mathbf{V}_c$ is simply equal to $-h_{fe}\mathbf{I}bR_c$. This leads to the expression for $\mathbf{I}_c$:

$$\mathbf{I}_c = [\mathbf{V} - (-h_{fe}\,\mathbf{I}_bR_c)]\,j\omega C = \mathbf{V}\left(1 + \frac{h_{fe}R_c}{h_{ie}}\right)j\omega C$$

The total current $\mathbf{I}$ is then equal to

$$\mathbf{I} = \mathbf{V}\left[\frac{1}{h_{ie}} + \left(1 + \frac{h_{fe}R_c}{h_{ie}}\right)j\omega C\right]$$

The admittance at port 1-1′ is therefore $\mathbf{I}/\mathbf{V}$, or

$$\mathbf{Y}_{in} = \frac{1}{h_{ie}} + \left(1 + \frac{h_{fe}R_c}{h_{ie}}\right)j\omega C \qquad (4\text{-}16)$$

The first term merely shows the effect of the shunt element h_{ie} that is connected across the 1-1′ port. The second term involves the expression $(1 + h_{fe}R_c/h_{ie})$, which we recognize as equivalent to $(1 + A_v)$, where A_v is the voltage gain of a common emitter amplifier.

Observe that the admittance given by Eq. (4-16) can be obtained from the circuit shown in Fig. 4-31. We therefore conclude that the effect of the circuit is to amplify the capacitance C by the factor $(1 + A_v)$ and to cause it to appear to be shunted across the input port. This effect is the Miller effect, and the capacitance C_M given by $C_M = (1 + A_v)C$ is sometimes called the *Miller capacitance*.

To carry out a rigorous calculation of Y_{in}, we could, for the network of Fig. 4-30, write node-voltage equations for the base and collector nodes. Simultaneous solution of these equations would eventually lead to an expression for Y_{in}. Instead of using that procedure, we shall find Z_{in} by using Blackman's equation, Eq. (4-12), and then invert the result to find Y_{in}. First of all, $Z_{11'}$ is simply $h_{ie} \| (R_c + 1/j\omega C)$. Short-circuiting the input port makes I_b zero, regardless of the fact that the controlled current source is replaced by an independent one of amplitude h_{fe} amperes. Hence $T_{sc} = 0$. With the input port open, the same procedure yields $I_b = -h_{fe}R_c/[R_c + h_{ie} + 1/j\omega C]$. Hence $T_{oc} = h_{fe}R_c/[R_c + h_{ie} + 1/j\omega c]$. Blackman's equation then yields

$$Z_{in} = \left[h_{ie} \| \left(R_c + \frac{1}{j\omega C} \right) \right] \frac{1}{1 + \dfrac{h_{fe}R_c}{h_{ie} + R_c + 1/j\omega C}}$$

$$= \frac{h_{ie}(R_c + 1/j\omega C)}{h_{ie} + R_c + 1/j\omega C} \cdot \frac{R_c + h_{ie} + 1/j\omega C}{R_c(1 + h_{fe}) + h_{ie} + 1/j\omega C}$$

Hence

$$Z_{in} = \frac{h_{ie}(R_c + 1/j\omega C)}{R_c(1 + h_{fe}) + h_{ie} + 1/j\omega C}$$

This seemingly perplexing result is more meaningful if we have the luck to carry out favorable, rather than unfavorable, algebraic manipulations. The trick is to multiply numerator and denominator by $j\omega C$ and to divide numerator and denominator by h_{ie}. This procedure yields

$$Z_{in} = \frac{1 + j\omega C R_c}{[R_c(1 + h_{fe})/h_{ie}] j\omega C + j\omega C + 1/h_{ie}}$$

At this point we again use the fact that the gain of a common-emitter amplifier A_v is approximately equal to $h_{fe}R_c/h_{ie}$. Hence the equation for Z_{in} becomes

$$Z_{in} = \frac{1 + j\omega C R_c}{(A_v + 1)j\omega C + 1/h_{ie}}$$

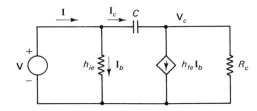

Fig. 4-30 Voltage source applied to find Z_{in}.

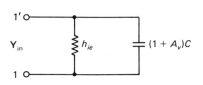

Fig. 4-31 Equivalent circuit.

The expression for $\mathbf{Y}_{in}$ is

$$\mathbf{Y}_{in} = \frac{1}{\mathbf{Z}_{in}} = \frac{(A_v + 1)j\omega C + 1/h_{ie}}{1 + j\omega C R_c}$$

Now we must look at the denominator of $\mathbf{Y}_{in}$. Suppose that C is 10 pF, R_c is 5 KΩ and the frequency f is 10^6 Hz. Then the denominator becomes $1 + j2\pi \times 10^6 \times 10^{-11} \times 5 \times 10^3 = 1 + j0.1\pi$. The magnitude of the denominator is $\sqrt{1 + (0.314)^2} = 1.05$, which suggests that $\mathbf{Y}_{in}$ can be approximated reasonably well (if the frequency is not too high) by

$$\mathbf{Y}_{in} \approx (A_v + 1)j\omega C + \frac{1}{h_{ie}} \qquad (4\text{-}17)$$

Eq.(4-17) is essentially the same as Eq. (4-16). This time, however, we see that the result is really an approximation; with the circuit parameters assumed above, $\mathbf{Y}_{in}$ would be seriously in error if f were 10 MHz instead of 1 MHz.

We can get some idea of the importance of the Miller effect by calculating the capacitance seen at the base of a common-emitter amplifier. If the internal capacitance from collector to base is 5 pF and $A_v = 50$, it is clear that $C(A_v + 1) \approx 255$ pF, a capacitance value whose effect upon a circuit may well be nonnegligible.

We are now equipped to consider an application of the Miller effect that we had previously avoided discussing. This application occurs in the circuit of the 741 operational amplifier. Figure 4-32 shows the pertinent part of the schematic diagram.

In Chap. 3 we determined that the voltage gain for this stage was about 1000 and that the input resistance was about 2.25 MΩ. The Miller capacitance resulting from the 30-pF capacitor that is connected from the output to the input is $C_M = (A_v + 1)C \approx 3 \times 10^{-8}$ F. This is an excellent example of the use of the Miller effect to produce a large effective capacitance with only a small physical capacitor.

Taken together with the input resistance, this capacitance results in a break frequency of $1/(2\pi \times 3 \times 10^{-8} \times 2.25 \times 10^6)$, or about 2.5 Hz. This is an example of lag compensation that is provided to make the gain fall off at -20 db per decade at frequencies well below those at which further breaks occur. In this way the gain is reduced to 0 db before further phase-shift influences occur. The purpose of this drastic compensation is to ensure that when the operational amplifier is so connected into an external feedback network that high values of $|\mathbf{T}|$ occur, the feedback system will be stable. This topic is discussed in detail in Chap. 5.

4-6 BOOTSTRAPPING

Another interesting and useful application of feedback principles is demonstrated by the bootstrap connection. The purpose of bootstrapping is to magnify the apparent resistance seen by the circuit because of the presence of a particular resistor. This is accomplished by connecting the low end of the resistor in question not to signal ground, but to a point having a signal voltage in phase with and almost equal

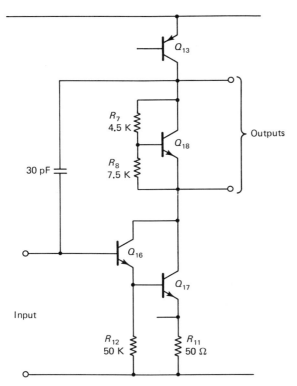

Fig. 4-32 Second stage of the 741 operational amplifier.

to that at the high end. Thus, since the voltage at the "low end" of the resistor varies nearly the same way as that at the "high end," the situation is as if the resistor were being "pulled up by its own bootstraps," as far as voltage is concerned. The result is to decrease greatly the resistor current, thus apparently magnifying the resistance. These ideas are illustrated by the circuits shown in Fig. 4-33.

Figure 4-33(a) shows the standard emitter-follower circuit. The Thevenin-equivalent resistance of the base-bias network is $R_x \| R_Y$. In the network of Fig. 4-33(b), the voltage divider comprising R_1 and R_2 establishes the same open-circuit voltage as resistors R_x and R_y. Bypass capacitor C_b fixes the tap voltage to the signal ground level. (We assume that the impedance of C_b is negligible.) Resistor R_b is adjusted so that $R_b + R_1 \| R_2 = R_x \| R_Y$. Thus the circuit of Fig. 4-33(b) is biased to the same quiescent point as that of Fig. 4-33(a).

In Fig. 4-33(c), the low end of the bypass capacitor is connected to the emitter of Q instead of to ground, thus accomplishing bootstrapping. The resistance looking into the base of Q is the same for all circuits; what is of interest is the effective value of resistor R_b in the circuits of Fig. 4-33(b) and 4-33(c). The effect of bootstrapping is shown in a simplified way in the circuit of Fig. 4-34.

We now wish to calculate the effective resistance R_{eff} due to R_b. This resistance must be equal to $\mathbf{V}_i / \mathbf{I}$, and the current $\mathbf{I}$ must be equal to $(\mathbf{V}_i - A\mathbf{V}_i)R_b$, so that

$$R_{\mathrm{eff}} = \frac{\mathbf{V}_i}{\mathbf{V}_i(1 - A)/R_b} = \frac{R_b}{1 - a}$$

Sec. 4-6   Bootstrapping   **147**

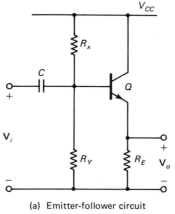

(a) Emitter-follower circuit

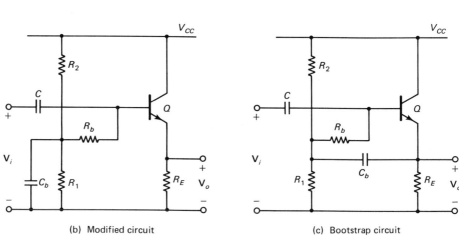

(b) Modified circuit   (c) Bootstrap circuit

Fig. 4-33 Emitter-follower circuit with bootstrapping.

Since the gain of an emitter follower is always less than 1, let us assume that A is equal to 0.9. Then we see that R_{eff} will be equal to $10R_b$. Thus the bootstrap connection has caused R_b to appear to the input voltage source as if it were magnified ten times.

    Another circuit that makes use of bootstrapping is shown in Fig. 4-35. In this circuit, transistors Q_2 and Q_3 operate so that over a full sine-wave variation of output current, Q_2 conducts on positive half-cycles, whereas Q_3 conducts on the negative ones. Thus Q_2 and Q_3, taken together, behave pretty much as a large-signal emitter follower having a gain somewhere near 1. Transistor Q_1 is a common-emitter driver that furnishes base signal current to Q_2 or Q_3. The collector resistor of Q_1 is R_c, and we see that it connects not to ground but to R_L, the load resistor. The value of R_L is low, such as 4 or 8 Ω, while R_c may be several hundred ohms. Therefore, this connection does not affect the DC biasing of Q_1. But, since the

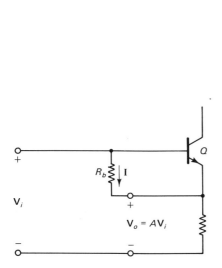

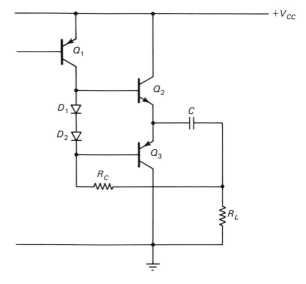

Fig. 4-34 Effect of bootstrapping upon R_b.

Fig. 4-35 Complementary-symmetry amplifier with bootstrapping.

signal voltages at the emitters and bases of Q_2 and Q_3 are almost equal, there is very little signal voltage across R_c. (Again, we are assuming that C is very large.) Thus the signal current passing through R_c is quite small, and therefore almost all the signal current produced by Q_1 is delivered to the base of Q_2 or the base Q_3.

REFERENCES

1. Robert Boylestad and Louis Nashelsky, *Electronic Devices and Circuit Theory*, 3rd ed. (Englewood Cliffs, N.J.: Prentice-Hall, Inc., 1982).
2. Adel S. Sedra and Kenneth C. Smith, *Microelectronic Circuits*, (New York: Holt, Rinehart and Winston, 1982).
3. Solomon Rosenstark, "A Simplified Method of Feedback Amplifier Analysis," *IEEE Transactions on Education*, E-7(1974): 192–8.
4. W. H. Williams and J. H. Parker, "An IC Medium-Power Voltage Regulator," *IEEE Spectrum*, 6(1969): 72–8.
5. John W. Chu and Richard D. Ricks, *The μA7800 Series Three-Terminal Positive Voltage Regulators*, Application Note 312, December 1971, (Mountain View, Calif.: Fairchild Semiconductor Corporation).

PROBLEMS

4-1. The transistor is connected as an emitter-follower.
   (a) Estimate β for this circuit at midband frequencies (where $X_c \approx 0$).
   (b) Find the value of A required to hold variations in A_{vf} to 2% if A varies by 40%.

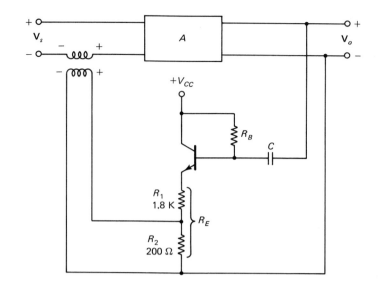

4-2. For the circuit of Prob. 1, an amplifier having a voltage gain A_v equal only to 100 is available. Find the new values of R_1 and R_2 that will maintain the same gain stability as in Prob. 1. R_E is to be kept equal to 2000 Ω.

4-3. Specify A_v and β required to yield an amplifier with overall gain of A_{vf} of 100 and to provide variations in A_{vf} not greater than 0.01% when A varies by 20%.

4-4. An amplifier has a midband gain of -100 and a bandwidth of 10^7 Hz.
(a) Find the value of β required to triple the bandwidth.
(b) What will be the midband gain using this value of β?

4-5. An amplifier has $A_v = 80$ db and a gain-bandwidth product $A_v f_h$ of 1 MHz. When negative feedback is applied, the bandwidth becomes 10 KHz. Find A_{vf} and β.

4-6. An audio-power amplifier produces an output voltage of 12 V across a 16-Ω load with a 1-V input signal. The output contains 3% harmonic distortion. We want to reduce the harmonic distortion to 0.5% by the use of feedback, as shown. Find K and β so that a 1-V input will produce a 12-V output with 0.5% distortion.

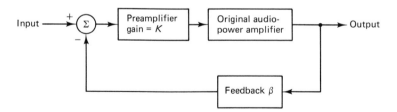

4-7. An amplifier has a gain of 2000 at room temperature and a gain of 1500 at $-30°C$. The amplifier is to be used in a feedback circuit that is to have a gain of 100 at room temperature. What gain will this circuit have at a temperature of $-30°C$?

4-8. Find the return ratio **T**. Assume all $C = \infty$. Also, $R_B > 500 \text{ K}\Omega$.

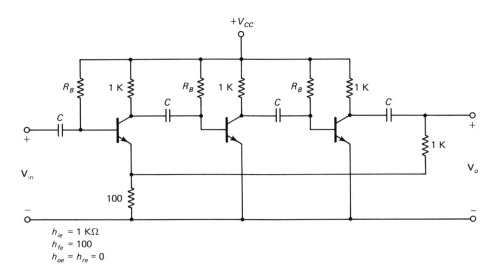

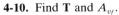

$h_{ie} = 1 \text{ K}\Omega$
$h_{fe} = 100$
$h_{oe} = h_{re} = 0$

4-9. For the circuit of Prob. 8, find the asymptotic gain K and the overall gain A_{vf}.

4-10. Find **T** and A_{vf}.

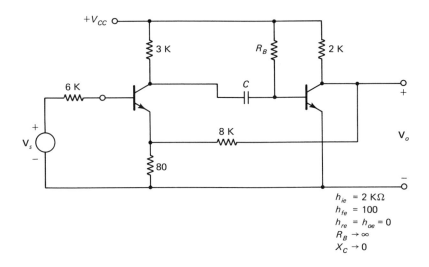

$h_{ie} = 2 \text{ K}\Omega$
$h_{fe} = 100$
$h_{re} = h_{oe} = 0$
$R_B \to \infty$
$X_C \to 0$

4-11. Assume all capacitors are short circuits at midband frequencies. Neglect the 300-KΩ bias resistor of Q_1 and calculate **T** and A_{vf}.

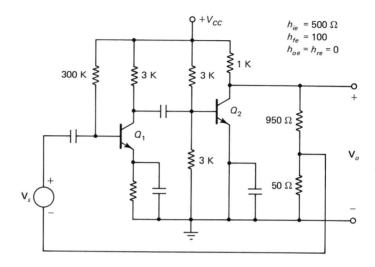

$h_{ie} = 500\ \Omega$
$h_{fe} = 100$
$h_{oe} = h_{re} = 0$

4-12. Bias resistors have been omitted for simplicity. Find (a) A_{vf}, (b) $\mathbf{Z}_i$, and (c) $\mathbf{Z}_o$. For an approximate result, begin by treating the three-transistor amplifier without the 50-KΩ feedback resistor.

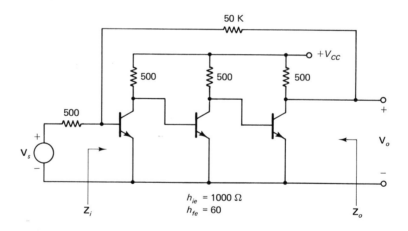

$h_{ie} = 1000\ \Omega$
$h_{fe} = 60$

4-13. (a) With the feedback resistor R_f open, calculate A_v for the amplifier.
(b) Calculate A_{vf}.

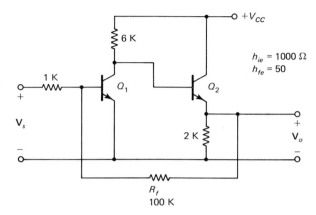

4-14. The differential amplifier can be treated as a feedback amplifier. In the circuit shown below, consider Q_1 to be an emitter follower driving Q_1, which should be taken as a common-base amplifier.
(a) Compute **T** with respect to Q_1.
(b) Find A_{vf}.

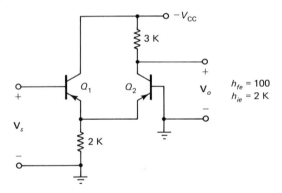

4-15. Treat the emitter follower as a feedback amplifier and find A_{vf}.

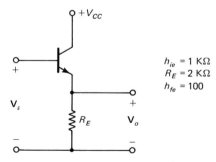

4-16. For the emitter follower of Prob. 15, use Blackman's equation to find R_i and R_o.

4-17.

    (a) Assume that the DC collector current of Q_2 is 1 mA. Determine I_{C1}, V_{CE1}, V_{B2}, V_{CE2}, and V_{E2}.

    (b) Suppose that the connection between the collector of Q_1 and the base of Q_2 is broken, but I_{C2} is maintained and caused to increase by 10%. Calculate the change in V_{C1}.

    (c) What effect would the result obtained in (b) have on I_{C2} if the collector-base connection is restored?

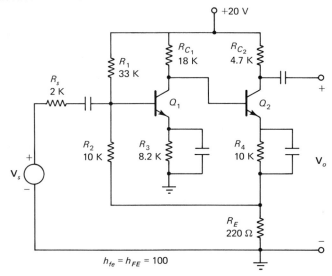

4-18. For the amplifier of Prob. 17, find A_{vf}. Assume capacitors are short circuits to AC.

4-19. **(a)** Use Blackman's formula to find the output resistance $R_{2\text{-}2'}$.

    (b) If $V_{ref} = 20$ V, what is the open-circuit voltage at terminals 22'?

    (c) If $V_1 = 40$ V and $R_L = 40$ Ω, what power is dissipated by Q_1? (Use reasonable approximation.)

    (d) *Percent voltage regulation* is defined as

$$\% \text{ regulation} = 100 \, \frac{V_{\text{no load}} - V_{\text{load}}}{V_{\text{load}}}$$

Calculate the percent regulation produced by switching the load R_L on and off.

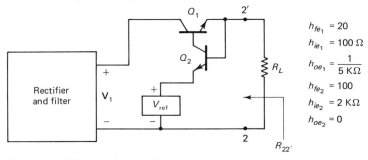

4-20. For the circuit of Prob. 19, calculate the deviation in output voltage that results from a change of 0.5 V in the output of the rectifier-filter circuit. Assume R_L is equal to 40 Ω.

4-21. Find the output resistance, r_o.

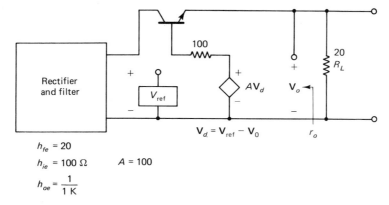

$h_{fe} = 20$
$h_{ie} = 100 \, \Omega$ $A = 100$
$h_{oe} = \dfrac{1}{1 \, K}$

4-22.

    (a) Assuming that the regulator circuit drives the input voltage to the differential amplifier (V) to zero, find V_o.

    (b) If $I_{SC} = 5$ A and $I_{L\text{max}}$ is to be 1 A and $R_1 = 1$ K, find R_S and R_2.

    (c) Sketch the foldback curve (see Fig. 4-27), labeling current and voltage values.

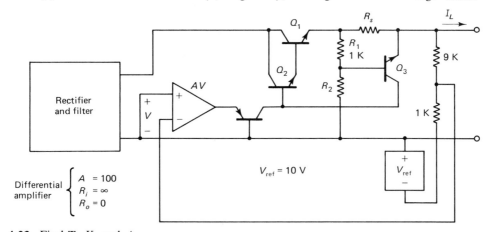

Differential amplifier $\begin{cases} A = 100 \\ R_i = \infty \\ R_o = 0 \end{cases}$

4-23. Find **T**, K, and A_{vf}.

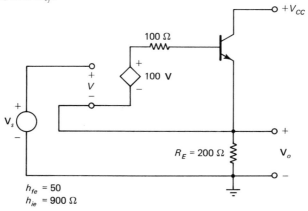

$h_{fe} = 50$
$h_{ie} = 900 \, \Omega$

4-24. (a) Calculate the approximate value of A_v for the circuit shown, assuming C_f had been removed. Assume the gain of Q_2 and Q_3 is unity.
   (b) Calculate the Miller capacitance that results when C_f is replaced in the circuit.
   (c) Find the bandwidth of the circuit.

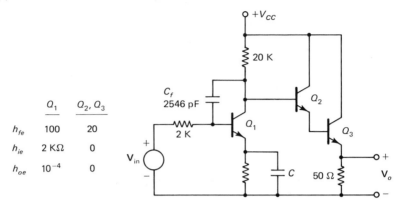

	Q_1	Q_2, Q_3
h_{fe}	100	20
h_{ie}	2 KΩ	0
h_{oe}	10^{-4}	0

Assume C is a short circuit.

4-25. Use the return-ratio concept to calculate the bandwidth of the circuit of Prob. 24.

4-26. The equivalent circuit shown occurs in the MC 1530 op-amp. Find A_{vf} using the return-ratio principle.

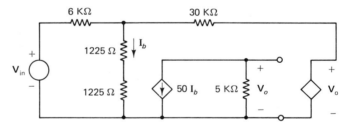

4-27. (a) Calculate the value of R_f required to set the quiescent level of $\mathbf{V}_o$ at $+10$ V.
   (b) Calculate A_{vf}. (See Prob. 15.)

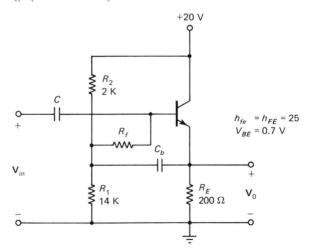

$h_{fe} = h_{FE} = 25$
$V_{BE} = 0.7$ V

Feedback Principles   Chap. 4

(c) Calculate the effective value of the bootstrapped resistor R_f.

(d) If the impedance of capacitor C_b must be at most 10% of the resistance it drives—that is, $R_1 \| R_2$—calculate the value for C_b if the signal frequency is 10 Hz.

4-28.

Shown is a simplified circuit based on the Fairchild μA706 integrated-circuit audio power amplifier. Transistors Q_8, Q_9 and Q_{10}, Q_{11} form a Darlington-connected complementary-symmetry output stage that functions as an emitter-follower with voltage gain of 0.98.

(a) When the input voltage is zero, the emitters of Q_9 and Q_{11} rest at $+7$ V. If the signal power delivered to R_L is 20 W, find the peak values of sinusoidal AC voltage and current in R_L.

(b) Find the quiescent collector current of Q_4.

(c) When Q_9 delivers its peak value of emitter current, what is the base current of Q_8? What is the collector current of Q_4 at this instant?

(d) If the connection of R_4 to V_{cc} is broken and R_4 is connected along the dashed line to R_L, what is the *apparent* value of R_4 viewed from the collector of Q_4?

(e) Compare the peak value of collector current in Q_4 as found in (c) with the value that would occur using the bootstrap connection of (d).

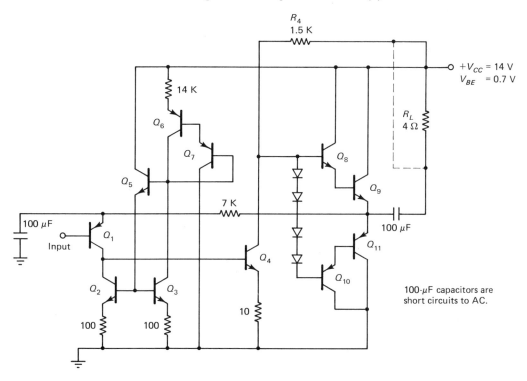

4-29. Find A_{vf}, r_i, and r_o.

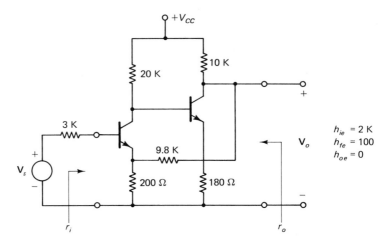

4-30. (a) For the circuit of Prob. 28, find the collector currents of Q_3 and Q_2.

    (b) With the input at 0 V and the emitters of Q_9 and Q_{11} at $+7$ V, what is the collector current of Q_1?

    (c) What is the base current of Q_4? If the quiescent collector current of Q_4 is 3.73 mA, what must be the minimum value of h_{FE} for Q_4?

4-31.

    (a) Find V_o.

    (b) Find r_i.

    (c) Find r_o.

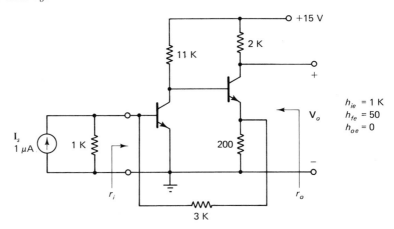

Feedback Principles   Chap. 4

4-32.

    (a) Find **T**.
    (b) Find A_{vf}.
    (c) Find r_i.
    (d) Find r_o.

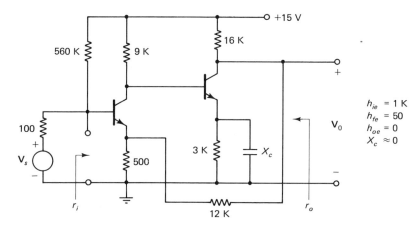

4-33. Find r_{of}.

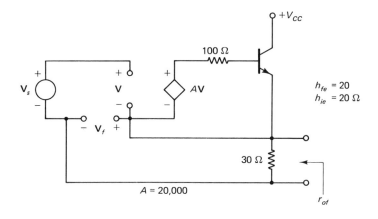

4-34. Find (a) A_{vf}, (b) $\mathbf{Z}_i$ looking into terminals 1-1′, and (c) $\mathbf{Z}_o$ looking into terminals 2-2′. Make reasonable approximations.

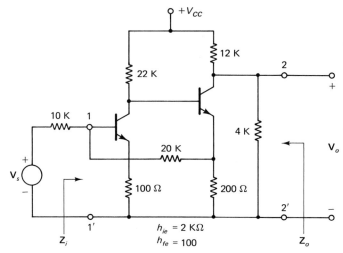

Bias circuitry omitted for simplicity.

CHAPTER 5

Dynamic Stability
of Feedback Amplifiers

5-1 INTRODUCTION TO THE CONCEPT OF STABILITY

Everyone has probably experienced the unpleasant phenomenon that sometimes occurs in a meeting hall when the microphone of the public-address amplifier is brought too close to the loudspeaker. "Feedback," people say, in trying to explain the irritating squeal. Indeed, an audio signal is fed back from the speaker to the microphone. But why does the squeal occur? Evidently the amplifying system is no longer functioning as an amplifier should but has instead begun to operate in some other (in fact, oscillatory) mode. That such a change can occur is evidence that the system has become unstable. What are the criteria that determine whether or not a feedback amplifier will be unstable? This chapter is concerned with the answer to that question.

To begin with, let us examine what happens when a feedback amplifier operates with positive feedback. If we look back to the feedback system shown in Fig. 4-1 and replace the feedback constant β by $-\beta$, it is easy to show that Eq. (4-1) changes to the form given by Eq. (5-1)

$$A_{vf} = \frac{A}{1 - A\beta} \tag{5-1}$$

If the denominator, $1 - A\beta$, approaches zero, then A_{vf} will approach infinity. We might suppose that achieving a large value of A_{vf} would be a good thing; sometimes this is true, though often not. Suppose, for example, that our amplifier has a *DC* supply voltage of 15 V, A_{vf} has become 100,000, and an input signal of 1 mV is applied. It looks as if v_o should equal $100{,}000 \times 10^{-3} = 100$ V. This is clearly impossible, since the supply voltage is only 15 V. What will happen is that one or more transistors will saturate or cut off, limiting the output to 15 V. Because of this severe clipping, the wave form of v_o will be grossly distorted in comparison to v_s. In effect, the amplifier has gone into another (nonlinear) mode of operation in contrast with its normal (linear) mode. Under certain conditions, this new mode of operation may give rise to an oscillatory output; the device we presumed to be an amplifier has unexpectedly become an oscillator.

In the case we have just discussed, we deliberately replaced β by $-β$ so that negative feedback became positive feedback. Of more interest is the case when negative feedback somehow becomes positive without our making deliberate changes in the circuit wiring. To answer the question of how this can occur, let us return to Eq. (4-1), which was derived under the assumption that the feedback is negative. From Eq. (4-1) we see that A_{vf} can approach infinity when $1 + Aβ \rightarrow 0$. Or, we can infer that a critical situation occurs when

$$Aβ = -1 \tag{5-2}$$

Recall that in deriving Eq. (4-1) we made use of the fact that $v_i = v_s - v_f$. In other words, 180° of phase shift was built in to the system. Hence, Eq. (5-2) tells us that if another 180° of phase shift occurs in the quantity $Aβ$ (which is the loop gain or, equivalently, the return ratio **T**), then instability occurs.

Since we shall be concerned with phase shifts from this point forward, it is appropriate to refer to voltages and currents in their phasor forms, **V**, **I**, and so on. To see how instability may occur, let us consider as an example, the circuit shown in Fig. 5-1.

To simplify things we shall assume that the bias resistors R_1 and R_2 are all large enough that they may be neglected. Similarly, we shall assume that the bypass and coupling capacitors C_E and C_c are all so large that at all frequencies of interest, we may neglect them and the emitter resistors R_E. We shall also assume that the input capacitor C_1 is so large that it, too, may be considered to be a short circuit

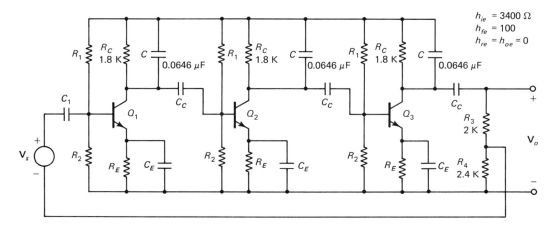

Fig. 5-1 Feedback amplifier to be analyzed.

to AC signals. What is left is an amplifier having three identical stages in which the parameters of importance are h_{ie} and h_{fe} of the transistors, collector resistors R_c, shunt capacitors C, and the load resistors R_3 and R_4. The resulting equivalent circuit is shown in Fig. 5-2.

To start the analysis we note first that $\mathbf{V}_i$ is given by

$$\mathbf{V}_i = \mathbf{V}_s + \mathbf{V}_f$$

This would seem to suggest that the feedback is positive rather than negative. However, we observe that since there are three common-emitter stages in this amplifier, at midrange frequencies there is an overall phase shift of 180°. Thus, the phase of $\mathbf{V}_o$, and hence of $\mathbf{V}_f$, is 180° with respect to $\mathbf{V}_i$. The feedback is therefore negative.

To simplify the analysis, we shall first find the transfer function that relates $h_{fe3}\mathbf{I}_{b3}$ to $\mathbf{I}_{b1}$. Starting with Q_1, we apply the current-division rule to find $\mathbf{I}_{b2}$ in terms of $\mathbf{I}_{b1}$. This yields

$$\mathbf{I}_{b2} = -h_{fe1}\mathbf{I}_{b1} \frac{R_c\|(1/j\omega C)}{h_{ie2} + R_c\|(1/j\omega C)}$$

Similarly,

$$\mathbf{I}_{b3} = -h_{fe2}\mathbf{I}_{b2} \frac{R_c\|(1/j\omega C)}{h_{ie3} + R_c\|(1/j\omega C)}$$

Substituting the first expression into the second and noting that $h_{fe1} = h_{fe2} = h_{fe}$ and $h_{ie2} = h_{ie3} = h_{ie}$, we obtain

$$\mathbf{I}_{b3} = h_{fe}^2 \, \mathbf{I}_{b1} \left(\frac{R_c\|(1/j\omega C)}{h_{ie} + R_c\|(1/j\omega C)} \right)^2$$

Let us denote the ratio $\mathbf{I}_{b3}/\mathbf{I}_{b2}$ by $\mathbf{A}_I$, where

$$\mathbf{A}_I = h_{fe}^2 \left(\frac{R_c\|(1/j\omega C)}{h_{ie} + R_c\|(1/j\omega C)} \right)^2$$

We can now redraw the circuit in the form shown in Fig. 5-3.

Next we find A_{vf}. To do this we must find T, a task that may be done conveniently with respect to the controlled source $\mathbf{A}_I\mathbf{I}_{b1}$. When the current source $\mathbf{A}_I\mathbf{I}_{b1}$ is replaced by an independent one of $\mathbf{A}_I$ amperes, $\mathbf{I}_{b3}$ is equal to $\mathbf{A}_I$. The source $h_{fe3}\mathbf{I}_{b3}$ therefore produces $h_{fe3}\mathbf{A}_I$ amperes. With $\mathbf{V}_s$ shorted, we find $\mathbf{I}_{b1}$ using the current-division rule twice. Thus we have

$$\mathbf{I}_L = -h_{fe3}\mathbf{A}_I \frac{R_c\|\mathbf{Z}_c}{R_3 + R_4\|h_{ie} + R_c\|\mathbf{Z}_c}$$

from which we obtain

$$\mathbf{I}_{b1} = \mathbf{I}_L \frac{R_4}{h_{ie} + R_4}$$

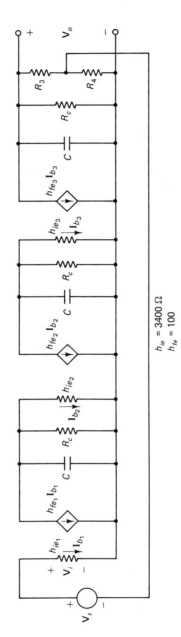

Fig. 5-2 Equivalent circuit of feedback amplifier.

$h_{ie} = 3400 \ \Omega$
$h_{fe} = 100$

164

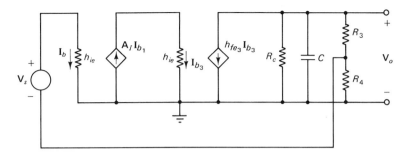

Fig. 5-3 Simplified equivalent circuit.

Since $\mathbf{T} = -\mathbf{I}_{b1}$, we have

$$\mathbf{T} = h_{fe3}\mathbf{A}_I \frac{R_c\|\mathbf{Z}_c}{R_3 + R_4\|h_{ie} + R_c\|\mathbf{Z}_c} \cdot \frac{R_4}{h_{ie} + R_4}$$

To find K we simply consider that as $\mathbf{A}_I$ becomes infinite, $\mathbf{I}_{b1}$ becomes zero and the output voltage $\mathbf{V}_o$ is maintained. With $\mathbf{I}_{b1} = 0$, no external current is drawn from the R_3-R_4 voltage divider, so by Kirchhoff's voltage law,

$$-\mathbf{V}_o \frac{R_4}{R_3 + R_4} - \mathbf{V}_s = 0$$

or

$$K = \frac{\mathbf{V}_o}{\mathbf{V}_s} = -\frac{R_3 + R_4}{R_4} \tag{5-3}$$

Let us now simplify the expression for $\mathbf{T}$. Writing out $\mathbf{A}_I$ with $R_c\|\mathbf{Z}_c$ expressed as $(R_c/j\omega C)/(R_c + 1/j\omega C)$, we have

$$\mathbf{A}_I = h_{fe}^2 \left(\frac{\dfrac{R_c/j\omega C}{R_c + 1/j\omega C}}{h_{ie} + \dfrac{R_c/j\omega C}{R_c + 1/j\omega C}} \right)^2$$

After some algebraic manipulation, this reduces to

$$\mathbf{A}_I = \left(\frac{h_{fe}R_c}{h_{ie} + R_c} \right)^2 \left(\frac{1}{1 + j\omega C \dfrac{R_c h_{ie}}{R_c + h_{ie}}} \right)^2$$

Since the term $R_c h_{ie}/(R_c + h_{ie})$ simply represents the parallel combination of R_c and h_{ie}, let us call this combination R_p. And let us define a radian frequency ω_1 by letting

$$\omega_1 = \frac{1}{[R_c h_{ie}/(R_c + h_{ie})]\,C} = \frac{1}{R_p C}$$

We can now write $\mathbf{A}_I$ in the simplified form

$$\mathbf{A}_I = \left(\frac{h_{fe}R_c}{h_{ie} + R_c}\right)^2 \left[\frac{1}{1 + j(\omega/\omega_1)}\right]^2$$

Let us now specify, for the sake of simplicity, that the resistance combination $R_3 + R_4\|h_{ie}$ shall be designed so that

$$R_3 + R_4\|h_{ie} = h_{ie}$$

This causes the complicated term in the expression for $\mathbf{T}$ to become simply

$$\frac{R_c}{h_{ie} + R_c}\left[\frac{1}{1 + j(\omega/\omega_1)}\right]$$

Hence $\mathbf{T}$ becomes

$$\mathbf{T} = \left(\frac{h_{fe}R_c}{h_{ie} + R_c}\right)^3 \left[\frac{1}{1 + j(\omega/\omega_1)}\right]^3 \frac{R_4}{h_{ie} + R_4} \tag{5-4}$$

As we discussed earlier, instability in A_{vf} occurs when $A\beta$ (or $\mathbf{T}$) approaches the value -1. There are a variety of ways to investigate how $\mathbf{T}$ behaves. All these are directed toward answering the question: Given the behavior of the *open-loop* gain (or return ratio $\mathbf{T}$), what will be the behavior of the *closed-loop* gain A_{vf}? The important point to note is that there is no need to calculate A_{vf} in order to answer this question.

5-2 NYQUIST DIAGRAMS

One way to deal with the problem is to make a plot of the complex quantity $\mathbf{T}$ as a function of frequency. The fact that $\mathbf{T}$ is complex suggests that the plot be made using polar coordinates. We can then observe how the resulting locus behaves with respect to the "magic number" -1, or, as it is represented in the complex plane, $1 \angle 180° = -1 + j0$. Such a plot is called a *Nyquist plot*.

Let us see how a Nyquist plot would look for the $\mathbf{T}$ function given by Eq. (5-4). For convenience, we shall first put the equation in polar form. That procedure yields

$$\mathbf{T} = M\left(\frac{1}{\sqrt{1 + (f/f_1)^2}}\right)^3 \angle -3 \tan^{-1}(b/b_1) \tag{5-5}$$

where

$$M = \left(\frac{h_{fe}\,R_c}{h_{ie} + R_c}\right)^3 \frac{R_4}{h_{ie} + R_4}$$

and f/f_1 replaces ω/ω_1 (which equals $2\pi f/2\pi f_1$). Rather than make a tedious point-by-point calculation of $\mathbf{T}$, we shall evaluate it at a few points where the calculation can be done by inspection. If a few more points are then needed, we can expend only as much effort as those few points require.

To give our work a more practical flavor, let us introduce numerical values. For M we have

$$M = \left(\frac{100 \times 1.8}{1.8 + 3.4}\right)^3 \left(\frac{2.4}{3.4 + 2.4}\right) = 1.72 \times 10^4$$

The frequency f_1 is equal to

$$f_1 = \frac{1}{2\pi R_p C} = \frac{1}{2\pi (R_c \| h_{ie})C} = \frac{10^6}{2\pi \times (1.8 \text{ K} \| 3.4 \text{ K}) \times 0.0676}$$

$$= 2000 \text{ Hz}$$

Table 5-1 shows values of $(1/\sqrt{1 + (f/f_1)^2})^3$ and $-3 \tan^{-1}(f/f_1)$ for certain values of f chosen to make the calculations trivial. Using just these three points we can begin to sketch the Nyquist plot. While much detail will be missing, we can at least begin to get some idea of the shape of the plot. The result is shown in Fig. 5-4. The three points are identified on the plot. Remember that all radii must be multiplied by $M = 1.72 \times 10^4$.

A critical question is whether the polar plot extends beyond the point $-1 + j0$. Put another way, the question is: Is there a frequency at which

$$M\left(\frac{1}{\sqrt{1 + (f/f_1)^2}}\right)^3 \angle{-3 \tan^{-1}f/f_1}$$

is real and negative, and has magnitude greater than 1? Let us answer this question by dealing with the angle first, for it is clear that under the condition postulated

$$-3 \tan^{-1}\left(\frac{f}{f_1}\right) = -180°$$

Hence $\tan^{-1}(f/f_1) = 60°$, or $f/f_1 = \sqrt{3}$. This value of f thus must be $\sqrt{3} \times 2000 = 3464$ Hz. At this frequency the magnitude of $\mathbf{T}$ becomes $1.72 \times 10^4 \times (1/\sqrt{1 + 3})^3 = 0.215 \times 10^4$, which is clearly much greater than 1. In any case, this calculation gives us a fourth point to add to the sketch.

At this point it is necessary to state certain principles that apply to the plotting and interpretation of Nyquist diagrams. The proof of these principles is beyond the scope of this book and will not be attempted. One of these concerns the use of negative frequencies; in plotting Nyquist diagrams, values of f are assumed to

TABLE 5-1

f (Hz)	$\left(\dfrac{1}{\sqrt{1 + (f/f_1)^2}}\right)^3$	$-3 \tan^{-1}\dfrac{f}{f_1}$
0	1	0°
2000	$\dfrac{1}{2\sqrt{2}}$	$-135°$
∞	0	$-270°$

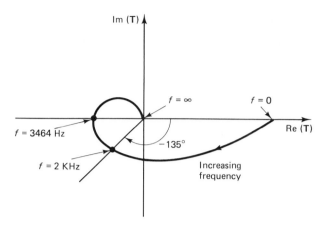

take on all values from $-\infty$ to $+\infty$. Notice that for any value of f—for example, f_k—there is a companion value, $-f_k$, which determines another point in the plane at which the magnitudes of the radii are equal and the angles are also equal but of opposite sign. This means that for every partial plot made for positive frequencies, such as the one we have sketched in Fig. 5-4, there is a second one for negative frequencies, which will be a mirror image of the first. The two branches will be exactly symmetrical with respect to the real axis. This idea is illustrated in Fig. 5-5. Note that the dotted-line branch results from considering negative frequencies.

It is one of the peculiarities of the Nyquist plot that the points representing $f = +\infty$ and $f = -\infty$ coincide. For this reason, the complete plot is always a closed figure of some sort.

How do we interpret the figure? To answer this question we state a second principle: Starting from the point corresponding to $f = -\infty$, imagine that a pencil traces over the figure in a direction corresponding to increasing values of f until it arrives at the point corresponding to $f = +\infty$. Suppose that as the pencil moves, a vector from the point $-1 + j0$ tracks with it. Monitor the total angle swept out by this vector; if this angle is at least 360°, then the point $-1 + j0$ is considered to be enclosed and the system is unstable. For the amplifier we have been consid-

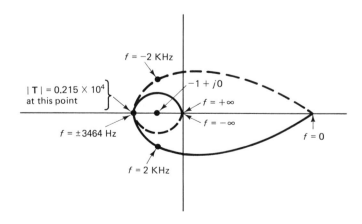

Fig. 5-5 Nyquist plot for all frequencies from $-\infty$ to $+\infty$.

ering, the point $-1 + j0$ is enclosed and the amplifier, though it incorporates negative feedback, actually develops positive feedback at a sufficiently high frequency.

In principle, the laboratory procedure suggested in Chap. 4 and illustrated in Fig. 4-7 should allow us to measure $\mathbf{T}$ both in magnitude and angle as a function of f. Subject to the practical limitations we discussed earlier, this is true; from such measurements, the positive-frequency branch of the Nyquist diagram could be plotted. Then the negative-frequency branch could be drawn by making use of symmetry.

Let us consider an additional example, which we shall derive from the first one. Suppose that in the circuit shown in Fig. 5-1, we discard Q_1 and Q_2 and couple $\mathbf{V}_s$ to the base of Q_3, leaving all circuit parameters unchanged. With this modification, it is easy to see that M now has the value

$$M = \left(\frac{100 \times 1.8}{1.8 + 3.4}\right)\left(\frac{2.4}{3.4 + 2.4}\right) = 14.3$$

The value of the frequency f_1 remains the same as in the last case, 2000 Hz, but the phase angle associated with $\mathbf{T}$ now becomes simply $-\tan^{-1}(f/f_1)$. Choosing easy values for f, we obtain Table 5-2.

The Nyquist plot corresponding to these values is shown in Fig. 5-6. It is clear that for this case, the point $-1 + j0$ is not enclosed, and the amplifier is stable at all frequencies.

5-3 BODE DIAGRAMS

We have seen that the Nyquist diagram provides a means of judging from open-loop gain and phase data whether the closed-loop amplifier will be stable. Unfortunately, it does not provide a convenient approach to the question of how to correct potentially unstable performance of an amplifier. A good way to do that is through the use of *Bode diagrams*.

We assume that you are already familiar with the essential principles of Bode diagrams. Accordingly, they will not be reviewed here. It is worthwhile to point out, however, that in a sense both the Nyquist plot and the Bode diagram present the same information about $\mathbf{T}$—that is, how its magnitude and phase vary with frequency. While the Nyquist plot shows the information on a single diagram, the

TABLE 5-2

f (Hz)	$\dfrac{1}{\sqrt{1 + (f/f_1)^2}}$	$-\tan^{-1}\dfrac{f}{f_1}$
0	1	$0°$
2000	$\dfrac{1}{\sqrt{2}}$	$-45°$
∞	0	$-90°$

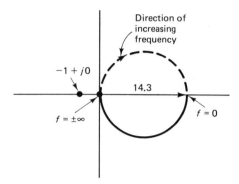

Fig. 5-6 Nyquist plot for single-stage amplifier.

Bode diagram does so on two separate graphs. On Nyquist plots, values of the independent variable, frequency, are hidden, whereas they are clearly shown on Bode diagrams.

Let us take the two examples we used in making Nyquist plots and construct Bode diagrams for them. For the three-stage amplifier, the magnitude of **T** for values of $f << f_1$ is equal to 20 log M = 20 log 1.72 × 10^4 = 84.5 db. The break frequency is f_1 = 2 KHz, from which the response rolls off at a slope of 3(-20) = -60 db per decade as frequency increases. The phase angle ϕ is equal to -3 tan^{-1}(f/f_1), as we saw earlier. The asymptotic and actual magnitude and phase plots are shown in Fig. 5-7.

For comparison we can plot the response curves for the single-stage amplifier. Here, M = 14.3, so that for $f << f_1$, 20 log M = 23.1 db. The break frequency remains equal to 2 KHz, whereas the roll-off is now at a slope of -20 db per decade. Phase angle ϕ is now equal to $-\tan^{-1}$ (f/f_1). The plots are shown in Fig. 5-8.

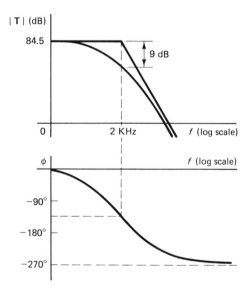

Fig. 5-7 Gain and phase plots of **T** versus f for the three-stage amplifier.

Dynamic Stability of Feedback Amplifiers   Chap. 5

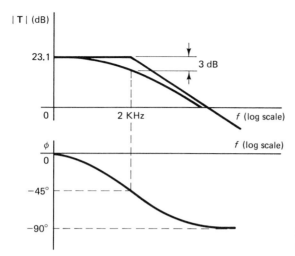

Fig. 5-8 Gain and phase plots of **T** versus f for the single-stage amplifier.

5-3.1 Gain and Phase Margins

As we saw earlier, the critical question is whether, at some frequency, **T** becomes equal to or worse than $1 \ \angle -180°$. What is meant by "worse than" was brought out in our discussion of Nyquist plots. When using Bode plots, we can approach this question in either of two ways. Either we can focus our attention on the frequency at which $|\mathbf{T}| = 1$ (or 0 db) and ask if the phase angle there is $-180°$ or less, or we can study the phase curve to see whether, for phase angles of $-180°$ or less, the magnitude of **T** has become 0 db or less.

To illustrate these ideas as well as to generate two additional definitions, let us consider the gain and phase plots shown in Fig. 5-9. First, we see that when the gain $|\mathbf{T}|$ equals 0 db, which corresponds to a numerical gain of unity, the corre-

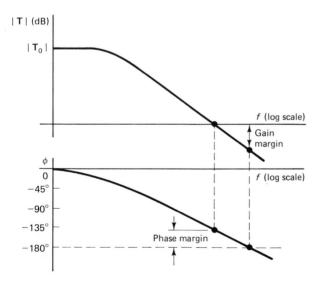

Fig. 5-9 Gain and phase plots showing gain and phase margins.

sponding phase angle ϕ is about $-150°$. Since the critical situation occurs when $|\mathbf{T}| = 1$ and $\phi = -180°$, we conclude that we are "safe" by 30°. Or, we say the *phase margin* is 30°.

Another measure of safety is that when the phase angle finally does reach $-180°$, the gain in decibels has reached some negative value. We call this value the *gain margin*. A phase margin of 50° and a gain margin of 10 db are considered comfortable. An amplifier designed to have such phase and gain margins would surely be stable despite changes in parameter values due to factors such as production variations.

5-4 COMPENSATING AMPLIFIER RESPONSE

The question that naturally arises is what remedy to seek if phase or gain margin requirements are not met. Examination of Fig. 5-9 suggests that bringing the gain down at a lower frequency (holding the phase curve constant) would increase the phase margin. Or, alternatively, causing the phase shift to reach $-180°$ at a higher frequency would increase the gain margin. To see how such effects can be achieved, we consider the two lag-type compensating networks shown in Fig. 5-10.

It is easy to show that the transfer functions for these networks are given by Eq. (5-6a) and (5-6b), corresponding to Figure 5-10(a) and 5-10(b), respectively:

$$\mathbf{A}_v = \frac{\mathbf{V}_2}{\mathbf{V}_1} = \frac{1}{1 + jf/f_o} \tag{5-6a}$$

where $f_o = 1/2\pi RC$, and

$$\mathbf{A}_v = \frac{\mathbf{V}_2}{\mathbf{V}_1} = \frac{1 + jf/f_1}{1 + jf/f_2} \tag{5-6b}$$

where $f_1 = 1/2\pi R_2 C$ and $f_2 = 1/2\pi(R_1 + R_2)C$. Bode diagrams for these networks are shown in Fig. 5-11.

We wish now to consider how we might apply one of these networks in compensating the open-loop response of an amplifier so that we can guarantee its closed-loop stability. Such a study is perhaps better done in terms of a numerical example. For that reason and to review the principles learned up to this point, we begin with the following example.

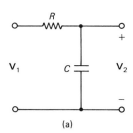 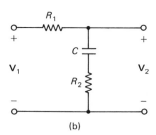

(a)   (b)

Fig. 5-10 Two lag-type compensating networks.

Dynamic Stability of Feedback Amplifiers   Chap. 5

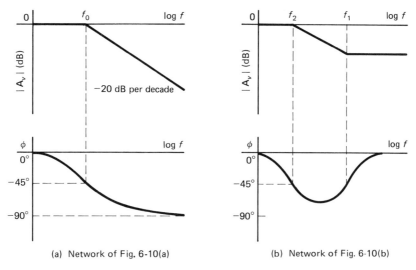

(a) Network of Fig. 6-10(a)   (b) Network of Fig. 6-10(b)

Fig. 5-11 Bode diagrams for compensating networks.

Example 5-1

We wish to determine whether the feedback amplifier shown in Fig. 5-12 will be stable. We begin by finding how **T** varies with frequency. We shall then plot Bode diagrams for the magnitude and phase angle of **T** and use these plots to discover whether or not stability exists.

We shall assume that at all frequencies of interest, the four coupling capacitors labeled C look like short circuits. To simplify the analysis, we shall treat the differential amplifier as a simple, controlled, voltage source of amplitude $A_{vd}(\mathbf{V}_f - \mathbf{V}_1)$, where $\mathbf{V}_f$ is the voltage that is fed back to the base of Q_2. The output resistance of this voltage source is 3 KΩ, whereas the input resistances seen at each base are $2h_{ie} = 5$ KΩ. The quantity A_{vd} is equal to $h_{fe}R_c/2h_{ie}$, or 100×3 K/5 K = 60.

The input resistance at the base of Q_3 is equal to $h_{ie} + (1 + h_{fe})R_E$, or approximately $h_{ie} + 100 \times 300$ Ω. Since the DC emitter current of Q_3 is about 10mA,

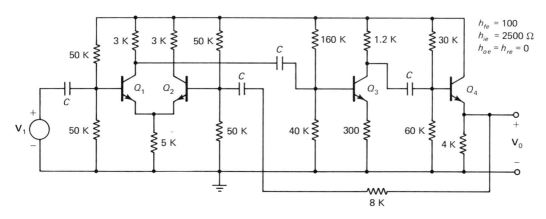

Fig. 5-12 Feedback amplifier.

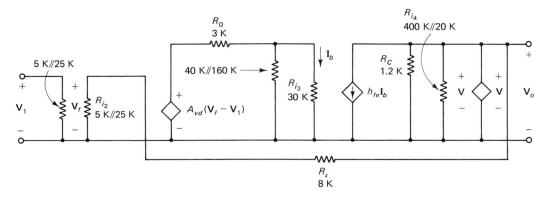

Fig. 5-13 Equivalent circuit of feedback amplifier.

h_{ie} for this stage is low so that R_{in} is approximated well enough by 30 KΩ. In a similar manner, we approximate the input resistance to the base of Q_4 as 100×4000, or 400 KΩ. Emitter follower Q_4 is treated as a voltage amplifier of gain equal to one with output resistance of zero.

Using these simplifications and approximations, we can represent the amplifier by the equivalent circuit of Fig. 5-13. Parallel resistor combinations occur on account of the biasing resistors.

To begin the analysis we calculate **T**, following the procedure just discussed. This results in the expression

$$\mathbf{T} = +h_{fe}(R_c\|R_{i4}) \frac{R_{i2}}{R_s + R_{i2}} \cdot \frac{A_{vd}}{R_o + 30\ \text{K}\|40\ \text{K}\|160\ \text{K}} \cdot \frac{40\ \text{K}\|160\ \text{K}}{30\ \text{K} + 40\ \text{K}\|160\ \text{K}}$$

$$= +100 \times 1.13\ \text{K} \cdot \frac{4.17\ \text{K}}{12.2\ \text{K}} \cdot \frac{60}{3\ \text{K} + 15.5\ \text{K}} \cdot \frac{32\ \text{K}}{62\ \text{K}} = 65.3$$

Expressed in decibels, this value of |**T**| corresponds to 36.3 db.

The asymptotic gain K is found by letting $\mathbf{I}_b$ become zero while the output voltage is maintained at $\mathbf{V}_o$. Since the requirement that $\mathbf{I}_b$ be zero means that the output of the differential amplifier $A_{vd}(\mathbf{V}_f - \mathbf{V}_1)$ must be zero, we conclude that $\mathbf{V}_f$ must equal $\mathbf{V}_1$. Since $\mathbf{V}_f$ equals $\mathbf{V}_o R_{i2}/(R_{i2} + R_s)$, we obtain

$$K = \mathbf{V}_o/\mathbf{V}_1 = (R_{i2} + R_s)/R_{i2} = \frac{12.2\ \text{K}}{4.17\ \text{K}} = 2.88$$

Therefore, A_{vf} must be equal to $2.88 \times 65.3/(1 + 65.3)$, or 2.84.

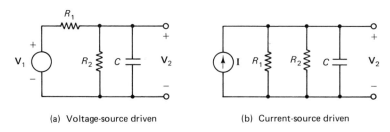

    (a) Voltage-source driven   (b) Current-source driven

Fig. 5-14 Shunt-capacitance circuit arrangements.

Our next task is to consider how the amplitude and phase angle of **T** vary with frequency. Suppose that the capacitance seen at the base of Q_2 is 58 pF, whereas those seen at the bases of Q_3 and Q_4 are 63 pF and 14.1 pF, respectively. It is not difficult to show that, for either of the circuit arrangements in Fig. 5-14, the transfer function $(\mathbf{V}_2/\mathbf{V}_1)$ or $(\mathbf{V}_2/\mathbf{I}_1)$ involves a multiplier of the form $1/(1 + jf/f_o)$. In fact, these networks turn out to be merely slight variations of the lag network shown in Fig. 5-10(a). The frequency f_o is given by $1/(2\pi \times R_p C)$, where R_p is the parallel combination of all the resistances. Such a multiplier causes the Bode diagram to have a break at f_o and a slope for $f > f_o$ of -20 db per decade. The phase shift due to this multiplier is $-\tan^{-1}f/f_o$.

When the shunt capacitances are taken into account, the circuit of Fig. 5-13 takes the form shown in Fig. 5-15.

The value of R_p for Q_2 is obtained from 4.17 K‖8 K and equals 2.74 KΩ. This number combined with the value of the shunt capacitance, 58 pF, yields a break frequency of 1 MHz. Similarly, R_p for Q_3 is 2.52 KΩ. Taken together with the 63-pF capacitance, this yields a break frequency that is also 1 MHz. The R_p value for Q_4 is 1.13 KΩ, and this number combined with the 14.1-pF capacitance yields a break frequency of 10 MHz.

You should recall that the phase angle corresponding to a multiplier of the sort produced by a shunt-capacitance circuit is zero at frequencies well below the break, is $-45°$ at the break frequency, and approaches $-90°$ for frequencies far above the break.

Let us now combine the information we have obtained thus far to construct a Bode diagram for **T**. This is shown in Fig. 5-16. Notice that 1 MHz is a "double-break" frequency; the slope for $f > 1$ MHz is -40 db per decade. Examination of Fig. 5-16 suggests that this amplifier will be unstable because, at the frequency where $\phi = -180°$, $|\mathbf{T}|$ seems to be above 0 db. We must exercise some caution in reaching this conclusion, however. First of all, we must remember that the magnitude part of the diagram is a straight-line approximation of the actual curve. Second, we cannot be too sure of the values of ϕ at the critical frequencies without actually calculating them. Nevertheless, the diagram does suggest that a more careful investigation is needed.

One way to approach the problem is to calculate the exact values of ϕ for frequencies above 1 MHz in order to find the frequency at which ϕ becomes $-180°$ Table 5-3 shows the results of such calculations. (Recall that $\phi = -2\tan^{-1}f/f_1 - \tan^{-1}f/f_2$.)

We see that at $f = 5$ MHz, ϕ has exceeded $-180°$. Therefore, we wish to

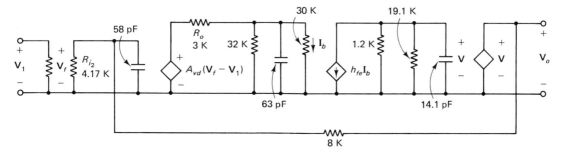

Fig. 5-15 Equivalent circuit of feedback amplifier including shunt capacitances.

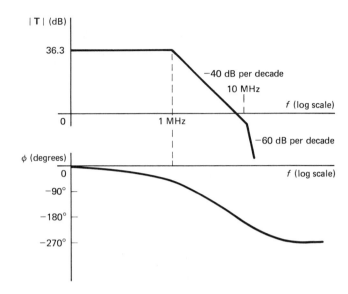

Fig. 5-16 Bode diagram of **T** versus f.

calculate the exact reduction in $|\mathbf{T}|$ that is produced by the three break factors at this frequency. This calculation is easily done if we remember that each factor (in terms of its *magnitude*) takes the form $1/\sqrt{1 + (f/f_o)^2}$. Hence the reduction of $|\mathbf{T}|$ at 5 MHz is given by

$$\left(\frac{1}{\sqrt{1 + (\frac{5}{1})^2}}\right)^2 \left(\frac{1}{\sqrt{1 + (\frac{5}{10})^2}}\right) = \frac{1}{26} \cdot \frac{1}{\sqrt{1.25}} = \frac{1}{29}$$

which corresponds to -29.2 db. Thus $|\mathbf{T}|$ at 5 MHz is $36.3 - 29.2 = 7.1$ db, and we see that our initial conclusion of instability is indeed verified.

We return now to the original question, which was how to compensate the response of this amplifier in order to make it stable. One obvious solution is simply to reduce the low-frequency level of $|\mathbf{T}|$. This could be done quite easily by reducing R_s to bring $|\mathbf{T}|$ (at low frequencies) below 29.2 db. But suppose that we are obliged to maintain the low-frequency value of $|\mathbf{T}|$. What can then be done to stabilize the amplifier?

One way to achieve stabilization is to "move" one of the pair of break factors by changing its break frequency to a value lower than 1 MHz. Suppose, for example, that we modify the amplifier network so that one of the three breaks occurs at 100 KHz. The actual circuit modification required to do this will be discussed a little later; at the moment, we shall consider only the effect of such a modification. This effect is shown in the Bode diagram of Fig. 5-17.

TABLE 5-3

f (MHz)	ϕ
1	$-95.7°$
2	$-138°$
3	$-163°$
5	$-183.9°$

Dynamic Stability of Feedback Amplifiers   Chap. 5

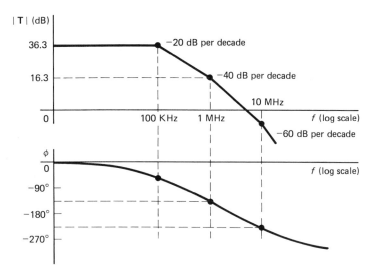

Fig. 5-17 Bode diagrams of modified network.

To see whether this proposed adjustment will result in a stable feedback amplifier, let us find the frequency at which ϕ becomes $-180°$. Then, as before, we can check whether the gain at that frequency is less than 0 db. This time ϕ is given by

$$\phi = -\tan^{-1}\frac{f}{100 \text{ KHz}} - \tan^{-1}\frac{f}{1 \text{ MHz}} - \tan^{-1}\frac{f}{10 \text{ MHz}}$$

Once again, the solution is found by trial and error as shown in the Table 5-4.

We see that at a frequency of about 4 MHz, ϕ becomes $-185.9°$. At this frequency the Bode diagram suggests that since 4 MHz is one decade plus two octaves higher than 100 KHz, then $|\mathbf{T}|$ must be $-20 + (-12)$, or -32 db below the 36.3-db level, or $+4.3$ db. This result suggests that the amplifier is still unstable. However, we must remember that the Bode (straight-line) diagram gives us only an approximation of the behavior of $|\mathbf{T}|$. To find out whether $|\mathbf{T}|$ truly exceeds 0 db, we must make an exact calculation. We do this by evaluating the break factors at 4 MHz:

$$\frac{1}{\sqrt{1 + (40)^2}} \cdot \frac{1}{\sqrt{1 + (4)^2}} \cdot \frac{1}{\sqrt{1 + (0.4)^2}} \approx \frac{1}{40} \cdot \frac{1}{4.13} \cdot \frac{1}{1.08}$$

or $1/179.7$, which corresponds to -45 db. Therefore, at 4 MHz, $|\mathbf{T}|$ is $36.3 - 45$, or -8.7 db, which is sufficient to insure stability—although this gain margin is not quite comfortable.

We have left unresolved the problem of what circuit modification to make in order to achieve this result. One such modification is to add shunt capacitance at the

TABLE 5-4

f (MHz)	ϕ
1	$-135°$
2	$-161.8°$
4	$-185.9°$

base of Q_2. Recall that the value of R_p for this part of the circuit was found to be 2.74 KΩ. Therefore, to produce a break frequency of 100 KHz, a value of C equal to $1/2\pi fR$ is required. Numerically, this is $1/(2\pi \times 2.74\ K \times 10^5) = 580$ pF. Since there is already a shunt capacitance of 58 pF present in the circuit at that point, all that is necessary to move the break frequency to 100 KHz is to add an additional capacitance equal to $580 - 58 = 522$ pF.

Compensation can equally well be accomplished by working with the base circuit of Q_3. Let us see how lag compensation might be done at this point in the circuit. Suppose that we add a series RC branch between the base of Q_3 and ground. This branch will appear in parallel with the 63-pF capacitance in the equivalent circuit of Fig. 5-15. If we replace the bias resistors, R_o, and the voltage generator representing the differential amplifier by a Thevenin-equivalent circuit, the resulting base network takes the form shown in Fig. 5-18.

The parallel combination of C_{in} and the RC network, together with R_{th}, form a voltage divider. Thus, in principle at least, it is easy to work out the expression for the transfer function of the network of Fig. 5-18. We shall omit the tedious details and state that the transfer function is given by

$$\mathbf{H}(\omega) = \frac{1 + j\omega RC}{1 - \omega^2 CC_{in}R_{th}R + j\omega C(R + R_{th})}$$

Using $f = \omega/2\pi$, $f_1 = 1/2\pi RC$, $f_2 = 1/2\pi C(R + R_{th})$, and $f_3 = 1/2\pi R_{th}C_{in}$, we can reduce the expression for $\mathbf{H}(\omega)$ to

$$\mathbf{H}(f) = \frac{1 + j(f/f_1)}{1 - (f^2/f_1 f_3) + j(f/f_2)} \tag{5-7}$$

Equation (5-7) appears to differ substantially from Eq. (5-6b), which represents the response of a lag network. The disagreement is caused by the troublesome $-f^2/f_1f_3$ term in the denominator. Let us see how important this term is by assuming some possible numerical values for f_1, f_2, f_3, and the variable frequency f.

In the circuit under consideration, the frequency f_3 is already determined. Since f_3 is given by $1/2\pi R_{th}C_{in}$, we observe that C_{in} is 63 pF and that R_{th} is the same as the resistance we have heretofore called R_p. For this stage, $R_p = 2.52$ KΩ. As we have already seen, these data fix f_3 at 1 MHz.

Suppose that we adjust the circuit parameters to cause f_1 to be 100 KHz and f_2 to be 10 KHz. If f (the independent variable frequency) equals 50 KHz, then the term $-f^2/f_1f_3$ equals $-(5 \times 10^4)^2/(10^5 \times 10^6)$, or -25×10^{-3}. Clearly the real part of the denominator, $1 - 25 \times 10^{-3}$, can be well approximated by 1. Even when $f = 10^5$ Hz, this term is $-(10^5)^2/10^{11} = -0.1$, and the error caused by considering the magnitude of the denominator to be $\sqrt{1 + (10^5/10^4)^2}$ instead of $\sqrt{(1 - 0.1)^2 + (10^5/10^4)^2}$ is negligible. The upshot of the matter is that the term $-f^2/$

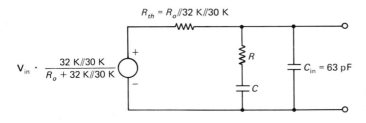

Fig. 5-18 Base circuit of Q_3 with lag network added.

$f_1 f_3$ can be neglected, and therefore the network of Fig. 5-18 can be treated as if the capacitance C_{in} were absent.

Even when f^2 is large compared to $f_1 f_3$, the annnoying term can be neglected. Suppose that with the values we have assumed for f_1 and f_2, the variable f has the value 10^6 Hz. Then $-f^2/f_1 f_3$ becomes $-10^{12}/10^{11} = -10$, which is certainly large compared to 1. However, the magnitude of the denominator of Eq. (5-7) now becomes $\sqrt{(1 - 10)^2 + (10^6/10^4)^2} \approx \sqrt{100 + 10^4}$. Once again we see that this troublesome term is negligible.

Using the values for f_1 and f_2 assumed above, let us see what effect this lag compensation will have upon the amplifier. Once again we plot the Bode diagrams for **T**, this time including the effect of the compensating network (see Fig. 5-19).

Figure 5-19 shows that the phase angle is asymptotic to $-180°$ but never exceeds that amount. Moreover, when |**T**| is at a level of 0 db, the phase angle is well above the $-180°$ level. To check this matter precisely, let us find the *crossover frequency*, that is, the frequency at which |**T**| reaches 0 db. Since the 0-db level corresponds to an absolute value of 1, we write the expression for |**T**| (taking into account the break factors) and set it equal to 1:

$$|\mathbf{T}| = 65.3 \frac{\sqrt{1 + (f/100 \text{ KHz})^2}}{\sqrt{1 + (f/10 \text{ KHz})^2}} \cdot \frac{1}{\sqrt{1 + (f/1 \text{ MHz})^2}} \cdot \frac{1}{\sqrt{1 + (f/10 \text{ MHz})^2}} = 1$$

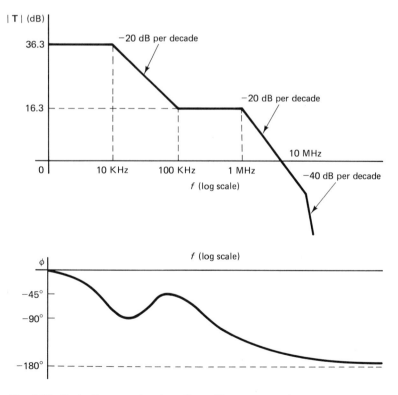

Fig. 5-19 Bode diagrams showing effect of lag compensation.

TABLE 5-5

Trial value of f	Calculated value of $|\mathbf{T}|$
4 MHz	1.47
6 MHz	0.922
5 MHz	1.14
5.7 MHz	0.99

It is, if not hopeless, certainly impractical to seek a "regular" analytical solution to this equation. But, guided by the Bode diagram of Fig. 5-19, it is not difficult to find the required value of f by trial and error. Table 5-5 shows such trials.

The value of 5.7 MHz for the crossover value of f is certainly acceptable. We shall use this value to calculate ϕ:

$$\phi = \tan^{-1}\frac{5.7}{0.1} - \tan^{-1}\frac{5.7}{0.01} - \tan^{-1}\frac{5.7}{1} - \tan^{-1}\frac{5.7}{10}$$

$$= +90° - 90° - 80.5° - 29.7° = -110.2°$$

This result shows that there is an ample phase margin of about 70°.

An alternative way to determine gain and phase margins is provided by Prog. 5-1. Note that this program assumes that $\mathbf{T}$ is given in the form

$$\mathbf{T} = K\frac{(1 + jf/z_1)(1 + jf/z_2) \cdots}{(1 + jf/p_1)(1 + jf/p_2) \cdots}$$

```
10   PRINT "COMPUTATION OF GAIN AND PHASE MARGINS"
11   REM   TRANSFER FUNCTIONS ARE CONSIDERED TO BE OF THE FORM
12   REM   H(S)=K((1+JF/Z1)*(1+JF/Z2)*...)
13   REM   /(1+JF/P1)*(1=JF/P2)*...)
20   INPUT "ENTER CONSTANT GAIN MULTIPLIER,K";K
25   PRINT
30   INPUT "ENTER NUMBER OF NUMERATOR BREAKPOINTS";NTP
35   PRINT
40   INPUT "ENTER NUMBER OF DENOMINATOR BREAKPOINTS";NBOT
42   PRINT
45   IF NTP = 0 THEN 80
50   FOR I = 1 TO NTP
60   INPUT "ENTER NUMERATOR BREAKPOINT VALUE IN HZ";Z(I): PRINT
70   NEXT I
80   FOR I = 1 TO NBOT
90   INPUT "ENTER DENOMINATOR BREAKPOINT VALUE IN HZ";P(I): PRINT
100  NEXT I
110  INPUT "GAIN MARGIN OR PHASE MARGIN?(G/P)";A$
111  PRINT
120  PRINT "SPECIFY FREQUENCY RANGE BY ENTERING LOW AND HIGH LIMITS."
121  PRINT
130  INPUT "LOW FREQUENCY LIMIT IN HZ?";FLO
131  PRINT
140  INPUT "HIGH FREQUENCY LIMIT IN HZ?";FHI
141  PRINT
142 F1 = FLO
143 F2 = FHI
150 G = 20 *  LOG (K) /  LOG (10)
151 DLST = 1
160 PLST = 1
165 X = 1
170  IF A$ = "P" THEN 210
180  GOTO 500
210  GOSUB 600
220  IF DB < .1 AND DB >  - .1 THEN 260
230  IF (DB * DLST) < 0 THEN X =  - X
240  GOSUB 300
```

Prog. 5-1.

```
250 DLST = DB
255  GOTO 210
260  GOSUB 400
270  PRINT "PHASE ANGLE AT GAIN CROSSOVER IS ";PHI;" DEGREES": END
300 TEMP = F1
310 F1 = F1 + X *  ABS (F2 - F1) / 2
320 F2 = TEMP
325  IF (FHI - F1) / (FHI - FLO) < .01 THEN  PRINT "NO ZERO CROSSING FOUND
    ": END
330  RETURN
400 PHI = 0
405  IF NTP = 0 THEN 440
410  FOR I = 1 TO NTP
420 PHI = PHI +  ATN (F1 / Z(I))
430  NEXT I
440  FOR I = 1 TO NBOT
450 PHI = PHI -  ATN (F1 / P(I))
460  NEXT I
461 PHI = (180 / 3.14159) * PHI
470  RETURN
500  GOSUB 400
510  IF (PHI + 180) < 0.1 AND (PHI + 180) >  - 0.1 THEN 560
520  IF ((PHI + 180) * (PLST + 180)) < 0 THEN X =  - X
530 PLST = PHI
540  GOSUB 300
550  GOTO 500
560  GOSUB 600
570  PRINT "GAIN AT PHASE CROSSOVER IS ";DB;" DB": END
600 DB = G
605  IF NTP = 0 THEN 640
610  FOR I = 1 TO NTP
620 DB = DB + 20 * ( LOG ( SQR (1 + (F(1) / Z(I)) ^ 2)) /  LOG (10))
630  NEXT I
640  FOR I = 1 TO NBOT
650 DB = DB - 20 * ( LOG ( SQR (1 + (F1 / P(I)) ^ 2))) /  LOG (10)
660  NEXT I
670  RETURN
```

Prog. 5-1 (cont.).

The price we pay in compensating the amplifier either by moving a break frequency lower or, as in this case, by lag compensation, is to reduce the bandwidth.

Having determined that lag compensation will stabilize the amplifier, we have yet to design the network to accomplish the task. The design equations for the frequencies f_1 and f_2 for the network of Fig. 5-18 are

$$f_1 = \frac{1}{2\pi\, RC}$$

and

$$f_2 = \frac{1}{2\pi(R + R_{th})C}$$

Since R_{th} is fixed, we must find R and C such that f_1 will be 100 KHz and f_2 will be 10 KHz. This means that we must require that $f_1 = 10f_2$. Hence we can write

$$\frac{1}{2\pi\, RC} = \frac{10}{2\pi(R + R_{th})C}$$

which leads to

$$\frac{1}{R} = \frac{10}{R + R_{th}}$$

Solving for R we get $R = R_{th}/9$. Since $R_{th} = 2.53$ KΩ, we find $R = 280$ Ω. Solving for C, we get $C = 1/2\pi \times 10^5 \times 280$, or $C = 5680$ pF.

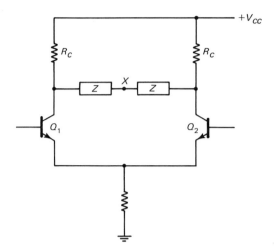

Fig. 5-20 Differential amplifier with impedance 2Z connected between collectors.

5-4.1 Compensating Networks Applied to Differential Amplifiers

There is another way to connect this *RC* compensating branch besides putting it between the base of Q_3 and ground. A glance at the schematic diagram of Fig. 5-12 should make it clear that this branch could equally well be connected between the collector of Q_1 and ground. A variation of this connection can be made by taking advantage of the symmetry of the differential amplifier. To understand how this is accomplished, let us consider the circuit shown in Fig. 5-20.

Since the collector voltages of Q_1 and Q_2 are 180° out of phase with each other and are of equal amplitude, the AC potential at point X is exactly zero. Hence the collector of Q_1 sees only the impedance **Z** connected (in effect) to ground. This means that we could connect an *RC* network having twice the impedance of the compensating network just designed between the collectors of Q_1 and Q_2. Since **Z** for the original network is equal to $R + 1/j\omega C$, 2**Z** must equal $2R + 1/j\omega(C/2)$. In other words, correct lag compensation can be obtained by connecting an *RC* network consisting of 560 Ω and 2840 pF between the collectors of the differential amplifier.

Some asymmetry exists in the circuit of Q_1 and Q_2 that is due to the fact that the collector of Q_1 is connected to the base circuit of Q_3, whereas the collector of Q_2 is floating. Since the resistance of the base circuit of Q_1 is about 15.5 KΩ but the collector resistance of Q_1 is 3 KΩ, the loading effect of Q_3 upon Q_1 is not large. However, we could correct this unbalance by reducing the collector resistance of Q_2 to a value equal to 3 K∥15.2 K = 2.52 KΩ.

REFERENCES

1 Jacob Millman and Christos C. Halkias, *Integrated Electronics: Analog and Digital Circuits and Systems* (New York: McGraw-Hill Book Company, 1972).
2 David C. Cheng, *Analysis of Linear Systems* (Reading, Mass.: Addison-Wesley Publishing Company, Inc., 1959).

3 Sundaram Seshu and Norman Balabanian, *Linear Network Analysis* (New York: John Wiley and Sons, Inc., 1959).

PROBLEMS

5-1. All capacitors are short circuits to AC except those equal to 0.01 μF. $g_m = 0.2$ mho, $R_D = 2$ KΩ, $r_d = \infty$. Assume R_B and R_G can be neglected. Find the return ratio **T** for this feedback amplifier.

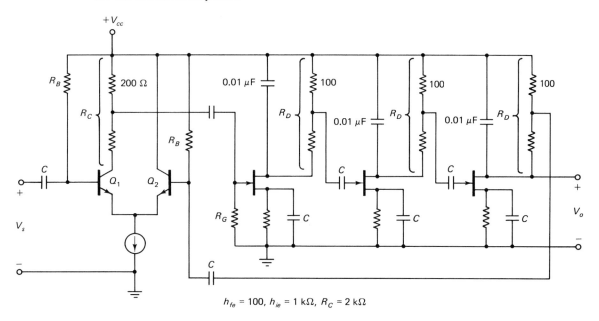

$h_{fe} = 100,\ h_{ie} = 1$ kΩ, $R_C = 2$ kΩ

5-2. **(a)** For the circuit of Prob. 1, sketch the Nyquist diagram.
  (b) Is the amplifier stable? If not, adjust the gains of each of the FET stages equally to produce stability.

5-3. Given

$$\mathbf{T} = \frac{20}{[1 + j(\omega/32.4)][1 + j(\omega/2.1)]}$$

  (a) Sketch the Bode plot of magnitude and phase.
  (b) Determine the phase margin of the system.

5-4.
  (a) Sketch the Bode gain and phase plots for this amplifier.
  (b) Find the gain and phase margins.

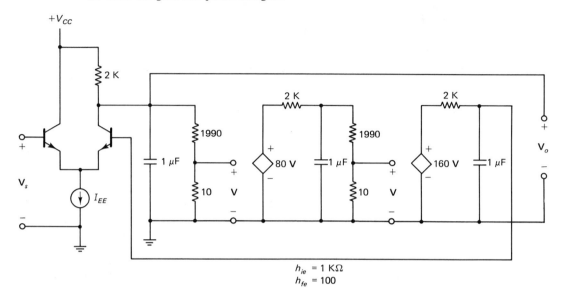

$$h_{ie} = 1 \text{ K}\Omega$$
$$h_{fe} = 100$$

5-5. (a) Find the maximum value of β for stable operation.
  (b) If $\beta = 0.01$, find the upper 3-db frequency for the closed-loop system.

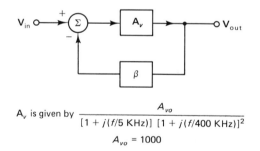

A_v is given by $\dfrac{A_{vo}}{[1 + j(f/5 \text{ KHz})] \, [1 + j(f/400 \text{ KHz})]^2}$

$$A_{vo} = 1000$$

5-6. An amplifier has an input resistance of 6 KΩ, an output resistance of 1 KΩ, and an open-circuit gain $\mathbf{A}_v$ that varies with frequency according to the asymptotic gain curve shown. The amplifier is connected in the feedback network shown. Assume that all capacitors labeled C are short circuits to AC. Resistors required for base bias of Q_1 and Q_2 are not shown but may be assumed to be so large as to allow them to be neglected.

   (a) Making reasonable approximations, calculate the value of **T** at low frequencies.

   (b) Assuming that $\phi = -\tan^{-1}(f/10 \text{ KHz}) - \tan^{-1}(f/40 \text{ KHz})$, calculate ϕ at the frequency at which $|\mathbf{T}| = 0$ db.

   (c) If lag compensation is used with $f_2 = 5$ KHz and $f_1 = 10$ KHz, estimate the gain-crossover frequency.

   (d) With this lag compensation, what is the value of ϕ at the gain-crossover frequency?

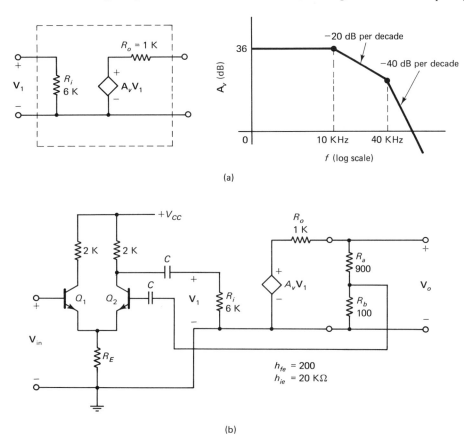

(a)

(b)

5-7. Design the lag compensation network for the circuit of Prob. 6, assuming that the network is to be connected between the collectors of Q_1 and Q_2.

5-8. The three-stage amplifier shown is to have a phase margin of 60°. The gain of stage 1 is given by

$$A_1 = \frac{-25}{1 + j(f/f_1)}$$

while stages 2 and 3 have gains given by

$$A_2 = A_3 = \frac{-20}{1 + j(f/10^6)}$$

(a) Find **T** for $f \to 0$.
(b) Find A_{vf} for $f = 0$.
(c) Determine the value of f_1 required to achieve the 60° phase margin.

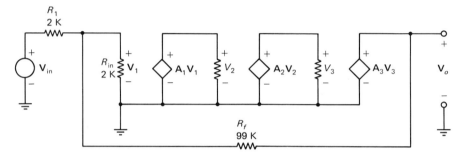

5-9. An amplifier has a return ratio $\mathbf{T} = \beta A_v/[1 + (f/f_1)^2]^{3/2} \angle -3 \tan^{-1}(f/f_1)$. If $A_v = 100$, find the maximum value of β beyond which the amplifier will become unstable.

5-10. Use the amplifier shown.
(a) Find the expression for V_4/V_1 in terms of K_a, K_b, C, and ω_1, where $\omega_1 = 1/RC$.
(b) Find an expression for **T**.
(c) If $\omega_1 = 10^4$ rad/s, find the frequency ω at which the phase angle of **T** equals $-180°$.

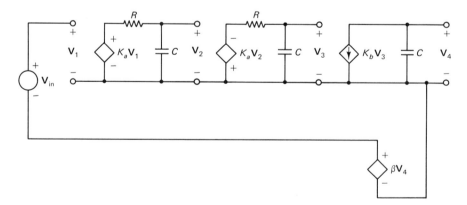

5-11. For the amplifier of Prob. 10, the value of C is 1 μF and $K_a = K_b = 10$.
(a) If $|\mathbf{T}|$ equals approximately 120 db when $\omega = 10$ rad/s, find β.
(b) Estimate the phase angle of **T** at a frequency ω that makes $|\mathbf{T}| = 0$ db.

5-12. (a) For the amplifier of Prob. 10, find the value of β required to make $\phi = -150°$ when $|\mathbf{T}| = 0$ db.

(b) Using this value of β, find the approximate value of $|\mathbf{T}|$ when $\omega = 10$ rad/s.

5-13. The network shown is a lag-lead compensating network. Show that, in terms of the Laplace variable S, $\mathbf{V}_2/\mathbf{V}_1$ is given by

$$\frac{\mathbf{V}_2}{\mathbf{V}_1} = \frac{(1 + SR_2C_2)(1 + SR_1C_1)}{R_1R_2C_1C_2S^2 + (R_1C_1 + R_2C_2 + R_1C_2)S + 1}$$

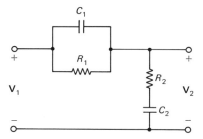

5-14. For the compensating network of Prob. 13, assume that $C_1 = C_2$ and $R_1 = R_2$. Also, $1/R_1C_1 = \omega_1$ radians per second.

(a) Determine the break frequencies in terms of ω_1.

(b) What is the ratio of break frequencies determined by the denominator of the expression found in Prob. 13?

5-15. For the compensating network of Prob. 14, let $\omega_1 = 1000$ rad/s.

(a) Sketch the straight-line Bode diagram for $|\mathbf{V}_2/\mathbf{V}_1|$ as a function of ω.

(b) By inspection of the network, determine decibel values of $|\mathbf{V}_2/\mathbf{V}_1|$ for $\omega = 0$ (DC) and $\omega \to \infty$.

5-16. For the network of Prob. 15, plot ϕ (the phase angle of $\mathbf{V}_2/\mathbf{V}_1$) as a function of ω. Discuss possible applications of this network in correcting a problematical phase margin in a feedback amplifier.

5-17. Shown below is the straight-line Bode diagram for $|\mathbf{T}|$ for a feedback amplifier.

(a) Find an expression for $\mathbf{T}$ for this amplifier in terms of the break frequencies and low-frequency amplitude.

(b) Find an expression for the phase angle of $\mathbf{T}$.

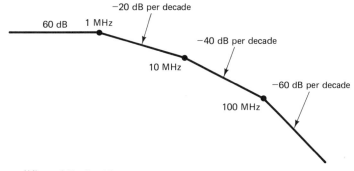

5-18. Use the amplifier of Prob. 17.

(a) Estimate the frequency at which $|\mathbf{T}| = 0$ db.

(b) Find the phase angle ϕ at this frequency.

(c) Find the frequency at which ϕ becomes $-180°$.

5-19. Assume that the compensating network of Fig. 5-10(a) is inserted into the amplifier circuit of Prob. 18 as shown. [The noninverting, controlled voltage sources keep the network from interacting with the rest of the amplifier circuit. Thus V_b/V_a is given by Eq. (5-6a).]

(a) If $1/2\pi RC = 1$ KHz, find the resultant $|T|$ when $f = 1$ MHz.

(b) Find the phase angle ϕ when $f = 1$ MHz.

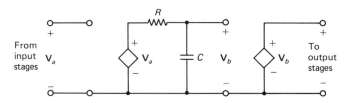

5-20. A feedback amplifier has a return ratio **T** whose magnitude varies as shown.

(a) Find **T** as a complex function of f, the frequency in Hertz.

(b) Calculate the frequency at which $|T|$ is 0 db. (Prog. 5-1 may be helpful.)

(c) Calculate the phase angle ϕ at this frequency.

(d) Is this amplifier stable? Explain.

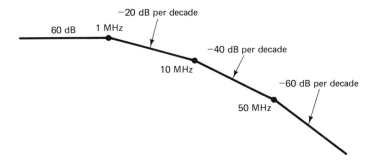

5-21. For the amplifier of Prob. 20, the compensating network of Fig. 5-10(b) is to be inserted according to the scheme used in Prob. 19.

(a) The compensating network is to use a capacitor whose value is 159 pF. Choose R_2 so that the break in the $|T|$ plot at 1 MHz will be eliminated.

(b) Determine R_1 so that a new break in $|T|$ will occur at 200 KHz.

5-22. Use the compensated amplifier of Problem 21.

(a) Find the phase angle at the frequency at which $|T| = 0$ db.

(b) For the frequency at which $\phi = -180°$, find $|T|$.

5-23. By how many decibels should the DC gain of the amplifier of Prob. 22 be reduced to provide a phase margin of 35°? At what frequency does gain crossover occur?

CHAPTER 6

Oscillators

6-1 CONDITIONS REQUIRED FOR OSCILLATION

The problem in designing an oscillator is the opposite of that encountered in designing a stable, feedback amplifier. To assure stability in a feedback amplifier, we saw that $\mathbf{T}$ could not be allowed to come close to the value $1 \underline{/180°}$. The design criterion that must be met to assure that a circuit *will* oscillate is that $\mathbf{T}$ (or $A\beta$) must *equal* $1 \underline{/180°}$. This requirement is known as the *Barkhausen criterion*. We have seen that the condition for stability of a feedback amplifier depended upon the *behavior* of $\mathbf{T}$ in the vicinity of $-1 + j0$ or upon the gain and phase *margins*. But in the case of an oscillator, we want $\mathbf{T}$ to equal exactly $-1 + j0$, or $1 \underline{/180°}$, expressed in polar form.

The reason for the Barkhausen requirement is not difficult to appreciate if we think a little about the way an oscillator works. What must happen in an oscillator is that while the amplifying part of the circuit develops an output signal, the feedback part of the circuit returns a portion of that output to the input port. This input signal has to have exactly the right amplitude and phase so that when it is amplified, the output that we assumed at first is produced. Clearly, the gain of the amplifier must be the reciprocal of the attenuation of the feedback network. Stated in terms of the return ratio $\mathbf{T}$, this means that $\mathbf{T}$ must be -1, or $1 \underline{/180°}$.

Linear oscillators that produce sinusoidal output wave forms make up one important category of oscillators. These all involve a frequency-dependent feedback network that acts to make $\mathbf{T}$ equal to -1 at the desired frequency but to have a magnitude less than one at all other frequencies. In designing an oscillator of this type, two basic problems must arise: (1) A frequency-dependent network must be

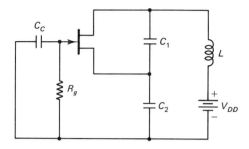

Fig. 6-1 Colpitts oscillator circuit.

provided with an attenuation that is a real number at some desired frequency, and (2) the amplifying element must provide a gain that will, in combination with the feedback network, cause **T** to equal -1.

There is another category of oscillator that operates in a nonlinear mode. An oscillator of this type has an amplifying element that is switched into saturation or cutoff. After some time elapses the device recovers, only to be switched again. The recovery time, which determines the frequency of oscillation, depends upon the time constants of the feedback network. Thus in this type of oscillator, the feedback network has a function in addition to providing a return path for signals transmitted from output to input.

There are also some types of switching-mode oscillators that generate sinusoidal wave forms. This is accomplished by using an *LC* filter (or tank) circuit at the output. Figure 6-1 shows one such oscillator, which is known as the *Colpitts oscillator*.

6-2 TAPPED-IMPEDANCE OSCILLATORS

The Colpitts circuit is one member of a class that uses a voltage-divider network formed by two branches having one type of reactance. This voltage divider is connected across a branch having the opposite type of reactance. Rather than analyze each of the oscillators of this type separately, we shall consider the general type shown in the equivalent circuit of Fig. 6-2. Note that we are considering the coupling capacitor C_c to be a short circuit and the gate resistor to be an open circuit.

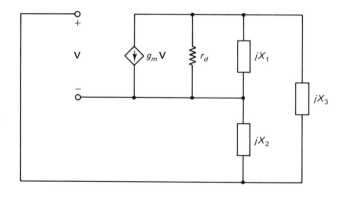

Fig. 6-2 Equivalent circuit of general tapped-impedance oscillator.

Oscillators   Chap. 6

We carry out this analysis by noting that, since this circuit is an oscillator, it has no input port in the usual sense. Or, viewed another way, the input of the circuit of Fig. 6-2 is short-circuited. That being the case, we have only to replace the current source $g_m V$ by an independent one of g_m amperes to find $\mathbf{T}$. Some advantage in computation is gained by exchanging the current source of g_m amperes in parallel with r_d for a voltage source of $g_m r_d$ volts in series with r_d. All these ideas are incorporated in the equivalent circuit shown in Fig. 6-3.

In terms of the loop currents shown in Fig. 6-3, a pair of simultaneous Kirchhoff's Voltage Law (KVL) equations can be written. These equations may be solved for the current $\mathbf{I}_2$ and the result multiplied by jX_2 to obtain $\mathbf{V}$. Since $\mathbf{V}$ is the variable that controls the controlled source $g_m \mathbf{V}$, then $\mathbf{T}$ must be the negative of $\mathbf{V}$. This means that the return ratio $\mathbf{T}$ is given by

$$\mathbf{T} = \frac{g_m r_d X_2 X_1}{X_1(X_2 + X_3) - jr_d(X_1 + X_2 + X_3)} \tag{6-1}$$

If the circuit is to oscillate, $\mathbf{T}$, as we have seen, must be equal to -1. Therefore, to begin with, $\mathbf{T}$ must be real. That means that the term $-jr_d(X_1 + X_2 + X_3)$ in the denominator of Eq. (6-1) must be zero. Consequently we see that

$$X_1 + X_2 + X_3 = 0 \tag{6-2}$$

This means that the reactances X_1, X_2, and X_3 must not be all of the same type; two may be capacitive, but then one must be inductive, and so on. This is true since $X_c = -1/\omega C$, whereas $X_L = \omega L$. Notice that Eq. (6-2) implies that the tank circuit must resonate at some frequency $\omega = 2\pi f$.

The result of the condition required by Eq. (6-2) is that Eq. (6-1) becomes

$$\mathbf{T} = \frac{g_m r_d X_1 X_2}{X_1(X_2 + X_3)} = \frac{g_m r_d X_2}{X_2 + X_3}$$

From Eq. (6-2) we see that $X_2 + X_3 = -X_1$, so this last equation becomes

$$\mathbf{T} = -\frac{g_m r_d X_2}{X_1}$$

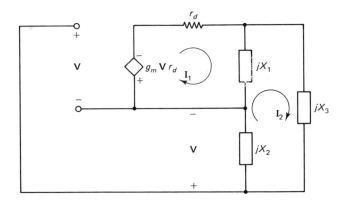

Fig. 6-3 Modified equivalent circuit.

Since **T** must equal -1, we obtain the relationship

$$\frac{X_2}{X_1} = \frac{1}{g_m r_d} \tag{6-3}$$

Equation (6-3) implies that the reactances X_1 and X_2 must be of the same type, be it inductive or capacitive. Furthermore, their ratio is the reciprocal of the gain that would be obtained if the FET could be operated with an infinite external drain resistor.

Example 6-1

Equations (6-3) and (6-2) are the two design equations we need to calculate the parameters for the Colpitts oscillator. Let us apply them for the case where $g_m = 4 \times 10^{-3}$ mho, $r_d = 20$ KΩ, and the frequency of oscillation is to be 10 MHz. The product $g_m r_d$ is equal to $4 \times 10^{-3} \times 2 \times 10^4 = 80$. Hence $(1/\omega C_2)/(1/\omega C_1) = 1/80$, or $C_1 = C_2/80$. Suppose that we have a coil whose inductance is 8.33 μH. Then, using Eq. (6-2), we have

$$\omega L = \frac{1}{\omega C_1} + \frac{1}{\omega C_2} = \frac{80}{\omega C_2} + \frac{1}{\omega C_2} = \frac{81}{\omega C_2}$$

Solving for C_2 we obtain

$$C_2 = \frac{81}{\omega^2 L} = \frac{81}{4\pi^2 \times 10^{14} \times 8.33 \times 10^{-6}}$$

which yields $C_2 = 0.246 \times 10^{-8} = 0.00246$ μF. Then $C_1 = 246 \times 10^{-11}/81 = 3.04 \times 10^{-11} = 30.4$ pF.

It is worthwhile to note that the series combination of C_1 and C_2 is equal to 30 pF, so that the effect of C_2 in determining the frequency of resonance is small. The function of C_2 is simply to allow 1/80 of the output voltage to be fed back to the gate of the JFET.

The preceding calculation is based on the idea that **T** must be precisely equal to -1. This is hardly a practical objective, since small changes occurring in g_m, r_d, or in the capacitance or inductance values could shift the value of **T** away from this number. For that reason it is desirable to design the network so that **T** will be about -1.05. In that way, one guarantees that the voltage fed back will be enough to maintain oscillation. However, the consequence of making |**T**| greater than 1 is that a gradual buildup in the level of output voltage can occur.

As the output-voltage level builds up, the voltage fed back to the gate of the JFET will also build up. Eventually, a level will be reached at which, on positive peaks of **V**, the gate-channel junction will be forward biased up to the threshold of forward conduction. This means that a very small current will flow into the gate, putting a small charge on the coupling capacitor C_c [see Fig. (6-1)]. This charge will be of a polarity that will make the gate side of C_c negative with respect to the tank-circuit side. On the negative swings of **V**, the instantaneous gate-channel voltage can now reach the pinch-off level. As this process continues, an average, negative, self-bias develops between the gate and the source of the JFET. Eventually, equilibrium is reached, so that the average charging current of C_c is balanced by a discharge current through R_g.

Since the drain current of the JFET is pinched off during a part of each cycle, the JFET is no longer operating in the linear mode. That is, the flow of drain-source current is actually interrupted for a part of each cycle of oscillation. Thus an oscillator of this type, although its initial operation may be in the linear mode, eventually operates as a nonlinear, or switching, device.

The self-biasing behavior of resonant-circuit oscillators of this type can actually be used as a means of verifying that oscillation is taking place. A high-impedance DC voltmeter connected between gate and source should indicate a small negative voltage if the circuit is oscillating. However, the impedance of the meter must be high compared to the resistance R_g if a meaningful result is to be obtained. Also, the instrument must show the DC level of the gate-source voltage, excluding the AC component. We must therefore be careful about interpreting the reading shown on, for example, a digital voltmeter used for this measurement.

If we exchange capacitors for inductors, and vice versa, in the circuit of Fig. 6-1, we obtain the circuit shown in Fig. 6-4, which is known as the *Hartley oscillator*.

The analysis of the generalized reactive-voltage-divider oscillator that resulted in Eqs. (6-1) and (6-2) applies also to the Hartley oscillator.

Example 6-2

We can carry out the design of a Hartley oscillator using the same parameters that were chosen for the Colpitts oscillator. That is, for $g_m = 4 \times 10^{-3}$ mho, $r_d = 20$ KΩ, $C = 30$ pF, and $f = 10$ MHz, we must find L_1 and L_2. Since $g_m r_d = 80$, then $X_2/X_1 = 1/80$. Since the total inductance $L_1 + L_2$ must resonate at 10 MHz with the 30-pF capacitor, we have

$$L_1 + L_2 = \frac{1}{4\pi^2 \times 10^{14} \times 30 \times 10^{-12}} = 8.33 \times 10^{-6} \text{ H}$$

Then $L_1 + L_1/80 = 8.33$ μH, so that $L_1 = 8.22$ μH, leaving a value of approximately 0.10 μH for L_2.

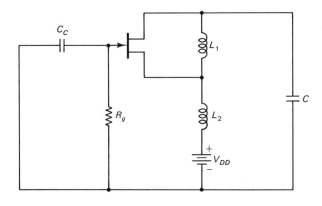

Fig. 6-4 Hartley oscillator.

The calculation in Example 6-2 was based on the assumption that any mutual-inductive coupling between the two inductors is negligible. While special steps could be taken to minimize such coupling (such as mounting the coils with their axes perpendicular to each other), this is rarely done. Hence, we must apply these design calculations with caution.

6-3 *RC* OSCILLATORS

The class of linear oscillators whose circuits do not involve resonant *LC* networks is sometimes referred to as the class of *RC oscillators*. Here the *RC* network must provide sufficient phase shift at some frequency to satisfy the Barkhausen criterion. One such circuit, known as the *phase-shift oscillator*, is shown in Fig. 6-5.

To simplify the analysis of this circuit, we make the following assumptions: (1) At frequencies of interest, capacitor C_s acts as a short circuit; (2) the drain resistance R_D has been so chosen that, in combination with the dynamic drain resistance r_d, a resistance equal to R results; (3) the gate resistor has also been made equal to R. Under these assumptions, the equivalent circuit shown in Fig. 6-6 results.

If we convert the current source $g_m V$ to a voltage source $g_m VR$ in series with resistor R, we can readily write loop equations and solve for the voltage V. Upon doing so, the following equation, written in terms of determinants, results.

$$
V = \frac{-g_m V R^4}{\begin{vmatrix} 2R + 1/j\omega C & -R & 0 \\ -R & 2R + 1/j\omega C & -R \\ 0 & -R & 2R + 1/j\omega C \end{vmatrix}}
$$

When the denominator is expanded and the controlled voltage source $g_m VR$

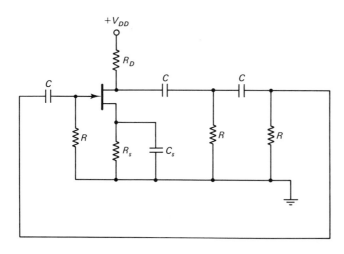

Fig. 6-5 Phase-shift oscillator.

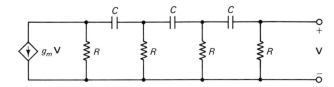

Fig. 6-6 Equivalent circuit.

is replaced with the independent source of $g_m R$ volts, we can find **T**:

$$\mathbf{T} = \frac{+g_m R_4 C^3 j\omega^3}{-j4(RC)^3\omega^3 - 10(RC)^2\omega^2 + j6RC\omega + 1}$$

Next we group the real and imaginary terms of the denominator together and then multiply the numerator and denominator by the complex conjugate of the denominator. This procedure yields

$$\mathbf{T} = \frac{+jg_m R^4 C^3 \omega^3 \{[1 - 10(RC)^2\omega^2] - j[6RC\omega - 4(RC)^3\omega^3]\}}{[1 - 10(RC)^2\omega^2]^2 + [6RC\omega - 4(RC)^3\omega^3]^2} \qquad (6\text{-}4)$$

This expression for **T**, according to the Barkhausen criterion, must equal $-1 + j0$. Accordingly, we must require that the imaginary term of the numerator, which is

$$+ jg_m R^4 C^3 \omega^3 [1 - 10(RC)^2\omega^2]$$

must equal zero. This will occur when

$$1 - 10(RC)^2\omega^2 = 0 \qquad (6\text{-}5)$$

from which we conclude that the frequency at which oscillation will occur must be

$$\omega = \frac{1}{\sqrt{10}RC} \qquad (6\text{-}6)$$

We must now examine the real part of Eq. (6-4) to discover the consequences of the requirement that the real part equal -1. Fortunately, we can now make use of the condition expressed by Eq. (6-5) to simplify Eq. (6-4). This results in

$$\mathbf{T} = \frac{+g_m R^4 C^3 \omega^3 [6RC\omega - 4(RC)^3\omega^3]}{[6RC\omega - 4(RC)^3\omega^3]^2} = -1$$

We can now see that

$$g_m R^4 C^3 \omega^3 = -6RC\omega + 4(RC)^3\omega^3$$

or

$$(g_m R^4 C^3 - 4R^3 C^3)\omega^2 = -6RC$$

Using the value of ω given by Eq. (6-6), we find

$$\frac{(g_m R - 4)R^3 C^3}{10R^2 C^2} = -6RC$$

from which it is easy to show that $-g_m R = 56$.

The quantity $-g_m R$ is the gain of a common-source, JFET amplifier with no external load. As we saw in Chap. 1, a gain of -56 is somewhat higher than typical. To examine this point in more detail, let us undertake the design of a phase-shift, JFET oscillator.

Example 6-3

Suppose we have a JFET whose data are as follows: $I_{DSS} = 9$ mA, $V_p = -3$ V, and $r_d = 20$ KΩ. We shall attempt to use this JFET in a phase-shift oscillator to operate at 5040 Hz. This rather odd frequency is chosen because it yields a value for $\omega = 2\pi \times 5040 = 3.16 \times 10^4$ rad/s. From Eq. (6-6) we find that $RC = 1/(\sqrt{10} \times 3.16 \times 10^4)$, or $RC = 10^{-5}$ s. Thus if we choose R equal to 10 KΩ, the capacitors must equal 1000 pF.

If we use a source resistor of 250 Ω, then we can bias the JFET to operate at $V_{GS} = -1$ V with a quiescent drain current of 4 mA. At this operating point g_m will be equal to 4×10^{-3} mho. Accordingly, the quantity $-g_m R$ will be equal to $-4 \times 10^{-3} \times 10^4 = -40$. Since, for this circuit configuration, the required value for A_v was found to be -56, we see that this design will not work!

To improve the situation we can bias the JFET so that the quiescent value of I_D will be closer to I_{DSS}, thus making g_m larger. In addition, we can also increase the value of R. Suppose then that we change R_s to 80 Ω, which will set the quiescent level of V_{GS} at -0.5 V. The quiescent drain current will then be 6.25 mA, and g_m will be 5×10^{-3} mho. At the same time we shall now choose a value of 12 KΩ for R. Under these conditions $g_m R$ will now become $5 \times 10^{-3} \times 1.2 \times 10^4$, or 60, which now meets the design requirement.

The circuit embodying this design is shown in Fig. 6-7. Notice that R_D has been chosen to be 30 KΩ. This is because the parallel combination of r_d and R_D yields 20 K‖30 K = 12 KΩ, which is the required value of R for this design. However, this large value of R_D gives rise to a very serious problem. With the quiescent value of I_D now set to 6.25 mA, the quiescent voltage across R_D must be 30 K $\times$ 6.25 mA, or 187.5 V! This high value implies that V_{DD} must be around 200 V.

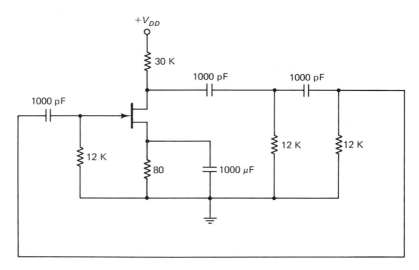

Fig. 6-7   Redesigned phase-shift oscillator.

This is not the only difficulty with this design. The fact that, with our best effort, we have just managed to achieve a value of A_v equal to -60—whereas a value of -56 is required—means that if a variation in parameters occurs, our design may be impossible to use. As we know, large parameter variations in JFETs are quite common. This would mean that to make this oscillator work, we might have to use only selected JFETs.

Example 6-4

Let us try a new design, this time using a BJT. The circuit for this design is shown in Fig. 6-8. Let us choose R to be 2 KΩ and then bias the transistor so that h_{ie} will have this same value. We shall neglect the influence of the base biasing network. For h_{ie} to be equal to 2 KΩ, I_c must be 1.3 mA, since h_{ie} will then be $(26/1.3) \times 100\ \Omega$. Therefore, let us set $R_2 = 20$ KΩ, $R_1 = 40$ KΩ, and $R_E = 1$ KΩ. If we make $V_{cc} = 6$ V, then the base voltage will be 2 V, the emitter voltage will be 1.3 V, and $I_c \approx I_E$ will be 1.3/1 K = 1.3 mA, as required. From Eq. (6-6) we know that C must be equal to $1/(\sqrt{10}\omega R)$. Since ω was chosen to be 3.16×10^4 rad/s, we find that C is equal to $1/(10 \times 10^4 \times 2 \times 10^3) = 0.5 \times 10^{-8} = 5000$ pF.

Instead of $g_m R$ we now must consider only A_v, which is now required to be at least -56. This requirement is easy to meet using a collector resistor of 2 KΩ.

6-4 AUTOMATIC GAIN CONTROL

We have not dealt with one matter in our discussion of the phase-shift oscillator— the fact that with $|A_v|$ in excess of 56, the amplitude of oscillations will build up to the point that the collector current of the transistor will be momentarily at the saturation level or cut off. Thus this oscillator will behave to some extent as the

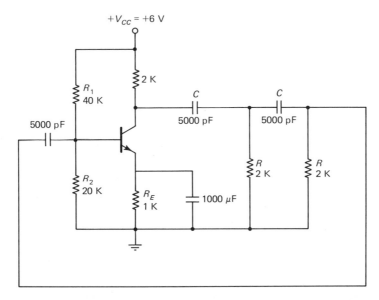

Fig. 6-8   Final design of BJT phase-shift oscillator.

resonant oscillators did. Unfortunately, however, *RC* oscillators have no tank circuits to filter the output and keep it sinusoidal. Therefore, some means must be found to control the gain of the amplifying element to keep it large enough to maintain oscillations but small enough to avoid clipping of the waveform.

One way to do this is to modify our transistor amplifier so that it includes an automatic gain-control feature. Since a common-emitter amplifier has $|A_v|$ equal to $h_{fe}R_c/h_{ie}$, and since h_{ie} is equal to $(0.026/I_E)h_{fe}$, we can control A_v by varying I_E. The circuit of Fig. 6-9 does this by incorporating a JFET in the emitter circuit of the BJT.

In this circuit the JFET is biased by means of a 250-Ω source resistor. (We are assuming that this JFET has the same parameters as the one used in Ex. 6-3.) Thus if no signal were present at the gate of the JFET, the drain current would have a quiescent value of 4 mA. This would also be the emitter current of Q_1, so that the quiescent value of h_{ie} for this transistor is about 650 Ω. The combination of bias resistors, the 1500-Ω series resistor, and h_{ie} causes capacitor C_4 to see a resistance roughly approximating 2000 Ω, as required by the oscillator design. The resistance looking into the emitter of Q_3 is only about 35 Ω. This means that a Thevenin-equivalent resistance of about 4 K‖4 K, or 2 KΩ, is seen looking back from the 5000-pF capacitor into the two 4-KΩ resistors. Thus the output resistance of the amplifier section is also held to the design value of 2 KΩ.

Capacitor C_1 bypasses Q_2 at the 5040-Hz frequency of oscillation. Thus transistor Q_1 operates, so far as AC signals are concerned, as a common-emitter amplifier. The gain of this stage is equal to $-h_{fe}R_c/h_{ie}$, or $-100 \times 150/650 = -230$ (assuming that h_{fe} is equal to 100).

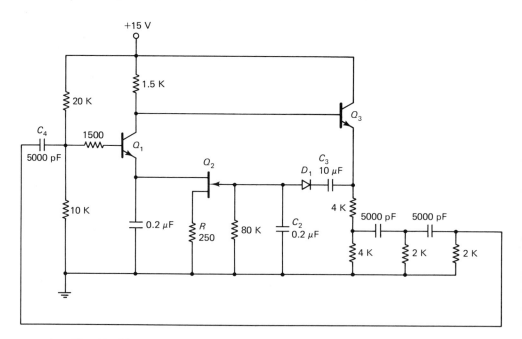

Fig. 6-9 Phase-shift oscillator with automatic gain control.

Transistor Q_3 operates as an emitter follower and thus has gain of 1 or less. The 5040-Hz output of Q_3 is connected via DC blocking capacitor C_3 to the rectifier-filter circuit consisting of D_1, D_2, and R. Notice that the time constant of R and C_2 is equal to 16 ms. This is quite large compared to the period of the 5040-Hz signal, which is about 0.2 ms. As a result, the bias voltage applied to the gate of Q_2 can not follow the variations of the signal cycle by cycle but instead responds only to slow variations in signal amplitude.

We might wonder whether the level of gate-bias voltage developed at Q_2 will be too large or too small for the amplitude of oscillation desired. We can make some estimate of the suitability of the bias-voltage level as follows: Since the quiescent voltage level at the collector of Q_1 is 12 V, the maximum swing that the collector voltage can have must be 6 V. That amount is just enough to bring the instantaneous collector voltage to the level of V_{cc}, which is $+18$ V. If we assume that the network comprising D_1, R, and C_2 operates to bring the voltage across C_2 to the peak negative excursion of the AC voltage wave, then a gate voltage of -6 V is applied to Q_2. Since V_p is only -3 V, Q_2 will be at pinch-off, no emitter current will flow in Q_1, and no oscillation will occur. Evidently, then, equilibrium must occur for some lesser amplitude of oscillation.

The voltage gain of the overall amplifier must include the effect of the tapped emitter resistor of the emitter-follower Q_3. This stage multiplies the voltage gain of Q_1 by one-half. To make $|A_v|$ (overall) equal to 56, we see that the gain of Q_1 must therefore be 112. Consequently, we require that h_{ie} be equal to $h_{fe}R_c/112$, or $100 \times 1500/112 = 1340\ \Omega$. This, in turn, requires that I_E be equal to $h_{fe} \times 0.026/h_{ie}$, or $100 \times 0.026/1340 = 1.94$ mA, or about 2 mA. Substituting this value for I_D in the JFET current equation, we find that the required value for V_{GS} is about -4.45 V. In other words, when the peak value of the 5040-Hz AC wave is about 4.5 V, the bias level of Q_1 will just produce the required value of A_v for sustained oscillation. Amplitudes greater than 4.5 V will cause the gain to fall below 56, thereby reducing the amplitude; for amplitude values below 4.5 V, the gain will be greater than 56, so that the oscillation amplitude will be increased.

6-5 THE WIEN-BRIDGE OSCILLATOR

Another widely used RC oscillator uses the Wien-bridge circuit. This configuration is shown in Fig. 6-10. For the moment we are representing the amplifying element of the circuit by a controlled voltage source that generates $A_v\mathbf{V}_i$ volts. This source has an input resistance equal to R_i and an output resistance R_o of 0 Ω. (Later we shall discard the unrealistic assumption that R_o is 0.) We are assuming that the output voltage of the source $A_v\mathbf{V}_i$ is in phase with the input voltage $\mathbf{V}_i$.

The Wien-bridge part of the circuit is said to be *balanced* when a voltage $\mathbf{V}_x$ is applied as shown and the voltage $\mathbf{V}_Q - \mathbf{V}_p$ becomes zero. It is clear that since $\mathbf{V}_p$ is determined simply by a resistive voltage divider, whereas $\mathbf{V}_Q$ depends upon the action of a voltage divider whose elements include reactances, balance can only occur when the phasor voltage $\mathbf{V}_Q$ is entirely real. Therefore, we begin our analysis by looking for the condition that will cause $\mathbf{V}_Q$ to be in phase with $\mathbf{V}_x$. We can

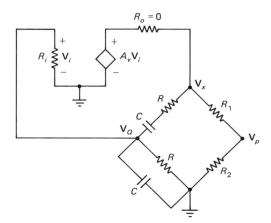

Fig. 6-10 Basic structure of the Wien-bridge oscillator.

write the expression for $\mathbf{V}_Q$ in terms of $\mathbf{V}_x$ as follows:

$$\mathbf{V}_Q = \frac{\dfrac{R(1/j\omega C)}{R + 1/j\omega C}}{R + \dfrac{1}{j\omega C} + \dfrac{R(1/j\omega C)}{R + 1/j\omega C}} \mathbf{V}_x$$

It is not difficult to show that

$$\mathbf{V}_Q = \frac{j\omega RC}{[1 - (\omega RC)^2] + j3\omega RC} \mathbf{V}_x \tag{6-7}$$

Now if the real term in the denominator could be made to vanish, then $\mathbf{V}_Q$ would become simply

$$\mathbf{V}_Q = \frac{j\omega RC}{j3\omega RC} \mathbf{V}_x = \frac{1}{3} \mathbf{V}_x \tag{6-8}$$

Then $\mathbf{V}_Q$ would be in phase with $\mathbf{V}_p$. If we set $R_1 = 2R_2$, then $\mathbf{V}_p$ would equal $\mathbf{V}_Q$ and the bridge would be balanced.

It is easy to see that the real term of the denominator will vanish when $\omega = 1/RC$. Obviously, this can occur at only one frequency. It is this property that we use in designing the Wien-bridge oscillator.

We have yet to see how the Barkhausen criterion is applied in making certain that the circuit of Fig. 6-10 will indeed oscillate. Before we do so, we need to make one more simplifying assumption. Notice that the input voltage $\mathbf{V}_i$ is the same as the voltage $\mathbf{V}_Q$ and that the resistor R between Q and ground is effectively in parallel with R_i. Our assumption is this: Let the lower bridge resistor R have some other value R', such that the parallel combination $R'\|R_i$ is equal to the original value R. Bearing in mind that R_o is zero, we can simplify the circuit to the form shown in Fig. 6-11.

It is clear that Eq. (6-8) applies to this modified circuit, yielding

$$\mathbf{V}_{\text{in}} = \tfrac{1}{3} A_v \mathbf{V}_{\text{in}}$$

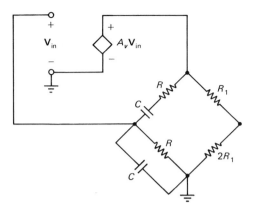

Fig. 6-11 Simplified circuit of Wien-bridge oscillator.

To calculate **T** we replace $A_v \mathbf{V}_{in}$ with A_v and take the negative of the resulting expression for **T**. This yields

$$\mathbf{T} = -\tfrac{1}{3} A_v$$

According to the Barkhausen criterion, **T** must equal -1. Therefore, we have

$$-1 = -\tfrac{1}{3} A_v$$

from which we conclude that A_v must equal at least $+3$.

If we look at the circuit of Fig. 6-11, we see that the resistors $2R_1$ and R_1 do nothing to produce oscillation, as they are not included in the feedback network. They are shown in the circuit at this point solely to remind you that the oscillator circuit has its origin in a standard, balanced-bridge structure. Since the oscillator will work perfectly well without resistors $2R_1$ and R_1, let us reconstruct the circuit of Fig. 6-10, omitting these resistors but restoring R_{in} and with R_o no longer required to be zero. The resulting circuit is shown in Fig. 6-12. The bridge resistors are now designated by R_a and R_b.

It should be clear that in the circuit of Fig. 6-12, we must require that $R_o + R_a = R$ and $R_{in} \| R_b = R$. Notice that the Wien-bridge oscillator requires an amplifying section whose output is in phase with its input. This is in contrast to the

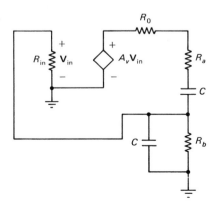

Fig. 6-12 Modified Wien-bridge oscillator circuit.

case of the phase-shift oscillator, where the amplifying element developed an output that was out of phase with its input.

One way to implement the Wien-bridge oscillator is by the use of a BJT differential amplifier. If the feedback signal is applied to one base while the other is at signal ground, either in-phase or out-of-phase output signals are available, depending on which collector is chosen for the output terminal. We can use the in-phase signal to drive the feedback network and the out-of-phase signal to drive the rectifier-filter circuit that controls the amplitude. These features are embodied in the circuit of Fig. 6-13.

The quiescent current passed by Q_3 is determined by the base current, which can be shown to be about 21 μA. That value of base current results in a collector current of about 2 mA, which sets the gain of the differential amplifier at a level of about 11.5, more than enough to ensure that oscillation will occur. The output resistance of the differential amplifier is very close to 300 Ω, so that, in series with the 500-Ω resistor, a bridge-resistance value of 800 Ω results. The resistance looking into the base of Q_1 is about 2600 Ω. This value in parallel with the 1150-Ω resistance yields 800 Ω, which agrees with the value just determined for the series resistor. The frequency of oscillation must therefore be equal to $1/2\pi RC$, or $1/(2\pi \times 800 \times 0.0636 \times 10^{-6}) = 3120$ Hz.

The output from the collector of Q_2 is coupled to the diode, which is polarized to develop an average voltage in a negative direction on the 0.2-μF capacitor, thus diverting base current from Q_3. This, in turn reduces the collector current of Q_3 and reduces the differential-amplifier gain until equilibrium occurs.

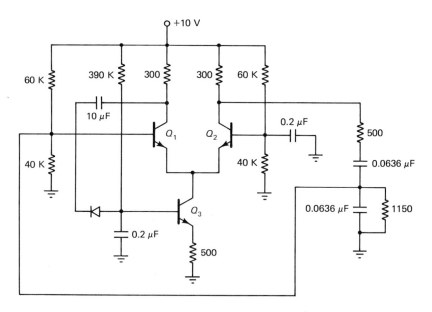

Fig. 6-13 Wien-bridge oscillator circuit.

REFERENCES

1. Jacob Millman and Christos C. Halkias, *Integrated Electronics: Analog and Digital Circuits and Systems* (New York: McGraw-Hill Book Company, 1972)
2. Lawrence B. Arguimbau, *Vacuum-Tube Circuits* (New York: John Wiley and Sons, Inc., 1948)

PROBLEMS

6-1. Find ω, the frequency of oscillation (radians per second) and the value of A required for stable oscillation.

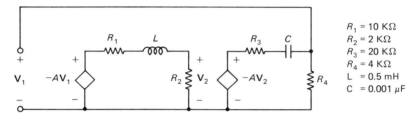

$R_1 = 10 \ K\Omega$
$R_2 = 2 \ K\Omega$
$R_3 = 20 \ K\Omega$
$R_4 = 4 \ K\Omega$
$L = 0.5 \ mH$
$C = 0.001 \ \mu F$

6-2. Amplifier K introduces 120° phase shift at all frequencies. Find the frequency of oscillation and required value of K.

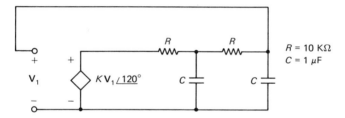

$R = 10 \ K\Omega$
$C = 1 \ \mu F$

6-3. Find the frequency of oscillation and the required value of A.

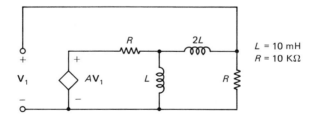

$L = 10 \ mH$
$R = 10 \ K\Omega$

6-4. What is the maximum value R can have that will just allow the circuit to maintain oscillation?

6-5. A proposed (simplified) oscillator circuit is given.

(a) Determine X_1 and X_2 so that the circuit will oscillate at a frequency of $\omega = 6 \times 10^6$ rad/s.

(b) Determine the values of the inductive or capacitive elements required. What modifications are required to make this circuit a practical one?

(c) What changes in C_1 and X_2 would be needed to maintain ω_o at 6×10^6 rad/s if a drain resistor of 2 KΩ were used?

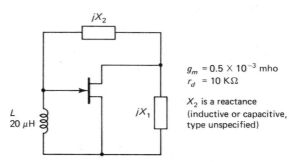

$g_m = 0.5 \times 10^{-3}$ mho
$r_d = 10$ KΩ

X_2 is a reactance
(inductive or capacitive,
type unspecified)

6-6. (a) Assume that the LC feedback circuit between points A and B is disconnected and write an expression for the voltage gain of the FET differential amplifier, V_B/V_A.

(b) With the LC circuit in place, find an expression for **T**.

(c) Determine the value for R that will just produce oscillation.

(d) Find the frequency of oscillation.

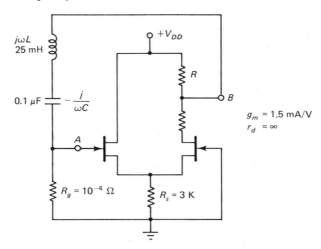

$g_m = 1.5$ mA/V
$r_d = \infty$

6-7. Modify the circuit diagram of the Colpitts oscillator shown in Fig. 6-1 to include the series resistance r of the inductor. Use this circuit to modify the expression for **T** given by Eq. (6-1) so that it takes into account the effect of r, thus showing that **T** is given by

$$\mathbf{T} = \frac{g_m r_d X_1 X_2}{X_1(\omega L + X_2) - r r_d - j[r X_1 + r_d(\omega L + X_1 + X_2)]}$$

6-8. (a) Use the expression for **T** given in Prob. 7 to determine the condition that makes $\phi = 0$.

(b) From the result in (a), show that the frequency of oscillation will be given by

$$\omega^2 = \frac{1}{L}\left[\frac{1}{C_2} + \left(1 + \frac{r}{r_d}\right)\frac{1}{C_1}\right]$$

6-9. For the circuit of Prob. 8, assume that $r/r_d \ll 1$ and show that the quantity $|g_m r_d|$ must satisfy

$$|g_m r_d| \geq \frac{C_2}{C_1} + \frac{(C_1 + C_2)}{L}rr_d$$

for oscillation to occur.

6-10. Design a Colpitts oscillator using the circuit of Fig. 6-1. The FET has $g_m = 10 \times 10^{-3}$ mho and $r_d = 5.5$ KΩ. For an inductor of 20 μH, choose values for C_1 and C_2 so that the circuit will oscillate at 5 MHz. Neglect coil resistance.

6-11.

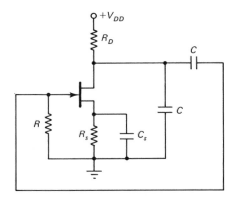

Shown above is a proposed oscillator circuit. The source resistor R_s is bypassed by capacitor C_s so that, for AC signals, it can be regarded as short-circuited. Drain resistor R_D is so chosen that $R_D \| r_d = R$.

(a) Show that **T** is given by

$$\mathbf{T} = \frac{g_m R}{3 + j(\omega RC - 1/\omega C)}$$

(b) Find an expression for the frequency of oscillation, ω.

(c) What is the minimum value of $g_m R$ required to produce oscillation?

6-12. A proposed oscillator circuit is given. Capacitor C_s bypasses R_s, so that at frequencies of interest, R_s appears to be short-circuited. R_D is so chosen that $R_D \| r_d = R$.

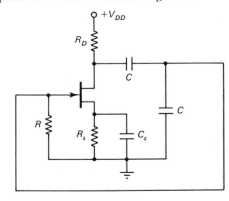

(a) Show that **T** is given by

$$\mathbf{T} = \frac{g_m R}{3 - j(\omega RC - 1/\omega RC)}$$

(b) Find an expression for the frequency of oscillation, ω.

(c) What minimum value of $g_m R$ will just permit oscillation?

6-13. For the oscillators of Probs. 11 and 12, the FET used has $g_m = 4 \times 10^{-3}$ mho and $r_d = 15$ KΩ. Oscillation frequency is to be 1326 Hz and capacitors of 0.024 μF are to be used.

(a) Determine R and R_D.

(b) Will oscillation occur?

6-14. For the circuit shown, assume that the emitter-follower has a voltage gain of approximately 1 and that its output resistance is negligible. The FET has a voltage gain equal to $-g_m(R_D\|r_d) = -10$.

(a) Find **T**.

(b) In terms of voltage gains, R and C, determine the conditions that must be satisfied to produce oscillation. Are these conditions realizable?

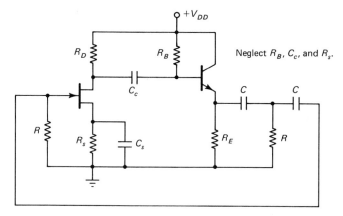

6-15. For the oscillator circuit shown, $R_D\|r_d = R$, C_s bypasses R_s to make it act as a short circuit, and R_g and C_c are so large compared to other circuit parameters that they can be ignored.

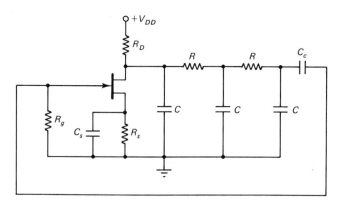

(a) Show that **T** is given by the expression

$$\mathbf{T} = \frac{g_m R}{[1 - 5(\omega RC)^2] + j[6\omega RC - (\omega RC)^3]}$$

(b) Find an expression for the frequency of oscillation.
(c) Determine the minimum numerical value for $g_m R$ in order that oscillation can occur.

6-16. For the oscillator circuit of Prob. 15, the FET has $g_m = 4 \times 10^{-3}$ mho and $r_d = 30$ KΩ. Determine R_D, R, and C so that the oscillator will operate at 5 KHz.

6-17. For the oscillator circuit shown, assume that the overall voltage gain between the gate of the FET and the emitter of the BJT is K, where $0 < K < 1$. Assume also that R_g and R_B are so large that they are negligible, that the coupling capacitor C_c is so large that it is negligible, and that the output resistance of the emitter follower is effectively zero.

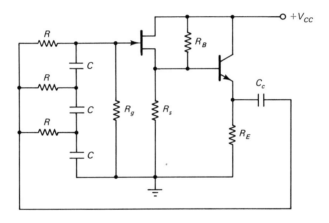

(a) Show that **T** for this circuit is given by

$$\mathbf{T} = K \frac{5/(\omega RC)^2 - j[1/(\omega RC)^3 - 6/\omega RC]}{1 - 5/(\omega RC)^2 + j[1/(\omega RC)^3 - 6/\omega RC]}$$

(*Hint:* Assume a current in the top resistor of the RC ladder. Since virtually no signal current enters the gate of the FET, this current must all pass through the top capacitor. By successive applications of KVL and KCL, work down to the bottom resistor and capacitor, expressing results in terms of the first current assumed.)

(b) Find an expression for the frequency of oscillation.
(c) What is the minimum numerical value of K that will produce oscillation?

6-18. For the oscillator of Prob. 17, $R = 4.7$ KΩ and $C = 0.028$ μF.
(a) Find the frequency of oscillation.
(b) If the voltage gain between the gate of the FET and the emitter of the BJT is K, determine the minimum numerical value of K to produce oscillation.

6-19. (a) Find an expression for **T** in terms of K, R_1, R_2, L, C and ω.
   (b) Find an expression for the frequency of oscillation.
   (c) Find an expression for the minimum value of K to produce oscillation.

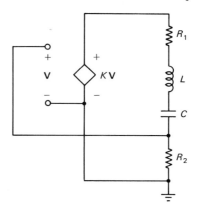

6-20. Assume C_s and C_c are short circuits to AC.
   (a) Determine the value of R_3 so that the circuit shown will just produce oscillation with L and C in series resonance.
   (b) Determine the frequency of oscillation.

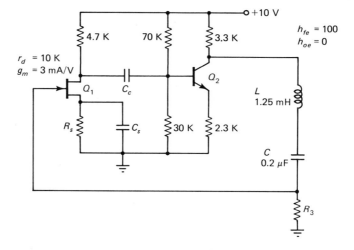

CHAPTER 7

Characteristics and Applications of Operational Amplifiers

7-1 PERFORMANCE CHARACTERISTICS

In Chap. 3 we studied the internal structure of operational amplifiers and learned that they have differential inputs, single outputs, and very high voltage gain. In Chap. 4 we found that an internal, compensating capacitor is often included in the circuit and that this capacitor resulted in a low value for the 3-db bandwidth of the amplifier. While these are important facts, operational amplifiers have such a wide variety of applications that it is necessary that we know still more about their characteristics.

Not a great deal has been said up to this point on the output capabilities of operational amplifiers. We have already seen examples in which the output section comprised a pair of transistors in a configuration resembling the totem-pole arrangement that we encountered in our study of logic-gate circuits. We avoided detailed analysis and treated these output circuits as if they were merely variations of the well-known emitter-follower circuit. This idea is essentially correct and will have to continue to serve until a later chapter, in which we shall deal with such power amplifiers in more detail. The important point for the present is that, like an ordinary emitter-follower, the typical output circuit of an operational amplifier has a low output resistance. It also has the capability of delivering or absorbing an output current that may be as much as 25 mA. The output-voltage range depends upon the supply voltage. Typical circuit arrangements call for both a positive and a negative voltage supply of 15 V. (Polarities are, of course, measured with respect to some reference point in the external circuit whose voltage is taken to be zero, or ground.) The output voltage can swing to values close to the positive or negative supply voltages. Thus for supply voltages of $+15$ V and -15 V, the output voltage can have a peak-to-peak value of almost 30 V.

An ideal operational amplifier would have, among other attributes, an infinite input resistance. In that way, the amplifier input resistance would have no loading effect upon any circuit connected to it but would merely sense the voltage applied. The situation is exactly analogous to the case of the ideal voltmeter, which would also have an infinite resistance. Obviously, a real operational amplifier has a finite input resistance. The best we can hope for is that this resistance will be large.

As we have seen, the input stage of an operational amplifier is invariably a differential amplifier. Thus if collector currents of 100 μA are used and if h_{fe} has a value of 100, we can expect a resistance from base to ground of about 50 KΩ. If each side of the differential pair consists of Darlington-connected transistors, we might expect that the input resistance could be as much as 1 MΩ. With FET transistors, input resistances as high as 10^9 Ω can be achieved.

At this point it is helpful to introduce a circuit model for the operational amplifier. Making use of the essential features that we know about up to now, we can symbolize the amplifier by the triangle shown in Fig. 7-1. An input applied to the positive terminal is amplified by a factor A_{voc} and produces an output that is in phase with the input. An input applied to the negative terminal produces an output that is 180° out of phase with the input but also multiplied by A_{voc}. These input terminals are referred to as the *noninverting* and *inverting* inputs, respectively.

A circuit model for the operational amplifier is shown in Fig. 7-2. The input and output resistances are included, and v_1 and v_2 are external input voltages.

All the considerations that applied to the differential amplifiers studied in Chap. 3 apply to the operational amplifier as well. Thus the quality of an operational amplifier depends upon its having a high CMRR and a constant DC output level. If we think back to the discussion regarding the use of differential amplifiers for instrumentation purposes, we can appreciate this last point. Suppose, for example, that the amplifier is used to connect the output of a temperature sensor to a pen recorder. If the temperature we are measuring is, for instance, the temperature of ocean water, which varies slowly over a day's time, then it is very important that the slowly varying amplifier output consist exclusively of the signal representing ocean temperature. Serious confusion would result if the output signal contained a component that was due to a slow drift of the amplifier's own quiescent operating point.

One possible source of such a slow drift is the thermal imbalance of transistor and resistor parameters in the input and intermediate stages of the integrated, operational-amplifier circuit itself. We know that the circuit is of monolithic construction on a tiny chip, so that temperature differences between one part of the chip and another are minimal. Also, as we learned in Chap. 3, the fabrication process for integrated circuits causes the parameters of transistors and resistors to be very closely matched. Nevertheless, small differences do exist.

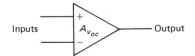

Inputs A_{voc} Output

Fig. 7-1 Symbolization for the operational amplifier.

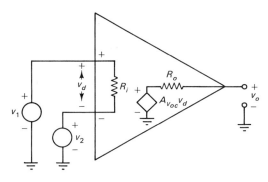

Fig. 7-2 Operational amplifier model.

There is not much that we can do about small temperature variations between points on a chip. Nor can we do anything about parameter variations in the intermediate or output stages of an operational amplifier. However, a good deal can be done about steady imbalances in the input stage, as we shall see in the following discussion.

Figure 7-3 shows the input stage of the 741 operational amplifier. (The complete circuit was shown in Fig. 3-31). Observe the 10-KΩ potentiometer connected

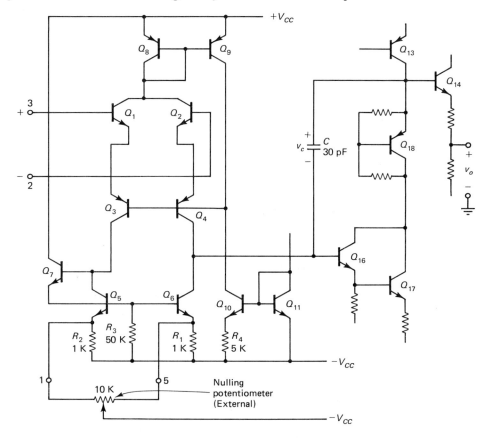

Fig. 7-3 Partial schematic of 741 operational amplifier.

Sec. 7-1   Performance Characteristics

externally between pins 1 and 5. With the slider of the potentiometer exactly centered, resistances of 5 KΩ are effectively placed in parallel with the 1-KΩ resistances of R_1 and R_2. If the input terminals (pins 2 and 3) are short-circuited and if all components of the operational amplifier are in exact balance, then the potentials at the emitters of Q_1 and Q_2 are exactly equal, the currents through the 5-KΩ added resistances will be equal, and no effect will be observed on the output to the second stage.

But if the emitter currents through Q_1 and Q_2 are in slight imbalance, an appropriate imbalance in the resistances in parallel with R_1 and R_2 can restore the balance of the emitter currents. This is the purpose of the 10-KΩ nulling potentiometer.

While the nulling potentiometer may "tease" the quiescent output voltage back to zero, it can have no effect on slight potential differences between the bases of Q_1 and Q_2 or upon the currents flowing into these bases. If the emitter currents in these transistors are slightly different, then the base-emitter voltages may differ, and a built-in potential difference will exist between the bases. On the other hand, even if the emitter currents are exactly equal, there may yet be differences in the h_{fe}s of these transistors, so that their base currents may be unequal. What all this means is that there may be two separate sources of unbalance in the input stage. These are referred to as the *input offset current* and the *input offset voltage*.

7-1.1 Input offset voltage

The input offset voltage V_{os} is the voltage required at one input (with the other grounded) that will bring the output voltage to a value of 0 V. We can represent the defining conditions by the diagram of Fig. 7-4. The input offset voltage varies with temperature. Manufacturers' data often include a number αV_{ID}, called the input-offset-voltage temperature coefficient. Typical values range from 5 to 10 μV/°C. Thus for αV_{ID} equal to 5 μV/°C, a temperature rise of 40°C would produce an offset voltage change of $40 \times 5 = 200$ μV.

7-1.2 Input bias and offset currents

In normal operation, there must be definite bias currents, I_1 and I_2, flowing into the respective input terminals even in the quiescent state, where the output is held at a level of 0 V. This flow is necessary because each base of the differential transistor pair in the input circuit requires a base bias current in order for the transistors to operate. The *difference* between these two currents is the input offset current, I_{os}. The defining conditions are illustrated in Fig. 7-5.

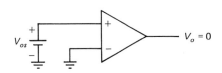

Fig. 7-4 Operational amplifier circuit to define V_{os}, the input offset voltage.

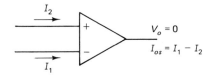

Fig. 7-5 Operational amplifier circuit to define I_{os}, the input offset current.

Each input has a total current flowing into it that is made up of the super-position of three component currents:

1. The bias component I_B needed merely to turn on the differential transistors.
2. The input offset current.
3. The current I that flows between the two input terminals, which is produced by the differential voltage input acting on the input resistance [or, $I = (V_2 - V_1)/R_i$].

The bias component is taken to be the average of the two currents I_1 and I_2; that is,

$$I_B = \frac{I_1 + I_2}{2}$$

If I_1 exactly equals I_2, then I_B will simply equal one of these currents.

If we simultaneously solve the equations for I_B and I_{os}, we can express the currents I_1 and I_2 in terms of I_B and I_{os}. This procedure yields

$$I_1 = I_B + \frac{I_{os}}{2} \tag{7-1}$$

and

$$I_2 = I_B - \frac{I_{os}}{2} \tag{7-2}$$

As we shall see later, it is often the case that operational amplifiers are connected in circuits where substantial amounts of resistance are placed in series with the input terminals. In such situations, the input currents can produce voltage drops such that a significant potential difference will be sensed at the input terminals. This difference, when magnified by the high gain of the amplifier, can result in serious displacement of the output voltage from its expected value.

Example 7-1

Suppose, for example, that an amplifier is connected as shown in Fig. 7-6. Clearly, the voltage $V' = V - I_1R$, or $V' = V - (I_B + I_{os}/2)R$. Typical values for the bias current I_B and the input offset current I_{os} are 300 nA and 50 nA, respectively. Thus we have

$$V' = V - 325 \times 10^{-9} \times 10^6 = V - 0.325 \text{ V}$$

Even if R were reduced to 10 K, there would still be a difference between V and V' of more than 3 mV.

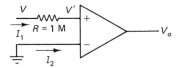

Fig. 7-6   Effect of input current.

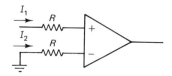

Fig. 7-7 Balanced input resistances.

The situation is greatly improved if the resistance R is placed in series with both input terminals. In that case, as illustrated in Fig. 7-7, the voltage deviation at the positive input would be the same, whereas that at the negative input would be equal to $(I_B - I_{os}/2)R$. For R equal to 1 MΩ, this latter figure would amount to $-275 \times 10^{-9} \times 10^6 = -0.275$ V. The resulting voltage *difference* between the input terminals would then be $-0.325 - (-0.275) = 0.05$ V. This result provides an argument for keeping the resistances seen from the two input terminals about equal.

Both the bias current and the input offset current vary with temperature. Typical variations of these quantities are shown in the curves of Fig. 7-8. Clearly, this and the temperature variation of offset voltage are potential sources of slow operating-point drift.

7-1.3 Slew Rate

There is one additional characteristic of operational amplifiers that remains to be discussed, the characteristic known as the *slew rate*. To explain this concept, it is helpful to make use of an actual circuit example. For this purpose we refer again to the partial schematic of the 741 operational amplifier shown in Fig. 7-3.

In Chap. 3 we saw that the current provided to the differential amplifier by the current-source transistor Q_8 was 30 μA. Call this current I_o. Suppose that the base voltage of Q_1 is suddenly driven in the negative direction so that Q_1 (and therefore Q_3) is cut off. The entire current I_o must therefore pass through Q_2 and Q_4 and be driven toward the base of the Darlington-connected amplifier Q_{16}-Q_{17}.

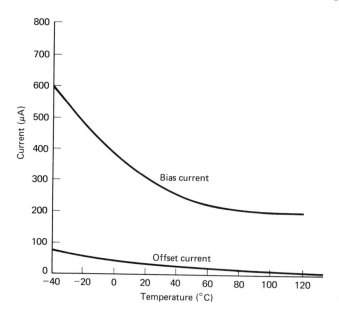

Fig. 7-8 Variation of bias and offset currents with temperature for type 741 operational amplifier.

These transistors present a large resistance compared to that seen looking into C, the 30-pF capacitor.

Recall that Q_{18} serves only to set bias-current levels in the output transistors and has negligible effect upon dynamic performance. Hence, in effect, the far end of C is connected to the collectors of Q_{16}-Q_{17}. The potentials of these collectors, the base of Q_{14}, and the output voltage v_o all vary together and differ from each other only slightly. Since the gain of the Darlington stage is about -1000, the *time-varying* potential across C can be well estimated by neglecting voltage variations at the near end of C. In other words, since v_c equals $v_{\text{far}} - v_{\text{near}}$ and we are neglecting v_{near},

$$v_c \approx v_{\text{far}} = v_o$$

Since the time-varying current through a capacitor always equals $C\, dv_c/dt$, we can write

$$I_o \approx C \frac{dv_o}{dt}$$

Because I_o and C are fixed, they set a maximum value on the derivative. In other words,

$$\left. \frac{dv_o}{dt} \right|_{\text{max}} = \frac{I_o}{C} \qquad (7\text{-}3)$$

This maximum value is called the slew rate. It expresses the speed with which the output of the operational amplifier can respond to a sudden change imposed upon the input.

In Sec. 4-5 we saw that the capacitor C determined the small 3-db bandwidth of the operational amplifier. As we shall see, this narrow bandwidth is needed to maintain stability. However, the point here is to recall the connection between the bandwidth of an amplifier and its response to a step-function input. Then we shall investigate the relationship between bandwidth and slew rate.

Suppose that the 3-db break frequency of an amplifier is f_o in Hertz or $\omega_o = 2\pi f_o$ in radians per second. If the DC gain of the amplifier is A_o, then the transfer function of the amplifier (output divided by input) is equal to

$$\frac{A_o}{1 + s/\omega_o}$$

Notice that we are using the Laplace variable s rather than $j\omega$. When the input is a step function, the transform output voltage $V_o(s)$ is given by

$$V_o(s) = \frac{1}{s} \cdot \frac{\omega_o A_o}{s + \omega_o} = \frac{A_o}{s} - \frac{A_o}{s + \omega_o}$$

Then the output voltage as a function of time must be

$$v_o(t) = A_o(1 - e^{-\omega_o t})$$

A plot of the output as given by this equation is shown in Fig. 7-9.

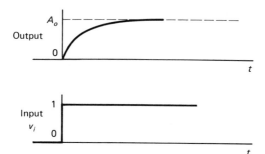

Fig. 7-9 Step response of low-pass amplifier.

The derivative dv_o/dt is given by

$$\frac{dv_o}{dt} = A_o\omega_o e^{-\omega_o t}$$

At $t = 0$, this equals $A_o\omega_o$.

The bandwidth, as we saw in Sec. 4-5, is determined by the Miller capacitance seen at the base of Q_{16}, which was equal to $C(A_v + 1)$ and the input resistance R_{in} to the base of Q_{16}. Here A_v refers to the gain of the Darlington stage. We thus have

$$\omega_o = \frac{1}{2\pi R_{in}C(A_v + 1)}$$

Hence

$$\frac{dv_o}{dt} = \frac{A_o}{2\pi R_{in}C(A_v + 1)} \tag{7-4}$$

Equation (7-4) is essentially a small-signal relationship that takes no account of the magnitude of the output voltage swing. For a step input of such tiny magnitude that only a small excursion of output occurs, the rate of change of v_o with time is given by Eq. (7-4). But if a large change in output is called for, then Eq. (7-3), which expresses the slew rate, sets a stricter limit on the speed of the response.

Example 7-2

Let us see how these ideas apply to the 741 operational amplifier. Using nominal values for ω_o and A_o (which differ a little from those we calculated in Chap. 4), we have $\omega_o = 2\pi \times 6$ and $A_o = 50,000$. As stated earlier, the values of I_o and C are 30 μA and 30 pF, respectively. Substituting in Eq. (7-3) we find the slew rate to be

$$\left.\frac{dv_o}{dt}\right|_{max} = \frac{30 \times 10^{-6}}{30 \times 10^{-12}} = 1 \text{ V/μs}$$

The small-signal derivative is simply

$$A_o\omega_o = 50,000 \times 2\pi \times 6 = 189 \times 10^4 = 1.89 \text{ V/μs}$$

Thus we see that the slew rate imposes a limit that is about half the limit that small-signal considerations alone would suggest.

7-2 COMPENSATION OF FREQUENCY RESPONSE

Another important performance characteristic is the unity-gain bandwidth. As we saw in the case of the 741 operational amplifier, the 3-db bandwidth is quite small, nominally 6 Hz. For frequencies above that point, the gain falls off at a rate of -20 db per decade. The unity-gain bandwidth is the frequency at which the gain becomes 1 or, expressed in decibels, 0 db. Since the nominal gain of the 741 is 50,000, or 94 db, almost 5 decades of roll off are required to bring the gain down to 0 db. Thus the unity-gain bandwidth must be something less than 600 KHz. The somewhat optimistic nominal value is 1 MHz.

The narrow 3-db bandwidth marks the frequency at which the roll off of -20 db per decade begins and thus determines the unity-gain bandwidth. As we saw in Sec. 4-5, the 3-db bandwidth was determined by the compensating capacitance C multiplied by $(A_v + 1)$. This yielded the Miller capacitance.

We might wonder why this capacitor (30 pF for the 741 operational amplifier) was chosen to restrict the 3-db bandwidth so severely. To answer this question, it is necessary to point out that other internal capacitances exist at other points in the circuit and that these create break frequencies of much higher values. Associated with these breaks at all times are the usual phase shifts. As we shall see, operational amplifiers are invariably operated with fairly large amounts of negative feedback. One example of this is the voltage-follower circuit in which the output is directly connected to the inverting input terminal. This connection makes $\mathbf{T} = \mathbf{A}$, and if $\mathbf{A}$ involves, for example, three break frequencies of a high value, the phase or gain margins may be inadequate to prevent oscillation.

The compensating capacitor creates a very low frequency break and thus brings the gain down to 0 db at a frequency below those of the troublesome, unwanted breaks. Some operational amplifiers do not have internal compensation but provide terminals to which external compensating capacitors may be connected.

The performance characteristics that we have discussed are summarized for several types of operational amplifiers in Table 7-1.

Operational-amplifier chips are enclosed in a variety of packages. One such package is the TO-99 case. Other common packages are the 8-pin and 14-pin DIP

TABLE 7-1 OPERATIONAL AMPLIFIER CHARACTERISTICS

Type	Input Offset Voltage (Max)	Input Offset Currrent (Max)	Input Bias Current (Max)	DC Voltage Gain (Min)	Slew Rate	Unity Gain Band-width	Input Offset Voltage Temp. Coeff.	Input Offset Current Temp. Coeff.
	mV	nA	nA		V/μs	MHz	μV/°C	pA/°C
741	6	200	500	20,000	0.5	1	7	—
770	10	10	30	35,000	2.5	1.3	10	—
308	7.5	1	7	25,000	—	—	30	10
8007	10	0.0005	0.002	50,000	6	1	75	—

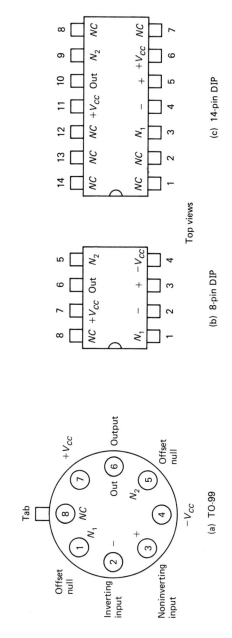

Fig. 7-10 Operational amplifier packages.

(dual in-line pin) cases. These, together with pin connections, are shown in Fig. 7-10.

Notice that while terminals are provided to connect the positive and negative supply voltages, there is no common return, or ground, terminal. This means that the entire operational amplifier circuit floats with respect to ground potential and that the ground reference potential has to be incorporated in some way in the external circuit into which the amplifier is connected.

7-3 OPERATIONAL AMPLIFIERS IN NEGATIVE-FEEDBACK CIRCUITS

Since the voltage gain of an operational amplifier can range from 10,000 to 100,000, it should be immediately clear that it is impracticable to operate one without negative feedback (that is, in the open-loop mode). Therefore, let us consider the use of operational amplifiers in negative-feedback circuits. The simplest negative-feedback arrangement is the voltage-follower circuit shown in Fig. 7-11. Notice that power-supply and offset-null connections are omitted for the sake of simplicity.

7-3.1 Voltage follower

We leave as an exercise to verify that for this circuit, $\mathbf{T} = A$ and $K = 1$. Using these values we obtain

$$A_{vf} = \frac{A}{1 + A} \approx 1$$

This result shows clearly that the output voltage v_o is very nearly equal to the input voltage v_1. Thus v_o follows v_1. The advantage of such a circuit is that it provides a high resistance to v_1 but develops v_o with a very low source resistance. Such a circuit could be useful as an electronic voltmeter, since the input side looks like a high resistance to the circuit whose voltage is measured, while the output is suitable for driving a relatively low-resistance meter movement.

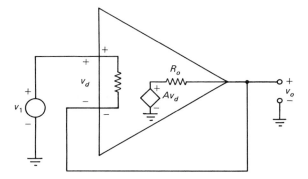

Fig. 7-11 Voltage follower circuit.

7-3.2 Finite-gain inverting amplifier

If a gain other than 1 is desired, the inverting amplifier circuit of Fig. 7-12 can be used.

What we wish to find is the gain with feedback. It is a simple matter to do this using the asymptotic gain formula. As usual, we replace the controlled source Av_d by an independent one of A volts in order to calculate the return difference $\mathbf{T}$. Since R_o is always small compared to R_f and R_1, we shall neglect it. We then obtain

$$\mathbf{T} = \frac{A(R_1\|R_i)}{R_f + (R_1\|R_i)}$$

The asymptotic gain K is found by assuming finite v_o with v_d equal to 0 and with v_1 in place. To perform this calculation, we redraw the circuit in the form shown in Fig. 7-13. Since v_d is required to be 0, no current flows through R_i. The component currents flowing into point P must then be simply $v_1/R_1 + v_o/R_2$, and these must have a sum of 0. Hence we have

$$K = -\frac{R_f}{R_1}$$

At this point we could, of course, combine the expressions for K and $\mathbf{T}$ and obtain an equation for A_{vf}. But rather than assemble such a messy equation, let us take into account typical numerical values to see what influence the parts will have upon the assembled whole. Suppose that $R_f = 50$ KΩ, $R_1 = 10$ KΩ, $R_i = 1$ MΩ, and $A = 10,000$. (These are all reasonable values. In fact, the one chosen for A is a little low). First of all, we see that $R_1\|R_i$ (or 10 K$\|$1 M) is almost equal to R_1 taken alone. Then $\mathbf{T}$ must be about $(10,000 \times 10$ K$)/(50$ K $+ 10$ K$)$, or about 1670. Hence A_{vf} will be given by

$$A_{vf} = \frac{K\mathbf{T}}{1 + \mathbf{T}} \approx \frac{(R_f/R_1) \times 1670}{1671}$$

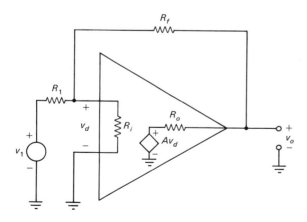

Fig. 7-12   Inverting amplifier circuit.

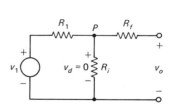

Fig. 7-13   Equivalent circuit to find K.

Clearly, 1670/1671 is so close to 1 that it can be replaced by 1. This leaves us with the excellent approximation

$$A_{vf} \approx K$$

There are other, equally good ways to obtain this result. One fact upon which all such approximate derivations depend is that v_d approaches zero. To repeat a phrase used in Chap. 4, we say that the node at which R_1, R_2, and R_i are connected is a virtual ground. *Negative-feedback operational amplifiers always operate to make v_d tend toward 0.* In the work that follows, it may sometimes be convenient to use that fact as a starting point in analysis without going through the full asymptotic-gain calculation.

7-3.3 Effects of input bias and offset parameters

An undesirable feature of the feedback circuit of Fig. 7-12 is the fact that the resistances seen by the inverting and noninverting inputs are quite different; as a result, the input bias current affects the two input voltages differently, thus introducing an anomalous voltage component in v_d. This problem can be overcome by equalizing the resistances in each input circuit, as is done in the configuration shown in Fig. 7-14. The resistance R_3 must be equal to $R_1 \| R_2$. Although the addition of R_3 compensates for the effect of the input bias current, it augments the problem created by the input offset current. To get some idea of the magnitudes of the quantities involved, we must undertake a more detailed analysis. For this purpose we make use of the diagram shown in Fig. 7-15, which includes current and voltage sources that represent the bias and offset currents and voltage. The current sources are chosen to conform with Eqs. (7-1) and (7-2). Notice that DC levels are chosen for input and output voltages V_1 and V_o.

Since the circuit of Fig. 7-15 is rather messy, we shall consider one offset effect at a time. Then, if we wish, we can combine the results using the principle of superposition.

If we short-circuit V_1 and remove all internal current sources, we are left with the circuit shown in Fig. 7-16. We are also neglecting the small resistance R_o for the sake of simplicity, and we are considering only the *deviation* of V_o from its quiescent value due to offset and bias effects. For that reason, the output in Fig. 7-16 is denoted by ΔV_o.

In the circuits of both Figs. 7-15 and 7-16, the controlling voltage V_d is shown as existing across R_i and not simply between the input terminals of the operational amplifier. At first glance it might seem that this arrangement excludes V_{os} from

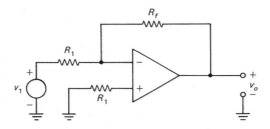

Fig. 7-14 Circuit to reduce effect of bias current.

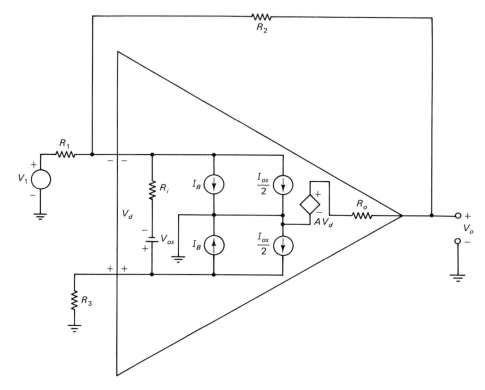

Fig. 7-15 Circuit including offset and bias effects.

any influence upon the output. This is certainly not the case, as we can see by considering what would happen if both operational-amplifier input terminals were grounded and no feedback was used at all. Such an arrangement is shown in Fig. 7-17. We can see at once that V_d must equal $-V_{os}$, and therefore that V_o must be $-AV_{os}$.

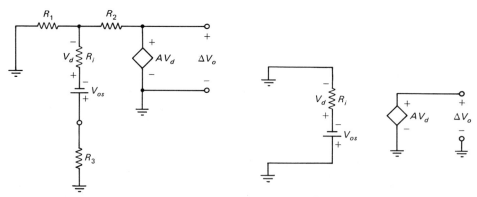

Fig. 7-16 Circuit showing effect of V_{os}.

Fig. 7-17 Grounded-input circuit without feedback.

Characteristics and Applications of Operational Amplifiers   Chap. 7

Returning to Fig. 7-16 we see that this circuit allows us to find the effect of V_{os} upon the output V_o when feedback is present. The voltage V_{os} acts as an input, and so to find **T**, we imagine V_{os} to be short-circuited. Since R_i is very large compared to R_1, R_2, and R_3, we shall assume it to be infinite. That procedure leads to an expression for **T**:

$$\mathbf{T} = \frac{AR_1}{R_1 + R_2}$$

The asymptotic gain K is found by keeping ΔV_o finite but assuming that V_d equals 0. This forces the current through R_i and R_3 to be 0 and therefore requires the voltage across R_1 to cancel $-V_{os}$ exactly. Therefore,

$$-V_{os} = \frac{\Delta V_o R_1}{R_1 + R_2}$$

Hence

$$K = \frac{\Delta V_o}{V_{os}} = -\left(\frac{R_1 + R_2}{R_1}\right)$$

The expression that shows the effect of V_{os} upon ΔV_o is

$$\Delta V_o = \frac{-AV_{os}}{1 + AR_1/(R_1 + R_2)}$$

If A is very large (as is usually the case), we can approximate this result by

$$\Delta V_o \approx -\frac{R_1 + R_2}{R_1} V_{os} \tag{7-6}$$

This result shows that the offset voltage, as it affects the output, is magnified by the ratio of $(R_o + R_2)/R_1$. Without feedback, V_{os} would be magnified by A, a much larger number.

The effect of the bias and offset currents can be accounted for by considering the two components together because, as we see in Fig. 7-15, I_B and $I_{os}/2$ appear as parallel current sources and may, therefore, be combined into a single source. As before, we disable those sources that are not of interest, V_1 and V_{os}. First considering only the bottom pair of current sources in Fig. 7-15, we can simplify the circuit to the form shown in Fig. 7-18.

If we convert the current source to a voltage source, the circuit has the same form as that of Fig. 7-16, so that we can make use of the calculations already performed. The circuit showing this conversion appears in Fig. 7-19. We can modify the results determined previously for V_o to obtain

$$\Delta V_o' = -\left(I_B - \frac{I_{os}}{2}\right) \frac{R_3(R_1 + R_2)}{R_1} \tag{7-7}$$

However, this is only half of the required result; we have yet to consider the top pair of current sources in Fig. 7-15. When these are combined into a single current source and then converted to a voltage source, the circuit shown in Fig. 7-20 results.

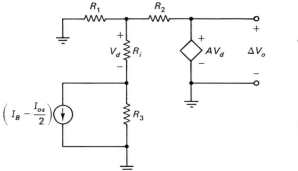

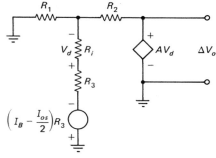

Fig. 7-18 Circuit showing effect of I_B and $I_{os}/2$.

Fig. 7-19 Circuit with conversion to voltage source.

This configuration is not quite the same as the one in Fig. 7-19 since R_i and V_d now appear in the leg of the circuit that contains no source. However, the expression for **T** remains the same; K is the parameter that is changed. We leave it to you to show that $K = R_2/R_1$. We now find that the contribution to V_o due to $(I_B + I_{os}/2)$ is equal to

$$\Delta V_o'' = \frac{[AR_1/(R_1 + R_2)] \cdot (R_2/R_1)}{1 + AR_1/(R_1 + R_2)} \left(I_B + \frac{I_{os}}{2} \right) R_1$$

Again, recognizing that A is very large leads to the approximation

$$\Delta V_o'' \approx R_2 \left(I_B + \frac{I_{os}}{2} \right) \tag{7-8}$$

Combining Eqs. (7-7) and (7-8), we obtain the total contribution to ΔV_o due to the bias and offset currents. This displacement of V_o, which we are calling ΔV_o, is given by

$$\Delta V_o = -I_B \left[\frac{R_3(R_1 + R_2)}{R_1} - R_2 \right] + \frac{I_{os}}{2} \left[\frac{R_3(R_1 + R_2)}{R_1} + R_2 \right] \tag{7-9}$$

Equation (7-9) shows the combined effect of the input bias and offset currents for the circuit of Fig. 7-15. If, as we suggested earlier, R_3 is chosen to be equal to $R_1 \| R_2$ or $R_1 R_2/(R_1 + R_2)$, then the I_B term vanishes and ΔV_o simply equals $I_{os} R_2$. Thus we can eliminate the effect of the bias current by judiciously choosing R_3, and the effect of the input offset current can be reduced but not entirely eliminated.

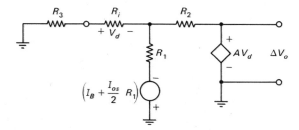

Fig. 7-20 Circuit with converted source.

The combined effects of input bias current, input offset current, and input offset voltage can be obtained by combining Eqs. (7-6) and (7-9).

It is not difficult to show that as long as R_i is large compared to R_1, R_2, and R_3, the gain of the circuit of Fig. 7-15 will be very nearly the same as that of the circuit of Fig. 7-12. That is,

$$A_{vf} \approx K \approx -\frac{R_2}{R_1}$$

7-4 DETERMINING OPEN-LOOP GAIN

Direct measurement of the open-loop gain A is not feasible because of the large values of A. Instead, measurement of A has to be carried out with the operational amplifier embedded in a negative-feedback circuit. Such an arrangement is shown in Fig. 7-21.

The only novelty in the circuit of Fig. 7-21 is that resistors R_2 and R_3 have been added to form an input attenuator. This is equivalent to considering $\mathbf{V}_x$ to be the voltage at the input terminal of an augmented amplifier (hence the dashed lines). The augmented amplifier has an input resistance that is equal to $R_2 + R_3$ and an open-loop gain equal to $-AR_3/(R_2 + R_3)$. Then $\mathbf{T}$ is easily shown to be given by

$$\mathbf{T} = \frac{R_1\|(R_2 + R_3)}{R_F + R_1\|(R_2 + R_3)} \cdot \frac{AR_3}{R_2 + R_3}$$

and K will be equal to $-R_F/R_1$.

Further algebraic treatment of this problem is not profitable. Instead, let us take some typical values for the resistances and see what result we get for A_{vf}. Suppose $R_F = 10$ KΩ, $R_1 = 1$ KΩ, $R_2 = 10$ KΩ, and $R_3 = 10$ Ω. Then $\mathbf{T}$ will be equal to

$$\mathbf{T} = \frac{1\text{ K}\|(10\text{ K} + 10)}{10\text{ K} + 1\text{ K}\|(10\text{ K} + 10)} \cdot \frac{A \times 10}{(10\text{ K} + 10)}$$

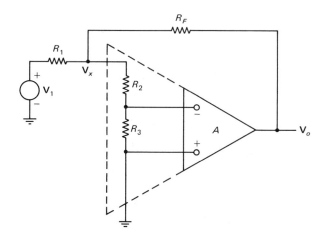

Fig. 7-21 Arrangement to measure open-loop gain.

and $K = -10 \text{ K}/1 \text{ K} = -10$. Hence, A_{vf} is approximately $-10A/(12 \text{ K} + A)$ or, for large values of A, $A_{vf} \approx -10$. This value is not of great importance in itself; its significance is to assure us that there is sufficient negative feedback so that reasonable values of $\mathbf{V}_1$ can be used without driving $\mathbf{V}_o$ to saturation levels.

What is important is the simple relationship between $\mathbf{V}_x$ and $\mathbf{V}_o$. This, as we have seen, is given by

$$\mathbf{V}_o = -\frac{AR_3}{R_2 + R_3}\mathbf{V}_x$$

Then A is given by

$$A = \frac{R_2 + R_3}{R_3}\frac{\mathbf{V}_o}{\mathbf{V}_x}.$$

It is a simple matter to measure $\mathbf{V}_x$ and $\mathbf{V}_o$ with an electronic voltmeter. Then A is readily calculated from the equation above.

Some additional caution has to be exercised in measuring A by this method. A problem arises in the choice of frequency at which the measurement is to be made. As we have seen, the open-loop break frequency f_o has a very low value, 10 Hz or less. To measure A correctly, we must choose a frequency low enough so that the value of A obtained will be from the plateau of the A-versus-f curve rather than on its roll off. A good way to assure that we have chosen the value of f sufficiently low is to display $\mathbf{V}_x$ and $\mathbf{V}_o$ in a Lissajou pattern on an oscilloscope. At frequencies above f_o, there will be significant phase shift between $\mathbf{V}_x$ and $\mathbf{V}_o$ so that an elliptical pattern will appear on the oscilloscope. When f is reduced to the point where no phase shift occurs, the ellipse will close to a straight line, assuring us that the measurement of A will yield the proper open-circuit, low-frequency value. Since the frequency at which this will occur will probably be only a few Hertz, it may be difficult to decide when the spot rotating on the CRT screen is describing a straight line instead of an ellipse. An oscilloscope with a long-persistent CRT phosphor should be helpful for this purpose.

7-5 DETERMINING OPEN-LOOP BREAK FREQUENCY

The measurement of the open-loop break frequency f_o presents the same problem that we encountered in the measurement of A. Just as we did in that case, once again we use a measurement made with feedback to infer the value of the open-loop quantity. This time the circuit arrangement is simpler; we merely use the inverting amplifier connection of Fig. 7-12 with both resistors equal. This means that the gain at low frequencies with feedback will be -1. Our procedure is to measure the frequency at which the gain with feedback decreases from a magnitude of 1 to a magnitude of $1/\sqrt{2}$. From this frequency we calculate the unity-gain bandwidth, from which, in turn, we can find the open-loop break frequency f_o. The basis for the necessary calculations is the subject of the discussion that follows.

To begin with let us repeat the now familiar expression for open-loop gain as a function of frequency:

$$\mathbf{A} = \frac{A_o}{1 + j\omega/\omega_0} \qquad (7\text{-}10)$$

where A_o is the DC, open-loop gain and ω_0 is the open-loop break frequency in radians per second. At the frequency ω_1 corresponding to the unity-gain bandwidth, $|\mathbf{A}|$ will equal 1. Therefore, we can write

$$1 = \left| \frac{A_o}{1 + j\omega_1/\omega_0} \right|$$

Since $\omega_1 \gg \omega_0$, $A_o \approx \omega_1/\omega_0$. Therefore, we can replace ω_o by ω_1/A_o in Eq. (7-10). This leads to an expression for $\mathbf{A}$ in terms of ω_1 instead of ω_o, which is

$$\mathbf{A} = \frac{A_o}{1 + jA_o\omega/\omega_1}. \qquad (7\text{-}11)$$

The circuit of Fig. 7-12 is shown with some modifications in Fig. 7-22. Modifications consist of assuming $R_i = \infty$ and the addition to the diagram of the voltage $\mathbf{V}_x$. Applying Kirchhoff's current law to the negative input terminal of the amplifier, we write

$$\frac{\mathbf{V}_1 - \mathbf{V}_x}{R_1} + \frac{\mathbf{V}_o - \mathbf{V}_x}{R_2} = 0$$

This equation leads to

$$\frac{\mathbf{V}_o}{R_2} = -\frac{\mathbf{V}_1}{R_1} + \mathbf{V}_x \left(\frac{1}{R_1} + \frac{1}{R_2} \right)$$

Since $\mathbf{V}_o = -A\mathbf{V}_x$, we can eliminate $\mathbf{V}_x$ to obtain

$$\mathbf{A}_{vf} = \frac{-1}{R_1/R_2 + (1/\mathbf{A})(1 + R_1/R_2)}$$

Since the circuit of Fig. 7-22 is set up to produce a gain of -1 at low frequencies, $R_1 = R_2$ and the last expression becomes

$$\mathbf{A}_{vf} = \frac{-1}{1 + 2/\mathbf{A}}$$

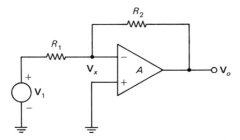

Fig. 7-22 Feedback circuit to measure f_o.

If we now substitute the expression for **A** from Eq. (7-11), we have

$$\mathbf{A}_{vf} = \cfrac{-1}{1 + \cfrac{2}{A_o/(1 + jA_o\omega/\omega_1)}} = \cfrac{-1}{1 + \cfrac{2(1 + jA_o\omega/\omega_1)}{A_o}}$$

Because A_o is very large, $\mathbf{A}_{vf}$ can be approximated by

$$\mathbf{A}_{vf} \approx \frac{-1}{1 + j2\omega/\omega_1}$$

The magnitude of $\mathbf{A}_{vf}$ will be reduced by a factor of $1/\sqrt{2}$ when $\omega = \omega_1/2$. This is the frequency that we find in our measurement procedure. Call this frequency ω_m. We then solve for the unity-gain frequency ω_1, which is twice ω_m. Then, knowing the open-loop gain A_o and using the straight-line asymptotic gain-frequency diagram shown in Fig. 7-23, we can find ω_o.

7-6 APPLICATIONS OF OPERATIONAL AMPLIFIERS

In the circuits and applications yet to be discussed, we shall not deal with bias and offset effects. These effects are always present, and you must keep that fact in mind. Whether or not these effects are troublesome has to be analyzed as the situation requires. Generally speaking, the smaller the resistances and the gain in the feedback circuit, the smaller will be the effect of the input-offset current and voltage.

Example 7-3

It is instructive to see, quantitatively, how the output of an operational amplifier deviates from its quiescent value when a feedback circuit is used. For this purpose, we shall assume that the operational amplifier considered in the last example is used again in the circuit of Fig. 7-15. We shall choose R_3 (as we discussed above) equal to $R_1 \| R_2$, or 8.33 KΩ. Suppose that the chip temperature varies from 20°C to 40°C. At 20°C, using the graph of Fig. 7-8, we see that I_{os} equals about 35 nA, whereas at 40°C, I_{os} decreases to about 30 nA. We shall take the input offset voltage V_{os} to be about 5 mV with a temperature coefficient equal to about 7 μV/°C. The combined effects will be to produce a value of ΔV_o at 20° equal to

$$\Delta V_o(20°C) = 35 \times 10^{-9} \times 5 \times 10^4 + \frac{60}{50} \times 5 \times 10^{-3} = 77.5 \text{ mV}$$

At 40°C the value of ΔV_o becomes

$$\Delta V_o(40°C) = 30 \times 10^{-9} \times 5 \times 10^4 + \frac{60}{50}$$

$$\times (5 \times 10^{-3} + 7 \times 20 \times 10^{-6}) = 76.7 \text{mV}.$$

The change in the level of ΔV_o due to the temperature change is thus seen to be less than 1 mV. Whether this is satisfactory or not depends upon the application.

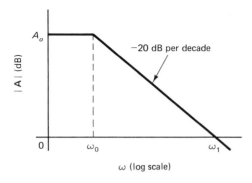

Fig. 7-23 Asymptotic diagram to find ω_o.

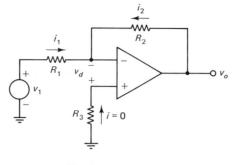

Fig. 7-24 Circuit of Figure 7-14 showing branch currents.

As we have seen, the voltage gain of the inverter circuit of Fig. 7-14 is very nearly equal to $-R_2/R_1$, or -5. This result can be achieved in an alternative and simpler way, which we shall now consider. The reason for bothering with another method is that its use will simplify some of the circuit analyses we shall undertake later.

7-6.1 Inverting amplifier

The key to the simplified method is the idea that we encountered earlier, namely, that the gain of the negative-feedback, operational-amplifier circuit is closely approximated by the asymptotic gain K. Furthermore, this asymptotic gain is readily calculated by making use of the idea that the negative feedback acts to drive the voltage v_d to zero. Also, we shall assume that R_i is infinite, and therefore no current flows into either input terminal. Let us now apply these ideas to the circuit of Fig. 7-14, which is redrawn in Fig. 7-24 to show the currents that are the focus of interest. Since instantaneous quantities are of interest, we use lowercase letters to represent such voltages and currents.

Because of the infinite input resistance, the current i in R_3 is zero. Therefore, the positive input terminal is at ground potential. Since v_d is zero, the potential of the negative terminal is also ground. We can therefore apply KCL to the negative terminal and write

$$i_1 + i_2 = 0$$

Using the appropriate voltages and resistances, this equation yields

$$\frac{v_1}{R_1} + \frac{v_o}{R_2} = 0$$

From this equation we obtain

$$A_{vf} = \frac{v_o}{v_1} = -\frac{R_2}{R_1}$$

which agrees with the result previously obtained.

You should note that, while the input resistance seen by v_2 is considered infinite, the source v_1 delivers a current i_1 to the circuit. Accordingly, the input resistance of the entire circuit must be v_1/i_1, or, simply, R_1.

7-6.2 Noninverting amplifier

Let us apply this technique to a few other interesting circuits. The circuit shown in Fig. 7-25 is that of a noninverting amplifier. Proceeding on the principle that v_d approaches zero, we see that the negative terminal must have a potential equal to v_1. (There is no virtual ground in this case.) Applying KCL to the negative terminal, we have

$$ i_1 - i_2 = 0 $$

or

$$ \frac{v_1}{R_1} - \frac{v_o - v_1}{R_2} = 0 $$

This leads to

$$ A_{vf} = \frac{R_1 + R_2}{R_2} $$

Notice that A_{vf} is a positive quantity, which means that there is no phase inversion between the input and the output of this circuit.

For this circuit, the input voltage source v_1 sees as input resistance just the very high input resistance of the operational amplifier. This contrasts with the case of the inverting amplifier of Fig. 7-24, where the input resistance was determined by one of the circuit resistors.

7-6.3 Differential amplifier

In the event that we might wish to use the operational amplifier as a differential amplifier, the circuit of Fig. 7-26 is useful. Of course, as we have seen, operational amplifiers always contain a differential amplifier as the input stage. However, most of the circuits in which operational amplifiers are used do not involve arrangements

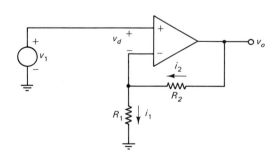

Fig. 7-25 Noninverting amplifier circuit.

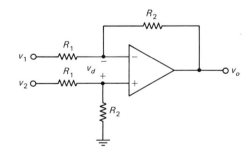

Fig. 7-26 Operational amplifier in a differential amplifier circuit.

in which two separate input signals are amplified differentially. Rather, one input terminal is usually used simply to accommodate the feedback signal derived from the output. The circuit of Fig. 7-26 provides both the means for negative feedback and the capability of developing an output voltage proportional to the *difference* of the two input signals. As we saw in Chap. 3, such a circuit is extremely useful in instrumentation applications.

Once again, the action of the circuit is to drive v_d to zero. Since the voltage at the positive terminal must be equal to $v_2 R_2/(R_1 + R_2)$, the voltage at the negative terminal must have this same value. Again applying Kirchhoff's current law to the negative terminal, we obtain

$$\frac{v_1 - v_2 R_2/(R_1 + R_2)}{R_1} + \frac{v_o - v_2 R_2/(R_1 + R_2)}{R_2} = 0$$

This equation yields

$$v_o = \frac{R_2}{R_1}(v_1 - v_2)$$

which shows that the output voltage is simply the difference of the two input voltages multiplied by R_2/R_1.

7-6.4 Summing amplifier

There are two operational-amplifier circuits that are found in analog computers. While analog computers are becoming rare except in research applications, these two circuits have wider uses and so deserve our consideration. The circuits in question are the *summing amplifier* and the *integrator*. Figure 7-27 shows the circuit of the summing amplifier; the integrator is shown in Fig. 7-28.

As usual, the action of the circuit is to drive the potential between input terminals to zero. In this case, this makes the negative terminal a virtual ground. Hence, application of Kirchhoff's current law yields

$$\frac{v_1}{R_1} + \frac{v_2}{R_1} + \frac{v_3}{R_1} + \frac{v_o}{R_2} = 0$$

It is easy to show that

$$v_o = -\frac{R_2}{R_1}(v_1 + v_2 + v_3).$$

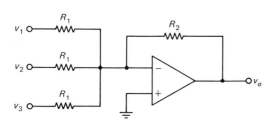

Fig. 7-27 Summing-amplifier circuit.

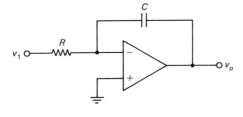

Fig. 7-28 Operational-amplifier integrator circuit.

Thus the output voltage is a negative constant times the sum of the input voltages. If the resistors in series with the input voltages are not identical, then v_o will be proportional to a sum of *weighted* voltage values.

7-6.5 Integrator

The integrator circuit of Fig. 7-28 operates on the same principle as the other circuits we have discussed. If we apply Kirchhoff's current law to the negative terminal (which is at virtual ground potential), we obtain

$$\frac{v_1}{R} + C\frac{dv_o}{dt} = 0$$

This equation can be rearranged to yield

$$dv_o = -\frac{1}{RC}v_1\,dt$$

We can integrate both sides to obtain

$$\int_0^{v_o} dv_o = v_o = -\frac{1}{RC}\int_0^t v_1\,dt$$

In general, both v_1 and v_o are functions of time.

By interchanging the positions of the resistor and the capacitor, a differentiator can be formed. However, this is rarely done because such a circuit enhances the effects of high-frequency noise.

7-7 INDUCTANCE SIMULATION

Operational amplifiers can be used to simulate inductors. Generally speaking, inductors are more expensive than the operational amplifiers used to simulate them, even when we include the cost of the few resistors and capacitors associated with these inductance-simulating circuits. To see how the simulation is accomplished, we consider the circuits shown in Fig. 7-29.

At first glance, there is nothing in the circuit of Fig. 7-29(a) to explain the equivalence claimed for the circuit of Fig. 7-29(b). The key to this equivalence is in the fact that the operational amplifier of Fig. 7-29(a) has an open-loop gain **A** that falls off with increasing frequency at a rate of -20 db per decade. The break frequency for a 741 operational amplifier, as we have seen, is less than 10 Hz. This

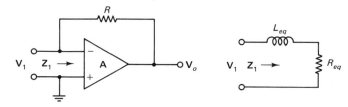

(a) Actual circuit   (b) Equivalent circuit   **Fig. 7-29** Inductance simulation.

is typical behavior for an operational amplifier that has internal compensation, as in the case of the 741.

As we did in earlier discussions, we shall represent the roll off of $\mathbf{A}$ by writing the expression

$$\mathbf{A} = \frac{A_o}{1 + j(\omega/\omega_o)}$$

where A_o is the DC gain (absolute, not in decibels) and ω_o is the break frequency in radians per second.

Since the input voltage $\mathbf{V}_1$ is applied to the inverting terminal of the amplifier, the output voltage $\mathbf{V}_o$ equals $-\mathbf{AV}_1$. The current into the resistor R must equal the potential difference across R divided by its resistance, or

$$\mathbf{I} = \frac{\mathbf{V}_1 - (-\mathbf{AV}_1)}{R} = \frac{\mathbf{V}_1(1 + \mathbf{A})}{R}$$

Upon substituting the expression for $\mathbf{A}$, we get

$$\mathbf{Z}_1 = \frac{R}{1 + A_o/[1 + j(\omega/\omega_o)]}$$

which, after a little algebraic manipulation, yields

$$\mathbf{Z}_1 = \frac{R}{1 + A_o + j(\omega/\omega_o)} + j\omega \frac{R}{\omega_o(1 + A_o)}$$

For values of $\omega/\omega_o \ll 1 + A_o$, we can approximate $\mathbf{Z}_1$ by the equation

$$\mathbf{Z}_1 \approx \frac{R}{1 + A_o} + j\omega \frac{R}{\omega_o(1 + A_o)}$$

This equation has the form

$$\mathbf{Z}_{eq} \approx R_{eq} + j\omega L_{eq}$$

so that we can identify the parts as follows, making use of the fact that $A_o \gg 1$.

$$R_{eq} = \frac{R}{1 + A_o} \approx \frac{R}{A_o} \tag{7-12a}$$

$$L_{eq} = \frac{R}{\omega_o(1 + A_o)} \approx \frac{R}{\omega_o A_o} \tag{7-12b}$$

If we use an operational amplifier with internal compensation, such as the 741, the break frequency ω_o is predetermined. However, to make use of this scheme for simulating inductance, we require some means of altering the break frequency. One way to do this is to use an uncompensated operational amplifier such as the 709 and connect a suitable capacitor to the compensating terminals.

There is another variable however, that cannot be controlled by the use of an uncompensated operational amplifier. That is the open-loop gain A_o. This quantity varies widely between one unit and another of the same type number. Since we require control of both A_o and ω_o, it seems natural to look for another circuit configuration that will accomplish this purpose. Such a circuit is shown in Fig. 7-30.

While this circuit looks more complicated, the additional number of components is well worth the trivial extra cost because the circuit allows us to control both the break frequency and the open-loop, DC gain regardless of operational-amplifier parameters. Of course, any break frequency we choose to introduce must be lower than the lowest one caused by the internal operational-amplifier compensation.

Operational-amplifier A_1 is simply a voltage follower that provides a gain of 1, high input resistance, and low output resistance. The resistor R_x and the capacitor C_x provide a transfer function equal to $1/(1 + j\omega/\omega_o)$. This causes a break frequency to occur at ω_o, where ω_o is equal to $1/R_x C_x$. To approximate ω_o reasonably closely, it is necessary that R_x be large compared to R_o of the amplifier, yet small compared to R_{in}. Typically, R_o is in the order of 100 Ω and R_{in} is in the order of 1 MΩ. Hence R_x could be set to a value of 10 KΩ, for instance.

Operational amplifier A_2 is also a voltage follower. Together, A_1 and A_2 isolate the $R_x C_x$ break-frequency network from the rest of the circuit. Amplifier A_3 is an inverting amplifier, whose gain is very nearly equal to $-R_2/R_1$. The entire cascade array of A_1, A_2, and A_3 corresponds to a single operational amplifier whose overall DC gain is $-R_2/R_1$ and whose break frequency is set at $(1/2\pi R_x C_x)$ Hertz. Equations (7-12a) and (7-12b) apply directly.

One word of caution must be offered. The internal break frequency of amplifier A_3 is not the nominal 6-Hz value (assuming that type 741 amplifiers are used) but, as we saw in Chap. 4, is instead equal to 6 Hz times $(1 + A\beta)$, where A and β refer to the gain and feedback factors associated with A_3. The factor $1 + A\beta$ is the return difference, equally well written as $1 + \mathbf{T}$. The value of $\mathbf{T}$ for an inverting amplifier is given by

$$\mathbf{T} = \frac{A R_1 \| R_i}{R_2 + R_1 \| R_i}$$

Assuming that A has the nominal value of 50,000, $R_i = 1$ MΩ, $R_1 = 10$ KΩ, and $R_2 = 10$ KΩ, then $\mathbf{T}$ has a value of about 25,000. Therefore, the break frequency now appears to be about $6 \times 25,000 = 150$ KHz. Even if A should be 10,000

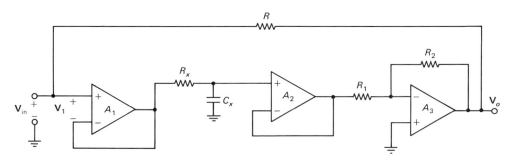

Fig. 7-30 Inductance-simulating circuit.

instead of 50,000 and R_1 and R_2 should be changed to 1 MΩ and 1 KΩ, respectively, the break frequency would still be about $6 \times 10{,}000 \times 500/(500 + 1) \approx 60$ KHz. What this means is that if R_x and C_x are set to produce a break frequency significantly lower than either of these values, proper operation would occur.

Example 7-4

Suppose we wish to simulate a 100-mH inductor at a frequency of 10 KHz. We can set A_o approximately to 1000 by making $R_2 = 100$ KΩ and $R_1 = 100$ Ω. Since the operating frequency is to be 10 KHz, we set the break frequency well below this value, such as at 1 KHz. That makes $\omega_o = 2\pi \times 10^3$ rad/s. For a value of R_x of 10 KΩ, C_x must equal $1/(10^4 \times 2\pi \times 10^3)$, or 0.0159 μF.

From Eq. (7-12b) we see that R must equal $\omega_o L_{eq} A_o$ or $2\pi \times 10^3 \times 0.1 \times 1000$, which yields $R = 628$ KΩ.

Unfortunately, Eq. (7-12a) shows that we get not only the desired simulation of L_{eq} but also an unwanted resistance R_{eq}. In this case, R_{eq} amounts to R/A_o, or $6.28 \times 10^5/1000 = 628$ Ω.

The situation can be improved by reducing ω_o. Suppose we set ω_o to 10 Hz by making C_x equal to 15.9 μF. We will now have R equal to 6.28 KΩ, and R_{eq} will become 6.28 Ω.

In Example 7-4, we could reduce R_{eq} by further increasing A_o. Rather than do this by increasing R_2 and decreasing R_1, we could perhaps connect A_2 as a noninverting amplifier with gain greater than 1.

Ultimately, however, it seems that some nonzero value of R_{eq} will always remain. What we wish to consider next is a technique to remove R_{eq} altogether. For this purpose we consider the circuit of Fig. 7-31.

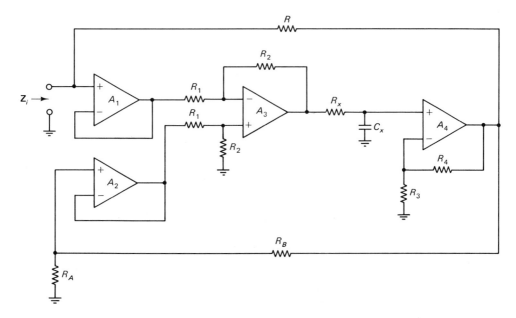

Fig. 7-31 Modified inductance-simulation circuit.

Amplifiers A_1 and A_2 are voltage followers that are used to take advantage of their high input resistances. Also, A_3 is a differential circuit, whose gain is R_2/R_1 times the difference of the two input signals coming from A_1 and A_2. Amplifiers A_3 and A_4 provide isolation for the $C_x R_x$ break-frequency-setting network. In addition, A_4 is a noninverting amplifier having gain of $(R_3 + R_4)/R_3$. The combined low-frequency gains of A_3 and A_4 produce a gain corresponding to A_o that is given by

$$A_o = \frac{R_2}{R_1} \cdot \frac{R_3 + R_4}{R_3}$$

We can represent the entire network by the simplified one of Fig. 7-32, where the four amplifiers and their associated circuitry are represented by the single amplifier A. Once again we seek an expression for $\mathbf{Z}_i$. The voltage at the positive terminal is equal to $\mathbf{V}_o R_A/(R_A + R_B)$. The voltage $\mathbf{V}_o$ is given by

$$\mathbf{V}_o = \mathbf{A}\left[\frac{\mathbf{V}_o R_A}{R_A + R_B} - \mathbf{V}_1 \right]$$

which leads to

$$\mathbf{V}_o = \frac{-\mathbf{A}\mathbf{V}_1}{1 - \mathbf{A}[R_A/(R_A + R_B)]}$$

The current into R is equal to

$$\mathbf{I} = \frac{\mathbf{V}_1 - \mathbf{V}_o}{R}$$

or

$$\mathbf{I} = \frac{\mathbf{V}_1}{R}\left\{ 1 + \frac{\mathbf{A}}{1 - \mathbf{A}[R_A/(R_A + R_B)]} \right\}$$

Hence

$$\mathbf{Z}_1 = \frac{\mathbf{V}_1}{\mathbf{I}} = \frac{R}{1 + \mathbf{A}/\{1 - \mathbf{A}[R_A/(R_A + R_B)]\}}$$

A little rearranging leads to

$$\mathbf{Z}_i = \frac{R}{1 + \mathbf{A}[1 - R_A/(R_A + R_B)]} - \frac{\mathbf{A}R[R_A/(R_A + R_B)]}{1 + \mathbf{A}[1 - R_A/(R_A + R_B)]}$$

If we make $R_A/(R_A + R_B) \ll 1$, then $\mathbf{Z}_i$ can be approximated by

$$\mathbf{Z}_i \approx \frac{R}{1 + \mathbf{A}} - \left(\frac{RR_A}{R_A + R_B} \right)\left(\frac{\mathbf{A}}{1 + \mathbf{A}} \right)$$

If we now substitute $A_o/(1 + j\omega/\omega_o)$ for $\mathbf{A}$ and carry out a few more manipulations,

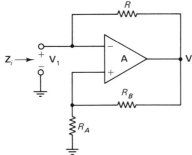

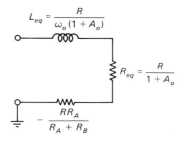

$$L_{eq} = \frac{R}{\omega_o(1 + A_o)}$$

$$R_{eq} = \frac{R}{1 + A_o}$$

$$-\frac{RR_A}{R_A + R_B}$$

Fig. 7-33 Equivalent circuit including negative resistance.

Fig. 7-32 Equivalent network.

we get

$$\mathbf{Z}_i \approx \frac{R}{1 + A_o + j\omega/\omega_o} + j\omega \frac{R}{\omega_o(1 + A_o)} - \frac{RR_A}{R_A + R_B} \cdot \frac{A_o}{1 + A_o + j\omega/\omega_o}$$

If $\omega/\omega_o \ll 1 + A_o$, we can make the further approximation that

$$\mathbf{Z}_i \approx \frac{R}{1 + A_o} + j\omega \frac{R}{\omega_o(1 + A_o)} - \frac{RR_A}{R_A + R_B} \tag{7-13}$$

Equation (7-13) leads to the same equation for L_{eq} that we had in Eq. (7-12b). However, the equation for R_{eq} given by Eq. (7-12a) is now modified to include a negative resistance term $-RR_A/(R_A + R_B)$. If we can adjust R_A and R_B so that this term is numerically equal to $R/(1 + A_o)$, we can cancel out the resistive term in $\mathbf{Z}_i$. The resulting situation is illustrated by the equivalent circuit of Fig. 7-33.

Example 7-5

Let us apply this negative-resistance circuit to the case we considered above, where $L_{eq} = 100$ mH, $A_o = 1000$, $R = 6.28$ KΩ, and $R_{eq} = 6.28$ Ω. Suppose we choose R_A to be 100 Ω. Then to cancel R_{eq} we need

$$R_B = \frac{R_A(R - R_{eq})}{R_{eq}}$$

or

$$R_B = \frac{100(6280 - 6.28)}{6.28} \approx 10^5 \ \Omega$$

As we might imagine, this scheme for simulating inductance has limitations. Perhaps the most important one is the fact that the voltage v_o (the instantaneous representation of the phasor voltage $\mathbf{V}_o$) cannot swing beyond the limits of the power-supply voltages. Thus if A_o is 1000 and v_o can have maximum excursions of from $+10$ V to -10 V, then the limits of the excursions of the input voltage v_1 are -10 mV and $+10$ mV.

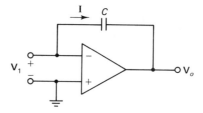

Fig. 7-34 Miller-capacitance circuit.

7-8 MILLER-CAPACITANCE CIRCUITS

At this point, we consider once again the use of the Miller effect to produce amplified capacitance. This subject was discussed in Chap. 4 in connection with the internal compensation of operational amplifiers; what we wish to examine now is the use of operational amplifiers as the amplifying devices in Miller-capacitance circuits.

The circuit of Fig. 7-34 shows a Miller-capacitance circuit using an operational amplifier. The current I must be equal to $(V_1 - V_o)j\omega C$. The output voltage V_o is, of course, equal to $-AV_1$. Therefore, I is given by

$$I = V_1(1 + A)j\omega C$$

The input admittance Y_i is I_1/V and so we have

$$Y_i = j\omega(1 + A)C$$

The effective capacitance is equal to $(1 + A)C$, which is consistent with results that we obtained earlier for other Miller circuits.

The circuit of Fig. 7-34 is hardly suitable for simulating a specified value of capacitance. This is because of the uncertain value of A. The circuit of Fig. 7-35 overcomes this problem.

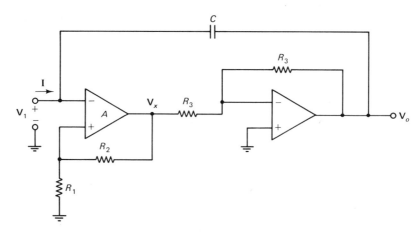

Fig. 7-35 Improved capacitance circuit.

Characteristics and Applications of Operational Amplifiers Chap. 7

For this circuit, $\mathbf{V}_x$ is given by

$$\mathbf{V}_x = A\left[\mathbf{V}_x \cdot \frac{R_1}{R_1 + R_2} - \mathbf{V}_1\right]$$

Solving for $\mathbf{V}_x$ yields

$$\mathbf{V}_x = \frac{-A\mathbf{V}_1}{1 - AR_1/(R_1 + R_2)} = \frac{-\mathbf{V}_1}{1/A - R_1/(R_1 + R_2)}$$

Since $A \gg 1$, we can approximate $\mathbf{V}_x$ by

$$\mathbf{V}_x \approx \mathbf{V}_1 \frac{R_1 + R_2}{R_1}$$

Notice that $\mathbf{V}_x$ is in phase with $\mathbf{V}_1$. As we saw in the circuit of Fig. 7-34, the Miller effect requires that voltage at the far end of the capacitor be out of phase with the voltage at the near end. For that reason, the second operational-amplifier circuit has been added. This stage has a gain of -1 and thus provides the needed phase inversion. As a result

$$\mathbf{V}_o = -\mathbf{V}_x = -\mathbf{V}_1 \frac{R_1 + R_2}{R}$$

The current $\mathbf{I}$ is equal to $(\mathbf{V}_1 - \mathbf{V}_o)j\omega C$, which, in turn, equals

$$\left(\mathbf{V}_1 + \mathbf{V}_1 \frac{R_1 + R_2}{R_1}\right)j\omega C$$

It follows that $\mathbf{Y}_i$ is given by

$$\mathbf{Y}_i = \left(1 + \frac{R_1 + R_2}{R_1}\right)j\omega C$$

Hence, the effective capacitance is equal to

$$C_{\text{eff}} = \left(1 + \frac{R_1 + R_2}{R_1}\right)C$$

It is important to consider the practical limitations that must be imposed on this capacitance-amplification scheme. Suppose, for example, that we wish to simulate a 10-μF capacitor using a 1-μF capacitor in the circuit of Fig. 7-35. To get the multiplying factor of 10, $(R_1 + R_2)/R_1$ must equal 9. This can be accomplished readily by using $R_1 = 10$ KΩ and $R_2 = 80$ KΩ. Presumably, we now have an effective 10-μF capacitor, so that if we apply an rms input voltage of 1 V at a frequency of 1000 Hz, the input current must be 62.8 mA rms, or ±88.8 mA (peak). We must now ask what path this current takes. Since the input current to the operational amplifier is negligible, this capacitive current must flow through C and into or out of the output terminal of the second operational amplifier. Since the typical output current capability of such an amplifier is to deliver or to absorb about 25 mA, we have clearly exceeded this capability. As a result, we shall either damage the amplifier or cause the current wave form to be seriously distorted.

There is another obvious limitation that, in this case, we have not exceeded. That is the limitation on output-voltage range. An rms input voltage of 1 V corresponds to a peak input voltage range of ± 1.41 V. With a gain of 9, the range of v_o will be ± 12.7 V. Thus, if supply voltages of $+15$ V and -15 V are used, this output-voltage range is safe.

There is yet another limitation that arises from a basic assumption that we tacitly made in developing the expression for the Miller capacitance. This assumption was that A is a constant. In fact, this is not true, as we know from our discussion of such matters as the unity-gain bandwidth. The value of A varies (rolls off) with frequency. If this behavior of A is taken into account, the result is that the expression for Y_i is not purely imaginary but contains a real term. As a result, we find that instead of obtaining a "pure" capacitance, we also get an unwanted resistance term. Fortunately, the effect is not serious if the gain is not set too high and if the operating frequency is well below the break frequency. Things work in our favor, since adjusting R_1 and R_2 to provide low gain adjusts the break frequency to a high value.

The limitation on current capability of the operational amplifier sets another limit on the performance of our simulated capacitor. That limit occurs in its transient behavior. Suppose that the simulated capacitor is charged to a voltage of 10 V. If it is suddenly connected across a resistor R, the initial current will be $10/R$. The current limitation of 25 mA requires that the minimum value of R be 400 Ω.

7-9 ACTIVE FILTER NETWORKS

To conclude this chapter we consider a more advanced application of operational amplifiers, which occurs in the construction of filter networks. Before we begin to discuss the principles involved, it is necessary to understand the nature of a filter network and for what reason we might wish to construct one.

To put the matter simply, a *filter network* is one that passes certain frequencies easily but inhibits the transmission of signals of other, unwanted frequencies. For example, the rules of the Federal Communications Commission require that AM broadcasting stations shall not transmit audio signals that are of frequencies higher than 5 KHz. Since audio frequencies normally extend to about 15 KHz, this upper limit of 5 KHz suggests that somewhere in the audio amplifier section of the transmitting system, a network must be included that will pass all frequencies up to 5 KHz and reject all higher frequencies. Such a network is called a *low-pass filter*. The range of frequencies to be transmitted is called the *pass band*, whereas the range of those to be rejected is called the *stop band*. The frequency that marks the boundary between the pass band and the stop band is called the *cutoff frequency*—in this case 5 KHz.

Filter networks have been studied for many years, and a variety of types has evolved. Some types provide a sharp boundary between the pass and stop bands but permit some variation in transmission in the pass band. Others have flatter characteristics in the pass band but a less well defined boundary between the pass

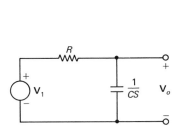

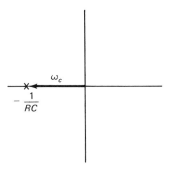

Fig. 7-36 Simple low-pass filter.

Fig. 7-37 Complex S-plane showing pole at $-1/RC$.

and stop bands. We shall restrict our interest in the discussion that follows to filters of the type known as *maximally flat*, or *Butterworth*, filters.

Perhaps the simplest filter is the RC network shown in Fig. 7-36. To verify that this circuit will behave as a low-pass filter, we need only consider two extreme cases. Suppose that the frequency is 0 (DC). Obviously, the capacitor will then charge up to the input voltage so that $\mathbf{V}_o$ equals $\mathbf{V}_1$, and therefore the gain equals 1 or 0 db. Now suppose that the frequency approaches infinity. This will certainly cause the capacitor to act as a short circuit, with $\mathbf{V}_o$ becoming 0. The gain then becomes 0 or $-\infty$ db. It is thus clear from physical reasoning that the circuit passes low frequencies farily well and high frequencies badly.

The question now arises: With respect to what value are frequencies considered high or low? One widely used criterion for this reference value is that it is the frequency at which the gain is equal to $1/\sqrt{2}$ times its zero-frequency value. Or, in an equivalent statement, at this frequency the gain is 3 db below its zero-frequency value. As we have seen before in connection with this form of RC network, this reference frequency (in radians per second) is equal to $1/RC$. This frequency is, of course, the cutoff frequency referred to earlier.

In Fig. 7-36 we have shown the network in terms of its Laplace-transformed impedance. That is, the impedance of the capacitor is shown as $1/CS$ rather than $1/j\omega C$. In terms of the Laplace frequency variable S, the gain function $\mathbf{V}_o/\mathbf{V}_1$ is equal to

$$\frac{\mathbf{V}_o}{\mathbf{V}_1} = \frac{1/CS}{R + 1/CS} = \frac{1}{RCS + 1} = \frac{1}{RC} \cdot \frac{1}{(S + 1/RC)}$$

The denominator of the last term of this expression contains the factor $(S + 1/RC)$. It is clear that when S takes on the value $-1/RC$, the factor becomes 0, and so $\mathbf{V}_o/\mathbf{V}_1$ becomes infinite. Values of S that cause $\mathbf{V}_o/\mathbf{V}_1$ to become infinite are called *poles*, and they are marked graphically in a plane in which values of S may be represented. (Since, in general, S may have complex as well as real values, a plane is needed to represent its possible values.) Fig. 7-37 shows the complex S-

plane and the pole where $S = -1/RC$. Note that the radius from the origin to this pole has a length of $1/RC$. We shall denote this radius by ω_c.

Now let us consider another, more complicated circuit—the *RLC* circuit shown in Fig. 7-38.

The gain function for this network is

$$\frac{\mathbf{V}_o}{\mathbf{V}_1} = \frac{1/CS}{R + LS + 1/CS} = \frac{1}{LC[S^2 + (R/L)S + 1/LC]}$$

Values of S that make the term in brackets become zero will cause $\mathbf{V}_o/\mathbf{V}_1$ to become infinite. These poles are found by setting the bracketed term to zero and solving the resulting quadratic equation. As is the case with all quadratic equations, the solution can yield real or complex conjugate roots. For the application we are considering, only complex conjugate roots (which will be the poles of $\mathbf{V}_o/\mathbf{V}_1$) are of interest. These are given by

$$S_1 = -\frac{R}{2L} + j\sqrt{\frac{1}{LC} - \left(\frac{R}{2L}\right)^2}$$

and

$$S_2 = -\frac{R}{2L} - j\sqrt{\frac{1}{LC} - \left(\frac{R}{2L}\right)^2}$$

As usual, we shall let ω_o^2 equal $1/LC$. For reasons that will shortly become clear, we are interested in the special case where $R = \sqrt{2L/C}$. With a little algebra, we can now write the expressions for S_1 and S_2 in the form

$$S_1 = -\frac{\omega_o}{\sqrt{2}} + j\frac{\omega_o}{\sqrt{2}}$$

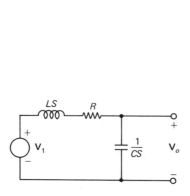

Fig. 7-38 *RLC* filter circuit.

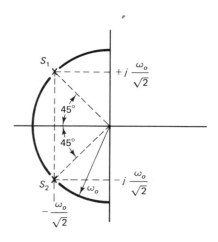

Fig. 7-39 Complex S-plane showing poles S_1 and S_2.

and

$$S_2 = -\frac{\omega_o}{\sqrt{2}} - j\frac{\omega_o}{\sqrt{2}}$$

In terms of ω_o and R (for this case), the expression for V_o/V_1 becomes

$$\frac{V_o}{V_1} = \frac{\omega_o^2}{(S^2 + \sqrt{2}\omega_o S + \omega_o^2)} \tag{7-14}$$

The poles given by S_1 and S_2 are plotted in the S-plane in Fig. 7-39. Notice that these poles lie on a semicircle of radius ω_o and that they stand $\pm 45°$ from the negative real axis.

To appreciate some of the significance of this special case, let us examine the characteristics of the filter network shown in Fig. 7-38. First of all, physical reasoning confirms that it is a low-pass filter because at zero frequency, the output voltage equals the input voltage, and at infinite frequency the output approaches zero.

Let us now find the cutoff frequency. To do that, we allow the variable S to take on values equal to $j\omega$. This is the mathematical equivalent of driving the network with a sinusoidal, variable-frequency oscillator. V_o/V_1 now becomes

$$\frac{V_o}{V_1} = \frac{\omega_o^2}{(-\omega^2 + j\sqrt{2}\,\omega\omega_o + \omega_o^2)}$$

Notice that when $\omega = 0$, V_o/V_1 equals 1. To find the cutoff frequency, we seek the value of ω at which $|V_o/V_1|$ becomes $1/\sqrt{2}$. The magnitude of V_o/V_1 is given by

$$\left|\frac{V_o}{V_1}\right| = \frac{\omega_o^2}{\sqrt{(\omega_o^2 - \omega^2)^2 + 2\omega^2\omega_o^2}}$$

and at the cutoff frequency ω_c, this quantity must equal $1/\sqrt{2}$:

$$\frac{1}{\sqrt{2}} = \frac{\omega_o^2}{\sqrt{(\omega_o^2 - \omega_c^2)^2 + 2\omega_c^2\,\omega_o^2}}$$

This leads to

$$(\omega_o^2 - \omega_c^2)^2 + 2\omega_c^2\,\omega_o^2 = 2\omega_o^4$$

from which we find that a real, positive solution for ω_c is $\omega_c = \omega_o$. From Fig. 7-39 we see that the radius of the circle on which the poles lie equals $\sqrt{(\omega_o/\sqrt{2})^2 + (\omega_o/\sqrt{2})^2}$, or ω_o. Thus we arrive at a simple and important conclusion: If the poles of the network are arranged in the pattern shown in Fig. 7-39, the radius of the circle on which they lie will equal the cutoff frequency.

Filter networks whose poles are complex conjugates and which lie in certain simple, symmetrical patterns on a circle in the S-plane belong to the class of maximally flat, or Butterworth, filters. The proof that these networks satisfy the condition of maximal flatness and the derivation of the patterns of the pole locations are beyond the scope of this book. Pole arrays for a few cases are shown in Fig. 7-40.

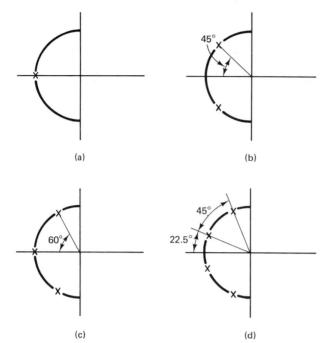

(a)

(b)

(c)

(d)

Fig. 7-40 Pole arrays for maximally flat, low-pass filters.

The number of poles that a filter network has is called its *order* and is designated by the letter *n*. By specifying the value of *n*, you can use Prog. 7-1 to generate the angles at which the poles must lie.

```
10   PRINT "CALCULATION OF POLE LOCATIONS FOR BUTTERWORTH LOW-PASS AMPLIFIE
     RS"
30   INPUT "ENTER ORDER OF FILTER";N
35   PRINT
52   PRINT "POLE NUMBER"; SPC( 10);"ANGLE(DEGREES)
53   PRINT
55   PHI = 180 / (2 * N)
60   FOR K = 1 TO N
65   THETA = (2 * K - 1 + N) * PHI
110   PRINT  TAB( 5);K; TAB( 23);THETA
130   NEXT K
140   END
```

Prog. 7-1.

Other arrays are possible and can be used to achieve some desired characteristic. For example, Tchebycheff filters are designed by making the poles lie on an ellipse instead of a circle. You will find such designs discussed in the literature.

One way to build low-pass filters is by using operational amplifiers instead of *RLC* networks. Advantages in using amplifiers are that we can eliminate inductors (always troublesome and expensive) and that the resulting networks are easy to adjust when precision performance is required.

We shall use an operational-amplifier network as a building block. Several building blocks can then be cascaded as needed to obtain a final design. In general, each building block produces a gain function having the form

$$\frac{\mathbf{V}_o}{\mathbf{V}_1} = \frac{ms^2 + cs + d}{s^2 + as + b}$$

Since the numerator of this expression is quadratic, as is the denominator, the building block that produces it is called a *biquad*. We are interested only in low-pass filters that have gain functions of the form that we developed above for the *RLC* network. The corresponding biquad, low-pass (LP) building block produces a gain function given by

$$\frac{\mathbf{V}_o}{\mathbf{V}_1} = \frac{d}{s^2 + as + b} \qquad (7\text{-}15)$$

A biquad LP building block producing this gain function is shown in Fig. 7-41.

The circuit of Fig. 7-41 forms a feedback network, and we shall analyze it in the usual way in order to find the relationships between the network parameter values and the required poles. Amplifier A_3 is connected as a unity-gain inverter and so provides merely a factor of -1 in the loop in which it is connected. Similarly, A_2 provides a factor of $-1/R_2C_2S$.

To compute $\mathbf{T}$, a simple procedure is to treat each operational amplifier and its associated circuitry as an ideal amplifier whose input resistance is infinite, whose output resistance is zero, and whose gain is -1, or $-1/R_2C_2S$, or the gain function associated with A_1. Since $\mathbf{T}$ is computed with $\mathbf{V}_1$ short-circuited, we can write the

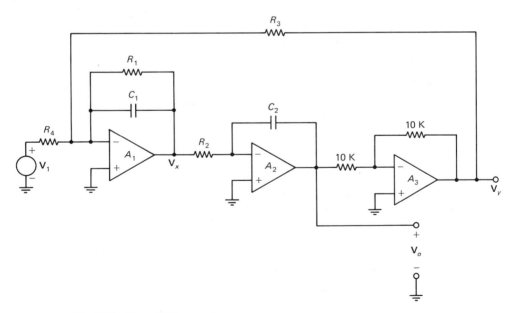

Fig. 7-41 Biquad LP network.

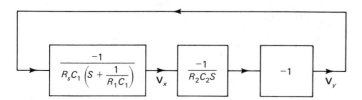

Fig. 7-42 Simplified network to determine T.

following equation to express current flows into the virtual-ground terminal of A_1:

$$\frac{\mathbf{V}_y}{R_3} + \frac{\mathbf{V}_x}{(R_1/C_1S)/(R_1 + 1/C_1S)} = 0$$

This equation leads to the following relation between $\mathbf{V}_x$ and $\mathbf{V}_y$:

$$\mathbf{V}_x = -\frac{1}{R_3C_1(S + 1/R_1C_1)}\mathbf{V}_y$$

We can thus represent the network in the simplified form shown in Fig. 7-42.

Focusing attention on the third amplifier, let us assume that it is an independent voltage generator producing -1 V and then compute what voltage is fed into it. It can be seen at once that $\mathbf{T}$ is equal to

$$\mathbf{T} = \frac{1}{R_2C_2R_3C_1S(S + 1/R_1C_1)}$$

To calculate K, we restore $\mathbf{V}_1$ and amplifier A_1 and require its input voltage to become zero, while we maintain $\mathbf{V}_o$ fixed. We shall leave the other two blocks as they were in Fig. 7-42. The modified network appears in Fig. 7-43.

We see that if the input voltage to A_1 becomes zero, $\mathbf{V}_x$ must be zero. Conditions at the input to A_1 are simply those shown in Fig. 7-44. It follows immediately

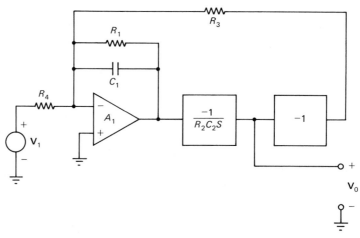

Fig. 7-43 Simplified network to find K.

Characteristics and Applications of Operational Amplifiers   Chap. 7

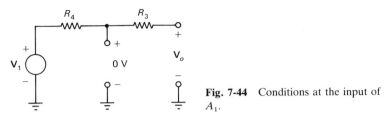

Fig. 7-44 Conditions at the input of A_1.

that

$$\frac{\mathbf{V}_1}{R_4} + \frac{-\mathbf{V}_o}{R_3} = 0$$

Hence

$$K = \frac{\mathbf{V}_o}{\mathbf{V}_1} = \frac{R_3}{R_4}$$

We now find $\mathbf{A}_{vf}$ to be equal to

$$\mathbf{A}_{vf} = \frac{\dfrac{1}{R_x C_2 R_4 C_1 S(S + 1/R_1 C_1)}}{1 + \dfrac{1}{R_2 C_2 R_3 C_1 S(S + 1/R_1 C_1)}}$$

After some algebraic manipulations, we arrive at

$$\mathbf{A}_{vf} = \frac{1}{R_2 R_4 C_1 C_2 (S^2 + 1/R_1 C_1 \, S + 1/R_2 C_2 R_3 C_1)}$$

Matching this expression to the biquad formula given by Eq. (7-15), we obtain the following relations between coefficients:

$$d = \frac{1}{R_2 R_4 C_1 C_2}$$

$$a = \frac{1}{R_1 C_1}$$

$$b = \frac{1}{R_2 C_2 R_3 C_1}$$

The following choices are now made:

C_1 is chosen arbitrarily.
C_2 is chosen arbitrarily.
R_2 is made equal to $1/\sqrt{b}\,C_1$.
R_3 is made equal to $1/\sqrt{b}\,C_2$.

Then R_1 must be equal to $1/aC_1$ and $R_4 = 1/dR_4 C_1 C_2$. There are alternative forms that can be used for R_4. We choose $R_4 = \sqrt{b}/dC_1$.

Example 7-5

Let us now apply the results we have developed to the design of a biquad low-pass filter having the following characteristics:

Low-frequency gain   0 db

Cutoff frequency   5 KHz

Maximally flat, second order   (2 poles)

We find ω_o to be $2\pi \times 5 \times 10^3 = 31{,}416$ rad/s. Comparing Eq. (7-14) and Eq. (7-15), we see that for this type of filter

$$a = \sqrt{2}\omega_o = \sqrt{2} \times 3.1416 \times 10^4 = 4.45 \times 10^4$$

$$b = \omega_o^2 \quad\quad = (3.1416 \times 10^4)^2 \quad = 9.89 \times 10^8$$

$$d = \omega_o^2 \quad\quad = 9.89 \times 10^8$$

Let us choose C_1 and C_2 to be 0.01 μF, or 10^{-8} F. Then

$$R_1 = \frac{1}{aC_1} \quad = \frac{1}{(4.45 \times 10^4 \times 10^{-8})} = 0.225 \times 10^4 = 2.25 \text{ K}\Omega$$

$$R_2 = \frac{1}{\sqrt{b}C_1} = \frac{1}{\omega_o C_1} = \frac{1}{(3.1416 \times 10^4 \times 10^{-8})} = 3.18 \text{ K}\Omega$$

$$R_3 = \frac{1}{\sqrt{b}C_2} = 3.18 \text{ K}\Omega$$

$$R_4 = \frac{\sqrt{b}}{dC_1} \quad = \frac{\omega_o}{\omega_o^2 C_1} = \frac{1}{\omega_o C_1} = 3.18 \text{ K}\Omega$$

By cascading two biquad sections, each producing a pair of complex-conjugate poles in the right arrangement, a fourth-order maximally flat filter can be designed.

Besides low-pass filters, biquad designs can be worked out for high-pass, band-pass, and band-elimination filters. Such designs are beyond the scope of this book. However, one category of band-pass filters does fall within our interest and will be discussed in connection with the design of tuned amplifiers.

7-10 WIEN-BRIDGE OSCILLATOR

The Wien-bridge oscillator using discrete elements was discussed in the preceding chapter. We now consider an oscillator of this type that uses an operational amplifier in its construction. Such a circuit is shown in Fig. 7-45.

To find the condition for oscillation, we must first find the return ratio **T**. This is easily shown to be equal to

$$\mathbf{T} = -A \left(\frac{j\omega RC}{-\omega^2 (RC)^2 + j3\omega C + 1} - \frac{R_1}{R_1 + R_2} \right)$$

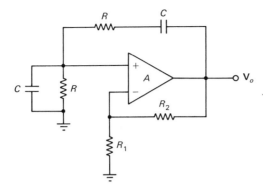

Fig. 7-45 Wien-bridge oscillator using operational amplifier.

The Barkhausen criterion requires **T** to equal -1. Hence we obtain

$$\frac{j\omega RC}{1 - \omega^2(RC)^2 + j3\omega C} - \frac{R_1}{R_1 + R_2} = \frac{1}{A}$$

Since A is very large, the right-hand side of this equation is very nearly zero. Therefore, we have

$$\frac{j\omega RC}{1 - \omega^2(RC)^2 + j3\omega C} \approx \frac{R_1}{R_1 + R_2}$$

Since the right-hand side of the equation above is real, the left-hand side must also be real. This can only be accomplished if ω is equal to

$$\omega = \frac{1}{RC}$$

We then obtain

$$\frac{R_1}{R_1 + R_2} = \frac{1}{3}$$

This result is essentially the same as that obtained in Chap. 6, but the circuit is simpler. Once again, the gain (with feedback) must be at least 3, whereas the frequency of oscillation (in Hertz) is given by

$$f = \frac{1}{2\pi RC}$$

Just as we did in Chap. 6, we can here employ an additional slow-speed, DC, feedback scheme using a JFET to control the gain and stabilize the amplitude of oscillation.

7-11 NOTCH FILTER

As a final example of the application of operational amplifiers, we consider another type of filter network, a *notch filter*. The name comes from the fact that a plot of gain versus frequency would show a sharp decline in gain in the vicinity of one

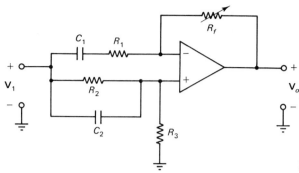

Fig. 7-46 Notch-filter network.

specific frequency, whereas the gains at other frequencies would be maintained at some more or less constant level. Such a filter would be useful in discriminating against an unwanted signal having a single frequency. For example, a troublesome 60-Hz signal could be reduced by means of a notch filter. (Of course, we ought to think twice before resorting to such a formidable way to cure problems that might be due simply to such factors as bad grounds.)

A notch-filter network is shown in Fig. 7-46. To simplify the analysis of this network, we first make use of the reduced form of the network shown in Fig. 7-47.

Referring to Fig. 7-47 and assuming that the operational amplifier has infinite input resistance, we can write KCL equations at the two input terminals:

$$\frac{\mathbf{V}_1 - \mathbf{V}_y}{\mathbf{Z}_1} + \frac{\mathbf{V}_o - \mathbf{V}_y}{\mathbf{Z}_f} = 0$$

and

$$\frac{\mathbf{V}_1 - \mathbf{V}_x}{\mathbf{Z}_2} - \frac{\mathbf{V}_x}{\mathbf{Z}_3} = 0$$

In addition to these equations, the input and output voltages of the amplifier are related by

$$\mathbf{V}_o = A(\mathbf{V}_x - \mathbf{V}_y)$$

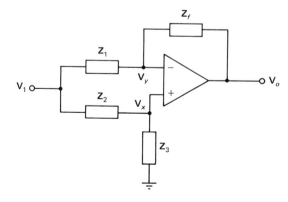

Fig. 7-47 Reduced network.

It is not too difficult to combine these equations, eliminate the voltages $\mathbf{V}_x$ and $\mathbf{V}_y$, and arrive at the equation

$$\mathbf{V}_o \left\{ \frac{1}{\mathbf{Z}_f} + \frac{1}{A}\left[\frac{1}{\mathbf{Z}_1} + \frac{1}{\mathbf{Z}_f} \right] \right\} = \mathbf{V}_1 \left\{ \frac{\mathbf{Z}_3}{\mathbf{Z}_2 + \mathbf{Z}_3}\left(\frac{1}{\mathbf{Z}_1} + \frac{1}{\mathbf{Z}_f} \right) - \frac{1}{\mathbf{Z}_1} \right\}$$

If we recognize that A is very large, then we can approximate the relationship between $\mathbf{V}_o$ and $\mathbf{V}_1$ by

$$\frac{\mathbf{V}_o}{\mathbf{V}_1} \approx \mathbf{Z}_f \left[\frac{\mathbf{Z}_3}{\mathbf{Z}_2 + \mathbf{Z}_3}\left(\frac{1}{\mathbf{Z}_1} + \frac{1}{\mathbf{Z}_f} \right) - \frac{1}{\mathbf{Z}_1} \right]$$

At this point we revert to Fig. 7-46 and replace $\mathbf{Z}_1$, $\mathbf{Z}_2$, $\mathbf{Z}_3$, and $\mathbf{Z}_f$ by the appropriate network elements, writing the impedances in terms of the Laplace variable S rather than $j\omega$. After some rather tedious algebraic manipulation, the following equation results.

$$\frac{\mathbf{V}_o}{\mathbf{V}_1} = \left(\frac{R_3}{R_2 + R_3} \right) \left[\frac{R_1 R_2 C_1 C_2 S^2 + \left(R_1 C_1 + R_2 C_2 - \dfrac{R_f}{R_3} R_2 C_1 \right)S + 1}{\left(\dfrac{R_2 R_3}{R_2 + R_3} C_2 S + 1 \right)(R_2 C_1 S + 1)} \right]$$

Notice that both the numerator and denominator of this equation are quadratic in S. The denominator is already in factored form, so that we can easily find its roots. To deal with the numerator we must resort to the well-known quadratic formula. And to simplify that procedure, we replace the messy coefficients by simple ones as follows. Let

$$a = R_1 R_2 C_1 C_2$$

and

$$b = R_1 C_1 + R_2 C_2 - \frac{R_f}{R_3} R_2 C_1$$

Then the roots of the numerator are given by

$$-\frac{b}{2a} \pm \sqrt{\left(\frac{b}{2a} \right)^2 - \frac{1}{a}}$$

Since b consists of terms that are both positive and negative, we can adjust the circuit elements to vary both the magnitude and the sign of b. This means that the roots of the numerator can be made to be real or complex conjugates, and when they are complex conjugates, the real part may be made negative, positive, or zero. We saw earlier that the roots of the denominator gave us the values of S that could make the $\mathbf{V}_o / \mathbf{V}_1$ function become infinite and were therefore called poles. Similarly, the roots of the numerator yield those values of S that cause $\mathbf{V}_o / \mathbf{V}_1$ to be zero and are called *zeros*. For the network under discussion, we see that there are two poles and two zeros. The poles are given by

$$P_1 = -\frac{1}{R_2 R_3 C_2 / (R_2 + R_3)}$$

and

$$P_2 = -\frac{1}{R_1 C_1}$$

Thus we see that they are negative and real. The zeros are found from the solution to the quadratic formula given above.

Rather than go through a complicated calculation to see what happens to the zeros as the circuit parameters change, we consider the only case that is of importance, the case in which the quantity b becomes zero. This case is the important one because when b becomes zero, the complex conjugate zeros of the $\mathbf{V}_o/\mathbf{V}_1$ function are purely imaginary. If we plot the poles and zeros in the complex S-plane as in Fig. 7-48, we can see what happens.

When a sinusoidal source, such as an oscillator, is applied to the input of the network and its frequency is increased from zero, the effect is the same as assigning a value to S that is purely imaginary (equal to $j\omega$) and increases from zero. At a frequency such that ω equals $\sqrt{1/a}$ or $\sqrt{1/R_1 R_2 C_1 C_2}$, the output of the network will fall to zero. That, of course, is what we want a notch filter to do.

Example 7-6

Let us consider an example. Suppose that we wish to construct a notch filter to eliminate an unwanted 60-Hz signal. (The radian frequency corresponding to 60 Hz is 377 rad/s.) To carry out the design we begin by making some choices that are entirely arbitrary. We finish by calculating the remaining quantities, which depend on the design requirements as well as upon our initial choices.

Suppose we choose the following parameters:

$$R_1 = 1 \text{ K}\Omega$$

$$C_1 = 2 \text{ }\mu\text{F}$$

$$C_2 = 1 \text{ }\mu\text{F}$$

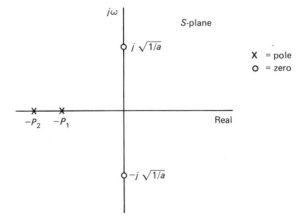

Fig. 7-48 Poles and zeros of $\mathbf{V}_o/\mathbf{V}_1$ in the S-plane.

Since the zeros must occur at $\pm j377$, we have

$$377 = \sqrt{\frac{1}{R_1 R_2 C_1 C_2}}$$

or

$$R_2 = \frac{1}{R_1 C_1 C_2 (377)^2} = \frac{1}{10^3 \times 2 \times 10^{-6} \times 10^{-6} \times (377)^2}$$

This yields $R_2 = 3.52$ KΩ. Next we must adjust b to equal zero. This means that

$$R_1 C_1 + R_2 C_2 - \frac{R_f}{R_3} R_2 C_1 = 0$$

or

$$10^3 \times 2 \times 10^{-6} + 3.52 \times 10^3 \times 10^{-6} - \frac{R_f}{R_3} \times 3.52 \times 10^3 \times 2 \times 10^{-6} = 0$$

Hence

$$\frac{R_f}{R_3} = \frac{2 + 3.52}{2 \times 3.52} = \frac{5.52}{7.04} = 0.784$$

If we let $R_3 = 5$ KΩ, then R_f will be 0.784×5 K, or 3.92 KΩ. This suggests that a 5-KΩ potentiometer will work quite well for R_f. An advantage of this circuit is that b can be tuned to zero by adjusting a single potentiometer, namely, R_f.

REFERENCES

1. Norman Doyle, "Some Useful Signal Processing Circuits Using FET's and Operational Amplifiers," *Application Note 243* (Mountain View, California: Fairchild Semiconductor Company, 1971).
2. Sanjit K. Mitra, "Synthesizing Active Filters," *IEEE Spectrum*, 6(1969): 47–63.
3. Moise Hamaoui, "Operational Amplifiers as Inductors," *Application Note 321* (Mountain View, California: Fairchild Semiconductor Company, 1973).
4. Lawrence P. Huelsman, *Theory and Design of Active RC Circuits*, (New York: McGraw-Hill Book Company, 1968).

PROBLEMS

7-1. **(a)** With its input terminals short-circuited, an operational amplifier has an output voltage of -4.5 V. Its gain A_{vo} is known to be 10,000. Find the input offset voltage.

(b) Using the nulling potentiometer, the output voltage observed in (a) is brought to zero. Then a 10-KΩ resistor is connected between the two input terminals and the output voltage becomes $+10$ V. Find the input offset current.

7-2. An operational amplifier has $A_{vo} = 10^4$. After the effect of V_{OS} has been nulled, the negative terminal is grounded. The positive terminal is connected to ground through a 1-KΩ resistor. The output voltage is found to be 3.6 V. If I_{os} is 10% of the bias current, find the bias current.

7-3. **(a)** A model of the operational amplifier showing offset current is given. Assuming R is 0 (short-circuited), find V_1 if $I_{os} = 50$ nA and $A_{vo} = 10,000$.
   (b) Find V_o.
   (c) With $R = 20$ K∥80 K repeat (a) and (b).

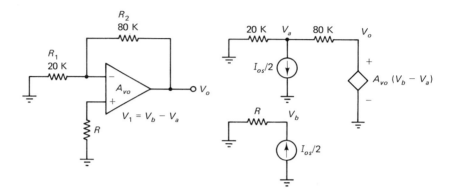

7-4. The circuit shown is to be used in adjusting the null potentiometer to cancel the effects of V_{OS}. Voltmeter V can register voltages ranging from -100 mV to $+100$ mV.
   (a) Use the model of Fig. 7-2 with $V_1 = V_{OS}$ and $V_2 = 0$ to estimate the range of values of V_{OS} that can be "nulled out."
   (b) After nulling V_{OS}, R_3 is short circuited. Meter V registers a change of 5 mV. Estimate I_B, assuming I_{os} is negligible.

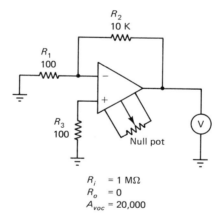

$R_i = 1$ MΩ
$R_o = 0$
$A_{voc} = 20,000$

7-5. Preliminary measurements have established that $V_{OS} = 3$ mV. The voltmeter V reads 5.8 mV. Estimate I_{OS}.

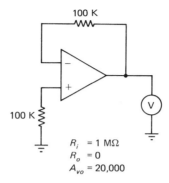

100 K

100 K

$R_i = 1$ MΩ
$R_o = 0$
$A_{vo} = 20{,}000$

7-6. Show that the asymptotic gain K for Fig. 7-20 is given by $K = R_2/R_1$. [Consider that the voltage source $(I_B - I_{OS}/2)R_1$ is replaced by a voltage source V_{in}.]

7-7. An empirical equation that relates the input offset current to the temperature for an integrated-circuit operational amplifier is

$$I_{OS} = 3.21 \times 10^{-3} T^2 - 0.621T + 50$$

where I_{OS} is in microamperes and T is in °C. If the amplifier is connected as shown, calculate the change in v_o as the temperature varies from 20°C to 80°C. Assume that $I_B = 8I_{OS}$ and that the effect of V_{OS} has been canceled.

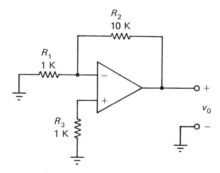

R_2
10 K

R_1
1 K

R_3
1 K

+

v_0

−

7-8. An experimenter attempting to determine the open-loop gain of an operational amplifier uses the circuit of Fig. 7-21. By mistake, only V_1 and V_o are measured, while V_x is omitted. Determine whether A (the open-loop gain) can be found from these measurements and, if so, calculate it. Assume resistance values are $R_F = 10$ KΩ, $R_1 = 1$ KΩ, $R_2 = 10$ KΩ, $R_3 = 10$ Ω, $V_o = -1.0$ V and $V_1 = 0.152$ V.

7-9. **(a)** Estimate the slew rate for the circuit shown.

(b) Estimate the 3-db bandwidth of this circuit assuming that, for Q_7 only, $h_{ie} = 3$ KΩ.

(c) Is the slew rate or the bandwidth the limiting factor in determining $dv_o/dt|_{max}$?

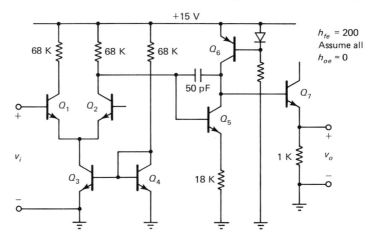

7-10. An operational amplifier has a 3-db bandwidth of 150 Hz, unity-gain bandwidth product equal to 6 MHz, and a slew rate of 4 V/μs. What is the maximum amplitude that a sinusoidal input voltage can have at a frequency of 60 KHz if the slew-rate limiting effect is to be avoided?

7-11. Show that for the voltage-follower circuit of Fig. 7-11, $\mathbf{T} = A$, $K = 1$, and $A_{vf} \approx 1$.

7-12. Using the circuit of Fig. 7-21, measurements are made to determine A. Circuit parameters are $R_F = 20$ KΩ, $R_1 = 5$ KΩ, $R_2 = 15$ KΩ, and $R_3 = 5\ \Omega$. The experimenter neglected to measure V_x but did measure V_o and V_1. Values found were $V_o = 1.4$ V and $V_1 = 100$ mV. Estimate V_x, and from this result calculate A.

7-13. Find an expression for i in terms of R_1, R_2, R_3, and v_s. Assume that the operational amplifier is ideal.

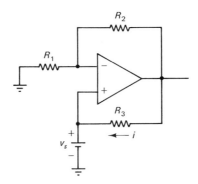

7-14. Find v_o/v_{in}. Assume that the operational amplifier is ideal.

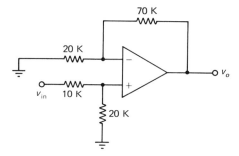

7-15. The operational amplifier shown is ideal. Find v_o.

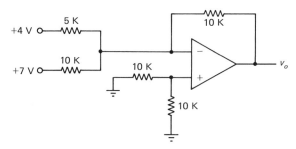

7-16. The operational amplifier is ideal.
    (a) Find the phasor voltage $\mathbf{V}_o$.
    (b) If $v_1 = 10 \cos 10t$, what is the expression for $v_o(t)$?

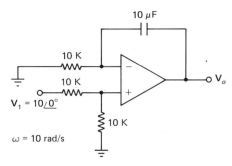

7-17. The operational amplifier is ideal. Find v_o.

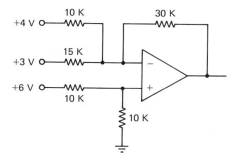

7-18. The operational amplifier is ideal. Find v_o/v_s.

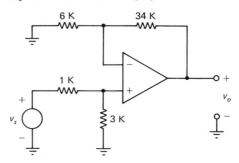

7-19. Both operational amplifiers are ideal.
  (a) Find v_o as a function of v_1, v_2, R_1, R_2, R_3, and R_4.
  (b) Equate two resistor *ratios* so that v_o depends *only* on $v_2 - v_1$.

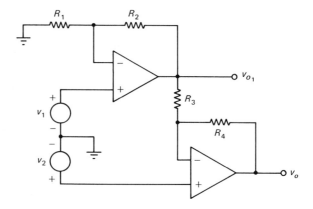

7-20. For the operational amplifier shown, $R_i \approx \infty$, $R_o \approx 0$, and $A_v \approx \infty$.

$$\Delta V_{os}/\Delta T = 15 \ \mu V/°C$$

$$\Delta I_{os}/\Delta T = 12 \ nA/°C$$

  (a) Find the value of R_3 required to eliminate the effect of I_B (bias current).
  (b) If voltage and current offsets were neglected, what would be the value of v_o?
  (c) Offset effects are initially canceled at 25°C. If the temperature varies from 25°C to 85°C determine the range of values of v_o that could result from this variation.

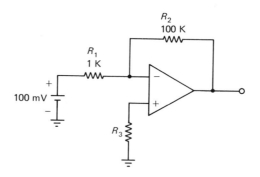

7-21. All operation amplifiers are ideal. Calculate v_o.

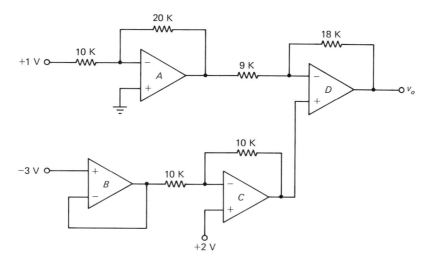

7-22. A student wishing to determine the open-loop break frequency ω_o of an operational amplifier connects it in the circuit shown in Fig. 7-22. By mistake, R_1 is chosen to be 20 KΩ, whereas R_2 is 10 KΩ. Using these parameter values, A_{vf} is found to be reduced from its low-frequency value by a factor of $1/\sqrt{2}$ when $f = 1.2$ MHz.
  (a) Find the unity-gain frequency f_1.
  (b) If the open-loop gain at frequencies approaching DC equals 106 db, find the 3-db open-loop frequency f_o.

7-23. The operational amplifier is ideal and $v_s(t)$ is shown. Sketch and dimension $v_o(t)$.

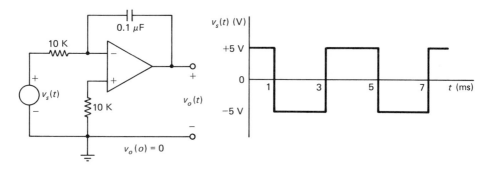

7-24. The *quality factor* Q of a coil having resistance R in series with inductance L is defined as $\omega L/R$. Show that Q_o, the value of Q for $\omega = \omega_o$ that is obtained using the inductance-simulating circuit of Fig. 7-29(a), is equal to 1.

7-25. The inductance-simulating circuit of Fig. 7-30 is to be used to simulate an inductance of 200 mH. Assume that $A = 50,000$, $R_i = 1$ MΩ, $R_2 = 10$ KΩ, and $R_1 = 1$ KΩ.
  (a) Using 6 Hz for the nominal value of the open-loop break frequency of amplifier A_3, find the closed-loop break frequency for this amplifier.
  (b) Choosing $R_x = 10$ KΩ, find C_x to make the break frequency 3 KHz.
  (c) Find R.
  (d) Determine Q at a frequency of 1 KHz ($Q = \omega L_{eq}/R_{eq}$).

7-26. In the circuit of Fig. 7-30, $R_1 = 1$ KΩ, $R_2 = 10$ KΩ, $R_x = 1590\ \Omega$, $C_x = 0.01\ \mu F$, $R_3 = 10$ KΩ, and $R_4 = 40$ KΩ.

    (a) Choose a value for R to make $L_{eq} = 20$ mH.

    (b) If the operating frequency must be at most one decade less than ω_o, specify the maximum operating frequency in radians per second.

    (c) If $R_A = 1$ KΩ, find the value required for R_B that will approximately eliminate R_{eq}.

7-27. The circuit of Fig. 7-35 is to be used to simulate a capacitance of 0.1 μF using a physical capacitor having a capacitance of 1000 pF.

    (a) Specify R_2 if $R_1 = 2$ KΩ.

    (b) If A (open loop) of the operational amplifier is 100 db and f_o (open loop) = 10 Hz, find the 3-db frequency of the amplifier section of the circuit.

    (c) At a frequency 1 decade below the 3-db frequency calculated in (b), a voltage of 1 V rms is applied to the simulating circuit. Calculate the *peak* current through the capacitor.

7-28. Design a biquad, low-pass, maximally flat, second-order filter network using the circuit of Fig. 7-41. Cutoff (3-db) frequency is to be 8 KHz. Choose C_1 and C_2 to be 0.01 μF.

7-29. A fourth-order, maximally flat LP network has poles arranged according to the pattern of Fig. 7-40(d). The radius of the circle on which the poles lie equals ω_o. The overall transfer function is obtained by cascading two biquad sections, each of which has a transfer function given by Eq. (7-15).

    (a) Find the values of a and b in Eq. (7-15) required to produce the pair of poles lying on radii that make angles of $\pm 22.5°$ with the negative real axis.

    (b) Find the values of a and b required to produce the pair of poles lying on radii at angles of $\pm 67.5°$ with the negative real axis.

7-30. For the fourth-order network of Prob. 29, specify circuit parameters for each biquad circuit for $f_o = 5$ KHz, using the values found. Choose $C_1 = C_2 = 0.05\ \mu F$. Set d in Eq. (7-15) equal to b and R_2 equal to R_3.

7-31. **(a)** For the circuit shown, find the cutoff (3-db) frequency f_o for the transfer ratio $\mathbf{V}_o/\mathbf{V}_1$.

    (b) To what value should R_1 be adjusted to produce a maximally flat response?

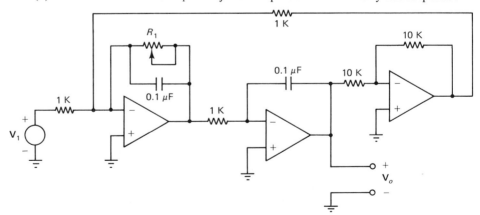

7-32. **(a)** Verify that the return ratio **T** for the circuit of Fig. 7-45 is given by

$$\mathbf{T} = -A\left(\frac{j\omega RC}{-\omega^2(RC)^2 + j3\omega RC + 1} - \frac{R_1}{R_1 + R_2}\right)$$

(b) Since **T** must equal $-1 + j0$, the imaginary part of **T** must be zero. Make use of this fact to find an expression for ω in terms of R and C.

(c) Use the result of (b) to determine the ratio $R_1/(R_1 + R_2)$.

7-33. For the circuit of Fig. 7-45, $C = 0.01 \mu F$ and $R_1 = 10 \ K\Omega$. Determine R and R_2 so that the circuit will oscillate at 10 KHz.

7-34. Using the circuit of Fig. 7-46, design a notch filter to remove an unwanted signal having a frequency of 400 Hz. Use C_1 and C_2 equal to 0.2 μF, $R_1 = 1 \ K\Omega$, and $R_3 = 10 \ K\Omega$.

CHAPTER 8

Wide-Band Amplifiers

8-1 THE NEED FOR WIDE-BAND AMPLIFIERS

Up to this point, most of the amplifier circuits we have dealt with have been intended to amplify frequencies extending over the audio range, that is, from about 20 Hz to 20,000 Hz. Of course, direct-coupled amplifiers, such as operational amplifiers, can amplify signals whose frequencies extend all the way down to zero (direct current). And, as we have seen, we can use negative feedback to extend the upper limits of the operating range up to frequencies in the order of hundreds of kilohertz. But what is of interest in this chapter is the class of circuits that can amplify signals whose frequency components range from zero to several megahertz.

To answer the question of how such signals arise, let us consider a hypothetical design for a television (video) amplifier. If we take as an example a simple black-and-white television set, we have the following items of data to use:

$$\text{Number of frames per second} = 30$$

$$\text{Number of lines per frame} = 525$$

We see that approximately $1/(30 \times 525) = 6.349 \times 10^{-5}$ s is required for a single horizontal line to be written on the picture tube. Now suppose that the picture tube has a face diameter of 38 cm. The question we must now answer is: Over what distance must a clear transition from light to dark occur if the picture we see is to have good resolution? Let us arbitrarily imagine that $\frac{1}{2}$ mm is the maximum distance we can tolerate for such a purpose. This means that the rise time of the video signal that controls the beam intensity must be $(\frac{1}{2})/380$ times the line-writing time determined earlier. In other words, the rise time t_r must equal $(0.5/380) \times 6.349 \times 10^{-5}$, or 8.35×10^{-8} s.

We have seen that there is a relationship between the rise time of a wave and the bandwidth of the network transmitting that wave. To understand this relationship quantitatively, we consider the circuit and the wave form shown in Fig. 8-1.

If the time-varying voltage v_1 were sinusoidal, we could replace v_1 and v_o by phasor voltages $\mathbf{V}_1$ and $\mathbf{V}_o$. Then, as we have often seen before, $|\mathbf{V}_o/\mathbf{V}_1|$ would be equal to $1/\sqrt{1 + (f/f_o)^2}$. The 3-db frequency f_o defines the bandwidth and is equal to $1/2\pi RC$.

If $v_1(t)$ is a unit step of voltage, $v_o(t)$ has the form shown in Fig. 8-1(b). A common engineering definition of rise time is the time interval during which the wave-form amplitude changes from 10% to 90% of its final level. The values of t when v_o achieves these levels are designated by T_1 and T_2, respectively. The rise time t_r is thus equal to $T_2 - T_1$. The equation for v_o for this case is $v_o = K(1 - e^{-t/RC})$. Upon substituting corresponding values of v_o and t, we obtain

$$0.1\,K = K(1 - e^{-T_1/RC})$$

and

$$0.9\,K = K(1 - e^{-T_2/RC})$$

These equations can be solved simultaneously to yield

$$t_r = T_2 - T_1 = RC \ln 9$$

We now make use of the fact that $RC = 1/2\pi f_o$ to find that

$$f_o = \frac{2.20}{2\pi t_r} = \frac{0.35}{t_r} \tag{8-1}$$

Equation (8-1) gives us the desired relationship between rise time and bandwidth. Returning to the example we were considering, we see that the required video-amplifier bandwidth must be given by

$$f_o = \frac{0.35}{8.35 \times 10^{-8}} = 4.19 \text{ MHz}$$

This number is in the range of acceptable figures for video-amplifier bandwidth. It clearly shows why wide-band amplifiers are required for video applications.

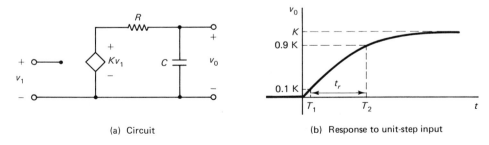

(a) Circuit   (b) Response to unit-step input

Fig. 8-1 To relate rise time and bandwidth.

8-2 TRANSISTOR HYBRID-π PARAMETERS

Up to this point we have ignored the fact that transistor parameters vary with frequency. Instead, we have considered all these quantities to be represented by real numbers that are taken to be resistances, conductances, or voltage or current ratios. This approach is good enough for circuits whose frequency ranges are restricted, but it will no longer serve in dealing with wide-band amplifiers.

One way to deal with this problem is to find empirical, frequency-dependent mathematical expressions for the various parameters. The trouble with such an approach is that it is abstract and dissociated from the circuit models that aid our thinking. A better way is to find new circuit models whose elements are all constant but which now include capacitors that account for transistor behavior at high frequency. One such model is called the *hybrid-π*.

Since the common-emitter connection is probably the most frequently used circuit arrangement, the only hybrid-π model used is based on this connection. A diagram showing the small-signal, hybrid-π circuit is shown in Fig. 8-2.

Since the circuit of Fig. 8-2 is supposed to represent the same *CE*-connected transistor that is represented by our usual *h*-parameter model, we would expect to find some kind of agreement between the two models. Such an agreement ought to occur at low frequencies, since it is only at such frequencies that the *h*-parameter circuit is valid. Of course, when the frequency is low enough, the capacitors in the circuit of Fig. 8-2 will have no effect. Since only the resistive elements will then be active, we can see that the following correspondences exist between parameters.

$$r_{bb'} = r_{b'e} = h_{ie} \tag{8-2a}$$

and

$$r_o = 1/h_{oe} \tag{8-2b}$$

An important difference between the *h*-parameter model and the hybrid-π model is the new node b' that appears in the latter. This node and its associated voltage $\mathbf{V}_{b'e}$ are inaccessible to the outside world. That is, b' represents a point somewhere in the transistor structure that is on the base side of the emitter-base junction. Only the external base terminal b is available for connection or measurement. Terminal b is separated from b' by some of the base material through which the base current must pass on its way to b', which is where the activity occurs. The resistance of this part of the base material is sometimes called the *base-*

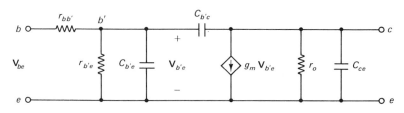

Fig. 8-2 Hybrid-π, small-signal, *CE* circuit.

spreading resistance and is referred to as the *ohmic resistance* r_{bb}'. Typical values for r_{bb}' range from 50 Ω to 100 Ω.

The resistance $r_{b'e}$ is the resistance associated with the base-emitter junction and therefore depends on the operating-point value of I_E. In fact, since we have heretofore ignored r_{bb}', the values that we used for h_{ie} ought really to have been assigned to $r_{b'e}$. In other words, we now recognize that

$$r_{b'e} = \left(\frac{0.026}{I_E}\right)(1 + h_{fe})$$

The base spreading resistance $r_{bb'}$ cannot be determined by direct measurement. If, however, we have a *measured* value for h_{ie}, then we can find $r_{bb'}$ by using the relationship

$$r_{bb}' = h_{ie} - \frac{0.026}{I_E}(1 + h_{fe}) \tag{8-3}$$

Again considering very low frequencies, if we short-circuit the collector and emitter terminals of the circuit of Fig. 8-2 and apply a current $\mathbf{I}_o$ into the base terminal, the short-circuit collector-emitter current will be simply $g_m V_{b'e}$. But $V_{b'e}$ will be equal to $r_{b'e}\mathbf{I}_b$. Therefore, the ratio of short-circuit collector-emitter current to base-emitter current will be $g_m V_{b'e}/\mathbf{I}_b$ and this current ratio is, by definition, h_{fe}. Thus we have

$$h_{fe} = g_m r_{b'e}$$

Since, at room temperature, $r_{b'e}$ is equal to $(0.026/I_E)(1 + h_{fe}) \approx (0.026/I_E)h_{fe}$, we have

$$h_{fe} = \frac{g_m \times 0.026}{I_E} h_{fe}$$

or

$$g_m = \frac{I_E}{0.026} \approx 40 I_E \tag{8-4}$$

If manufacturer's data give us the values of h_{fe}, h_{oe}, and h_{ie} for some stated quiescent operating level of I_E, we can determine $r_{b'e}$, r_{bb}', and r_0 from Eqs. (8-2), (8-3), and (8-4). We are then left with the problem of finding the capacitances $C_{b'e}$, $C_{b'c}$, and C_{ce}.

As we stated earlier, the purpose of the capacitors in the circuit of Fig. 8-2 is to account for the high-frequency performance of the transistor. The main thing that happens as the frequency increases is the reduction in the value of h_{fe}—the short-circuit current gain. Let us see how this circuit accounts for this behavior.

When the collector-emitter terminals in Fig. 8-2 are short-circuited so as to satisfy the conditions under which h_{fe} is defined, the circuit takes the form shown in Fig. 8-3. Note that a base input current $\mathbf{I}_b$ is assumed. With the short circuit in place, C_{ce} and r_o have no effect, and the current $\mathbf{I}_c$ is simply equal to $g_m V_{b'e}$. The capacitance $C_{b'c}$ is effectively placed in parallel with $C_{b'e}$, so that the expression

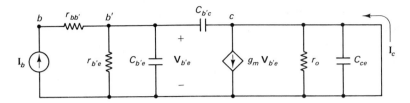

Fig. 8-3 Hybrid-π with output short-circuited.

for $\mathbf{V}_{b'e}$ is

$$\mathbf{V}_{b'e} = \mathbf{I}_b \frac{r_{b'e}[1/j\omega(C_{b'e} + C_{b'c})]}{r_{b'e} + 1/j\omega(C_{b'e} + C_{b'c})}$$

This expression is easily simplified to the form

$$\mathbf{V}_{b'e} = \mathbf{I}_b \frac{r_{b'e}}{1 + j\omega r_{b'e}(C_{b'e} + C_{b'c})}$$

We can now write

$$\mathbf{h}_{fe} = \frac{\mathbf{I}_c}{\mathbf{I}_b} = \frac{g_m r_{b'e}}{1 + j\omega r_{b'e}(C_{b'e} + C_{b'c})}$$

When ω becomes zero the value of h_{fe} yielded by this equation will clearly be the usual DC value. The magnitude of $\mathbf{h}_{fe}$ will be reduced by a factor of $1/\sqrt{2}$ when $\omega = 1/r_{b'e}(C_{b'e} + C_{b'c})$. We designate this value of ω by the symbol ω_β or, to express this frequency in Hertz, by f_β. Thus we have

$$f_\beta = \frac{1}{2\pi r_{b'e}(C_{b'e} + C_{b'c})}$$

As we have seen, we can estimate $r_{b'e}$ from a knowledge of h_{fe} and the quiescent level of I_E. The capacitance $C_{b'e}$ arises from the juxtaposition of bound charges at the base-emitter junction as well as from diffusion effects in the base material. If we knew the dimensions, doping concentrations, and so on, we could, at least in principle, calculate this capacitance. However, the principles involved go beyond the scope of this book. Furthermore, such a calculation would give us only a fair approximation of $C_{b'e}$ at best. Yet, if we knew $C_{b'e}$ and $C_{b'c}$ (which is very small compared to $C_{b'e}$), we could combine that value with the one for $r_{b'e}$ and find f_β.

Instead of finding $C_{b'e}$ and $C_{b'c}$ and then calculating f_β, we shall reverse the procedure and use the values of f_β and $r_{b'e}$ to find $C_{b'e}$ and $C_{b'c}$. It is clear that we could fairly easily measure the 3-db frequency f_β in order to do this. Common practice is to find the unity-gain frequency f_T and to use that figure to find f_β. The short-circuit unity-current-gain frequency f_T is defined as that frequency at which $|\mathbf{h}_{fe}|$ becomes equal to 1. Using this definition we can write

$$1 = \frac{g_m r_{b'e}}{\sqrt{1 + (f_T/f_\beta)^2}} \approx \frac{g_m r_{b'e}}{f_T/f_\beta}$$

This leads to

$$f_\beta = \frac{f_T}{g_m r_b'_e}$$

Since the DC value of $\mathbf{h}_{fe}$, or h_{fe}, is equal to $g_m r_b'_e$, we can write

$$f_\beta = \frac{f_T}{h_{fe}} \text{ or } f_T = h_{fe} f_\beta \tag{8-5}$$

Manufacturers' data usually include f_T. Thus we can easily find f_β. Then, knowing $r_b'_e$, we can easily calculate the *total* capacitance $C_b'_e + C_b'_c$. For the values of the capacitances $C_b'_c$ and C_{ce}, we must again depend upon manufacturers' data. The capacitance $C_b'_c$ is not usually specified using that symbol. Instead, it is commonly given as C_{ob}. The subscripts ob suggest the notation used for common-base h-parameters. That is, C_{ob} is the capacitance associated with the output (collector) circuit on a common-base basis. If we look back to Fig. 8-2 and imagine that the base terminal is the grounded (reference) terminal and that we look into the collector, we see that, except for the small resistance r_{bb}', the capacitance we observe is $C_b'_c$.

Example 8-1

A typical set of transistor data might look like the following:

$$h_{fe} = 150 \qquad\qquad C_{ob} = 3 \text{ pF}$$

$$h_{oe} = 2 \times 10^{-5} \text{ mho} \qquad C_{ce} = 2 \text{ pF}$$

$$I_E = 2 \text{ mA} \qquad\qquad f_T = 100 \text{ MHz}$$

$$r_{bb}' = 50 \ \Omega$$

Using these data we now calculate the parameters for the hybrid-π circuit:

$$r_b'_e \approx \frac{0.026}{2 \times 10^{-3}} \times 150 = 1950 \ \Omega$$

$$f_\beta = \frac{f_T}{h_{fe}} = \frac{100 \times 10^6}{150} = 6.67 \times 10^5 \text{ Hz}$$

$$(C_b'_e + C_b'_c) = \frac{1}{2\pi r_b'_e f_\beta} = \frac{1}{2\pi \times 1.95 \times 10^3 \times 6.67 \times 10^5} = 122 \text{ pF}$$

Since $C_b'_c = 3$ pF, we see that $C_b'_e = 119$ pF.

$$g_m = \frac{h_{fe}}{r_b'_e} = \frac{150}{1950} = 0.077 \text{ mho}$$

$$r_o = \frac{1}{h_{oe}} = 50 \text{ K}\Omega$$

Since we now know how to find all the necessary parameters to use in the equivalent circuit of Fig. 8-2, we might suppose that we need only to proceed with the usual circuit analysis to find the gain or whatever other information we desire. Of course, such an approach will certainly work. However, even so simple a task as finding the voltage gain of a single-stage amplifier turns out to be messier than we would like because the capacitor $C_{b'c}$ connects the b' and c nodes, so that a rigorous solution of the circuit requires us to solve at least two simultaneous node-voltage equations.

What we would like to find is a way to uncouple the b' and c nodes. The principle of the Miller effect provides us with a means of doing so. To see how this procedure works, let us consider the circuit of Fig. 8-4, which represents a common-emitter amplifier with a collector resistor R_c. Base-bias resistors are assumed to be large enough to neglect, and we also assume $r_o \gg R_c$.

At low frequencies $C_{b'c}$ and C_{ce} can be neglected, so that the output voltage $\mathbf{V}_o$ is given by

$$\mathbf{V}_o \approx -g_m \mathbf{V}_{b'e} R_c$$

The current through $C_{b'c}$ is equal to

$$\mathbf{I}_{b'c} = j\omega C_{b'c}(\mathbf{V}_{b'e} - \mathbf{V}_o) = j\omega C_{b'c}(\mathbf{V}_{b'e} + g_m R_c \mathbf{V}_{b'e})$$

Therefore, the admittance looking into $C_{b'c}$ from b' is equal to $\mathbf{I}_{b'c}/\mathbf{V}_{b'e}$, or

$$\mathbf{Y} = j\omega C_{b'c}(1 + g_m R_c)$$

This is the same admittance that would be observed if a capacitance equal to $C_{b'c}(1 + g_m R_c)$ were connected between b' and e. Thus we see that we can represent the circuit of Fig. 8-4 by the simplified one of Fig. 8-5.

To get some idea of the effect of $C_{b'c}$, let us suppose that the transistor whose hybrid-π parameters were determined in Example 8-1 is connected with a collector resistor of 2 KΩ. The effective capacitance at the input circuit is

$$C_{b'e} + C_{b'c}(1 + g_m R_c) = 119 + 3(1 + 0.077 \times 1950) = 572 \text{ pF}$$

Note that the seemingly insignificant 3-pF capacitance has by far the greater effect in determining input-circuit capacitance.

Suppose that the voltage source has an internal resistance R_s equal to 2 KΩ. Let us now find the voltage gain of the circuit of Fig. 8-5 and see the effect of frequency upon it. Considering first the collector circuit, we see that because of R_c and C_{ce}, there is a break frequency f_2 equal to $1/(2\pi \times 2 \times 10^3 \times 2 \times 10^{-12})$ = 39.8 GHz. Since this frequency is far beyond the value of f_T, we can ignore it altogether. (In general we must be careful about rushing to such a conclusion. If

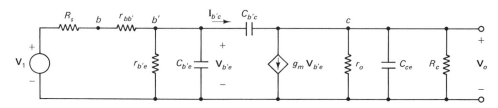

Fig. 8-4 Hybrid-π circuit with load resistor R_C.

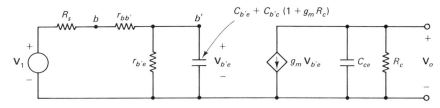

Fig. 8-5 Simplified circuit.

there is a substantial capacitance associated with the load, a very different result could occur.) We turn next to the base circuit, where we see that a break frequency f_1 occurs that is given by

$$f_1 = \frac{1}{2\pi[C_{b'e} + C_{b'c}(1 + g_m R_c)]r_{b'e}\|(R_s + r_{bb'})}$$

The value of f_1 is

$$f_1 = \frac{1}{2\pi \times 572 \times 10^{-12} \times 1950\|2050} = 278 \text{ KHz}$$

This, of course, will be the 3-db frequency of the amplifier. Low-frequency gain is given by

$$A_v = \frac{-r_{b'e}}{r_{b'e} + r_{bb'} + R_s} \cdot g_m R_c$$

$$= \frac{-1950}{4000} \times 0.077 \times 2000 = -75$$

For an audio amplifier this design would be more than adequate. However, the bandwidth of 278 KHz is far short of what is needed for a video amplifier. What can be done to increase the bandwidth? Some improvement can be had through the use of feedback. As we recall from Chap. 4, reducing the gain through the use of negative feedback will increase the bandwidth by the same factor. One simple way to do this is by adding an unbypassed emitter-resistor to modify the circuit, along the lines of the feedback circuit of Fig. 4-11.

It is not too difficult to show that with an emitter-resistor equal to R_E, the return ratio **T** at low frequencies is given by

$$\mathbf{T} = g_m \frac{r_{b'e} R_E}{R_E + R_S + r_{bb'} + R_S}$$

Hence the factor by which gain is reduced or bandwidth increased is $1 + \mathbf{T}$ which, for an emitter-resistor equal to 200 Ω, is

$$1 + \mathbf{T} = 1 + \frac{0.077 \times 1950 \times 200}{4200} = 8.15$$

Thus the gain now becomes -9.2, while the bandwidth expands to 8.15×278, or 2.26 MHz.

This last result is still not adequate to meet the bandwidth requirements of the video amplifier we are planning. Since we have just successfully exercised the now familiar trade-off of gain versus bandwidth, we are tempted to reduce the gain still further in order to attain the desired bandwidth. That, of course, would leave us with a low figure for gain. It is therefore tempting to see whether we can do better by using another transistor.

Let us see what results we can achieve using the 2N2616, for which we have the following data:

$$h_{fe} = 20 \qquad C_{ob} = 2.8 \text{ pF}$$

$$I_E = 50 \text{ mA} \qquad f_T = 600 \text{ MHz}$$

No data are listed for h_{oe}, h_{ie}, C_{ce}, or r_{bb}' in the 1977 D.A.T.A. transistor data book. Let us again assume R_c and R_s to be 2 KΩ. Of course, with I_E equal to 50 mA (the data point at which the other parameters are measured), this value of R_c would require a supply voltage V_{cc} greater than 100 V! For the moment, let us put aside this impractical requirement and see what results we might expect for gain and bandwidth.

First of all, $r_{b'e}$ is approximately $(26/50) \times 20 = 10.4 \ \Omega$. The frequency $\hat{f}_\beta$ is equal to $600/20 = 30$ MHz. Therefore, the total capacitance $C_{b'e} + C_{ob}$ is equal to $1/(2\pi \times 30 \times 10^6 \times 10.4) = 510$ pF. Since C_{ob} is so small compared to this figure, we need not subtract it. The figure for g_m is equal to $h_{fe}/r_{b'e}$, or $20/10.4$, which is 1.92 mho. Therefore, the low-frequency gain of the amplifier will be $g_m R_c$, or 3840 (a surprisingly large number). The Miller capacitance due to C_{ob} will be about $3840 \times 2.8 \times 10^{-12}$, or 10.75×10^{-9}, which is 10,750 pF. Notice how this Miller capacitance nearly swamps $C_{b'e}$. The effective base-circuit capacitance now turns out to be 510 pF + 10,750 pF = 11,260 pF. This, together with $r_{b'e}$, yields a bandwidth of $1/(2\pi \times 11.26 \times 10^{-9} \times 10.4)$, or 1.36 MHz. Once again it appears that we have failed to meet our design objective.

The trouble seems to arise from the high gain associated with the quantity $g_m R_c$ because the Miller capacitance, which is the chief factor reducing the bandwidth, is equal to $(1 + g_m R_c)C_{b'c}$. It seems reasonable to suppose that if we reduce g_m—and perhaps R_c as well—that we ought to be able to reduce the Miller capacitance and thus to increase the bandwidth.

Since g_m is directly proportional to I_E, it is a simple matter to adjust the bias current for the transistor in order to reduce g_m. There is, as we saw above, an additional reason to reduce I_E, the inordinately high supply voltage needed when I_E is held at 50 mA. If we operated the transistor with I_E at a level of 5 mA, the V_{CC} supply would need to be only 15 V, for instance, a much more reasonable value than we had before.

It turns out that h_{fe} is quite sensitive to changes in the level of I_E. Thus if we reduce I_E to 5 mA, h_{fe} will probably decrease to about 15. The value of $r_{b'e}$ then becomes $(26/5) \times 15$, or about 78 Ω. The frequency f_β will then be $600/15$, or 40 MHz. Therefore, the combined capacitance $C_{b'e} + C_{b'c}$ will be $1/(2\pi \times 40 \times 10^6 \times 78)$, or 510 pF, as before. This time, however, g_m is $15/78$, or 0.192 mho. Hence the Miller capacitance is $(1 + 0.192 \times 2 \times 10^3) \times 2.8 \times 10^{-12}$, or $385 \times 2.8 \times$

10^{-12}, which equals 1070 pF. The bandwidth is therefore $1/(2\pi \times [510 + 1070] \times 10^{-12} \times 78)$, or 1.29 MHz. This bandwidth is about the same as the figure last calculated. By reducing R_c to 1.5 KΩ, for instance, somewhat more bandwidth can easily be obtained. Notice that the low-frequency gain with R_c equal to 2 KΩ is about equal to $g_m R_c$, or 384, a far better result than we had for the preceding example.

8-3 INTEGRATED-CIRCUIT, WIDEBAND AMPLIFIERS

There are integrated circuits that are specifically designed to function as video amplifiers. One of these is the RCA CA3001, whose schematic diagram is shown in Fig. 8-6. The circuit is structured around the differential pair comprising transistors Q_3 and Q_4. Terminals 1 and 6 are the input nodes, whereas terminals 11 and 8 deliver double-ended output. (The CA3002 is structurally similar to this unit but provides only a single-ended output.) The bases of Q_3 and Q_4 are not driven

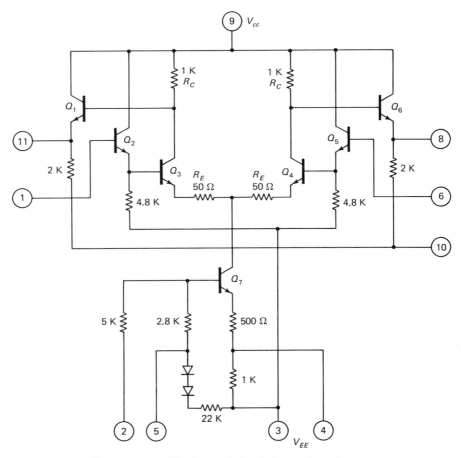

Fig. 8-6 The RCA CA3001 video amplifier. (Courtesy RCA)

directly from the input terminals but are driven by transistors Q_2 and Q_5, which are emitter-followers that are used to make the input resistance high. Transistors Q_1 and Q_6 are also emitter-followers and are used to provide low-output resistances. Transistor Q_7 functions in the now-familiar role of current source for the differential amplifier.

A simple form of negative feedback is introduced by the 50-Ω emitter-resistors. These modify the voltage-gain equation by adding a term equal to $2(1 + h_{fe})R_E$ to the usual denominator term of $2h_{ie}$. As a result, the voltage-gain equation becomes

$$A_v = \frac{h_{fe}R_c}{2[h_{ie} + (1 + h_{fe})R_E]}$$

To judge the relative weight of the two denominator terms, we need to know h_{ie}. And, as usual, to know h_{ie} we have to analyze the DC operating conditions for Q_3 and Q_4.

Suppose that V_{CC} and V_{EE} are $+6$ V and -6 V, respectively. Let us assume that terminals 4 and 5 are left unconnected. (These terminals are provided to allow adjustment of the bias of Q_7 by permitting the diodes and the 1-KΩ resistor to be either included in the circuit or short-circuited, as required.) With terminals 2 and 10 connected to ground, or 0 V, the voltage at the base of Q_7 is equal to $-(6 - 1.4) \times 5$ K/(5 K $+$ 2.2 K $+$ 2.8 K), or -2.3 V. The emitter voltage of Q_7 is therefore $-2.3 - 0.7$, or -3 V. Hence the emitter current of Q_7 is equal to $(-3 - [-6])/$ (1 K $+$ 500), or 2 mA. Therefore, the emitter currents of Q_3 and Q_4 are each equal to 1 mA. At room temperature, h_{ie} for these transistors must equal approximately $26h_{fe}$. Thus

$$A_v \approx \frac{h_{fe}R_c}{2[26h_{fe} + R_E h_{fe}]} = \frac{R_c}{2(26 + R_E)}$$

Since R_c is 1 KΩ and R_E is 50 Ω, we find that A_v equals about 6.6, or 16.4 db. (We are assuming that the voltage gains for the emitter-followers are equal to 1.)

The figure for A_v that we have just calculated is based on the assumption that we are driving a single input and observing only the output-voltage signal at a single output terminal. If two inputs are driven with equal, out-of-phase signals, then the figure for A_v would have to be doubled, or increased by 6 db. And if the output is measured between the output terminals rather than from a single output to ground, then A_v would have to be doubled yet again.

Increasing the emitter currents of Q_3 and Q_4 reduces the base-emitter dynamic resistance (the 26-Ω factor) and thus increases A_v. This is easily accomplished by connecting terminals 3 and 4. When this connection is made, the emitter current of Q_7 becomes $(-3 - [-6])/500$, or 6 mA. The emitter currents of Q_3 and Q_4 are then 3 mA each, and A_v becomes

$$A_v \approx \frac{R_c}{2(26/3 + R_E)} = \frac{1000}{2(8.67 + 50)} = 8.53$$

This figure corresponds to 18.6 db.

These calculations highlight one of the features of IC video amplifiers—the fairly high levels of quiescent current at which the differential amplifiers operate compared to the differential amplifiers used in operational amplifiers. Perhaps a more striking feature is the relatively low gain compared to that obtained from operational amplifiers. This low gain is, of course, consistent with the fact that we are trading off gain for bandwidth.

To permit interconnection of the CA3001 with other devices with minimal interaction, the input resistance is made high and the output resistance is made low. As we have already noted, this is done through the use of emitter-followers. We can get a good picture of what the input and output resistances are by the following line of analysis.

First we consider the output resistance as seen looking into terminal 8. To find this resistance, we need to know the quiescent emitter current of Q_6. For that purpose we assume that Q_7 is biased so that Q_3 and Q_4 are passing 1 mA. This situation brings the collector of Q_4 to a potential $6 - 1$ mA $\times$ 1 K, or $+5$ V. The voltage at the emitter of Q_6 is therefore $5 - 0.7$, or 4.3 V. With terminal 10 grounded, the emitter current of Q_6 must be 4.3/2 K, or 2.15 mA. This makes h_{ie} for Q_6 equal to $h_{fe} \times 26/2.15$, or about $13h_{fe}$ Ω. The resistance looking into the emitter is equal to $[h_{ie}/(1 + h_{fe})]\|2000$, which is about 13 Ω. This is an optimistic value, though it is certainly reasonably close; manufacturer's data specify a typical figure of 45 Ω.

To find the input resistance, we need first to know something about the input resistance to the base of Q_4. This resistance is simply equal to $2[h_{ie} + (1 + h_{fe})R_E]$. If we assume that h_{fe} equals 50, then, using a value of $26h_{fe}$ for h_{ie}, we get a value of about $2 \times (2600 + 50 \times 50)$, or about 10 K$\Omega$. This resistance is seen to be in parallel with the 4.8-KΩ resistor when viewed from the emitter of Q_5. Hence the input resistance at terminal 6 must be about $50 \times (10$ K$\|4.8$ K$)$, or roughly 160 KΩ. This figure compares well with the manufacturer's typical value of 140 KΩ.

The CA3001 has a 3-db bandwidth of 29 MHz according to manufacturer's data. This bandwidth is achieved by forming the transistors so that they have small capacitances due to junction and storage effects. Furthermore, integrated circuits that are intended for wide-band operation are packaged in structures that allow lead inductances and capacitances between leads to be minimized.

If higher gain and somewhat greater bandwidth are desired, then we could use the CA3040. This unit, whose schematic diagram is shown in Fig. 8-7, has a single-ended voltage gain of 31 db and a bandwidth of 55 MHz. The increased bandwidth is achieved by the differential cascode connection of transistors Q_3, Q_4, Q_5, and Q_6.

Calculation of quiescent operating conditions for the circuit of Fig. 8-7 is straightforward but laborious. However, it can be shown that with $+12$ V applied to terminal 2, terminals 5 and 11 grounded, terminal 9 connected to terminal 7, and terminals 8, 3, and 1 left unconnected, the emitter current of Q_9 will be 4.42 mA and the potential of terminal 1 will be 5.92 V above ground. If we assume that h_{fe} is about 50, then h_{ie} for Q_3 and Q_4 will be about 590 Ω. The single-ended voltage gain will then be about 110, or 41 db. When we take into account the fact that the input and output emitter followers multiply the differential-amplifier gain

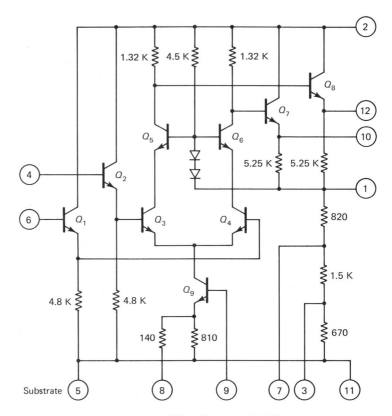

Fig. 8-7 The CA3040 video amplifier. (Courtesy RCA)

by factors that are something less than 1, we see that the 41-db figure is not too far out of line with the nominal 31-db value.

With both the CA3001 and the CA3040, the input and output terminals operate at DC potentials that may be removed from ground, which means that some sort of capacitive coupling must be used in circuits that utilize these devices. A typical circuit connection using the CA3040 is shown in Fig. 8-8.

Double bypass capacitors are used (0.1 μF in parallel with 0.001 μF) because the two capacitors are of different construction. The 0.1-μF unit might be a Mylar-insulated type, in which long strips of metallized plastic are rolled up to form the capacitor. Such a unit has some inductance associated with it, which, at sufficiently high frequencies, may be troublesome. This effect is overcome by the use of the small 0.001-μF capacitor, which may be a silvered mica unit having minimal inductance.

From the diagram of the CA3040 unit in Fig. 8-7 we see that the potential of the emitters of Q_3 and Q_4 must be positive with respect to ground. That being the case, the DC potentials at the bases of the input transistors Q_1 and Q_2 must also be positive. This bias requirement is satisfied in the circuit of Fig. 8-8 by the 1-KΩ resistors that connect terminals 4 and 6 to terminal 1, which—as we noted before—was at a potential of 5.92 V above ground.

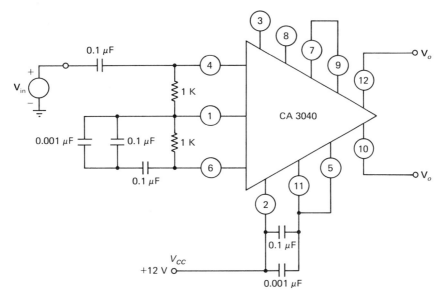

Fig. 8-8 Connection of the CA3040 video amplifier.

8-4 THE GAIN-BANDWIDTH TRADE-OFF

If we want to increase the bandwidth of an amplifier, the best procedure we have found so far is to trade off gain for bandwidth. There is obviously a limit to how far we can go with this approach. If the trade-off is accomplished by a feedback circuit external to the basic amplifier circuit, then the gain-bandwidth product is essentially a constant, as we saw in Chap. 4. But if, for a single-transistor amplifier, gain variation is obtained by changing either the source resistance R_s, the collector resistance R_c, or both, then the gain-bandwidth product is no longer a simple constant and the trade-off is a little more complicated.

To see how this works out, let us consider the transistor whose hybrid-π parameters we found earlier. For convenience, we show the approximate circuit in Fig. 8-9. The circuit is approximate because it uses the Miller capacitance, thus uncoupling the b' and c nodes.

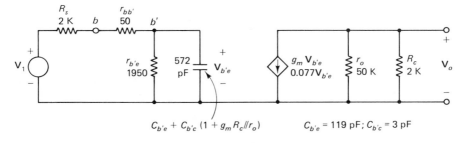

Fig. 8-9 Hybrid-π circuit with Miller capacitance.

The low-frequency gain of this circuit is given algebraically by

$$|A_{vo}| = \frac{r_{b'e}}{r_{b'e} + r_{bb'} + R_s} \cdot g_m(r_o \| R_c)$$

The 3-db bandwidth will be determined by the Thevenin resistance seen by the Miller capacitance, which is $(R_s + r_{bb'}) \| r_{b'e}$. Hence the bandwidth is simply

$$f_1 = \frac{1}{2\pi[(R_s + r_{bb'}) \| r_{b'e}][C_{b'e} + C_{b'c}(1 + g_m R_c \| r_o)]}$$

If we combine these two expressions to form the gain-bandwidth product, we get

$$|A_{vo}|f_1 = \frac{r_{b'e}}{r_{b'e} + r_{bb'} + R_s} \cdot \frac{r_{b'e} + r_{bb'} + R_s}{2\pi r_{b'e}(R_s + r_{bb'})} \cdot \frac{g_m R_c \| r_o}{C_{b'e} + C_{b'c}(1 + g_m R_c \| r_o)}$$

After cancellation of like terms in the numerator and denominator, we obtain

$$|A_{vo}|f_1 \approx \frac{g_m R_c \| r_o}{2\pi(R_s + r_{bb'})[C_{b'e} + C_{b'c}(1 + g_m R_c \| r_o)]}$$

In the usual case, where extended bandwidth is our main concern, $R_c \ll r_o$. Hence we may simplify the expression further to

$$|A_{vo}|f_1 \approx \frac{g_m R_c}{2\pi(R_s + r_{bb'})[C_{b'e} + C_{b'c}(1 + g_m R_c)]} \tag{8-6}$$

We can get some idea of how the gain-bandwidth product varies with the quantity $g_m R_c$ by plotting $|A_{vo}|f_1$ versus $g_m R_c$. For convenience, we define

$$x = g_m R_c$$

$$a = 2\pi(R_s + r_{bb'})(C_{b'e} + C_{b'c})$$

$$b = 2\pi(R_s + r_{bb'})C_{b'c}$$

Using these definitions, Eq. (8-6) takes the form

$$|A_{vo}|f_1 \approx \frac{x}{a + bx}$$

A plot of this function is shown in Fig. 8-10 with a few corresponding values of $|A_{vo}|f_1$ and x indicated.

For the transistor whose parameters are shown in Fig. 8-9, let us assume g_m constant at the value of 0.077 mho and find the values of $|A_{vo}|f_1$ and R_c corresponding to the point $(a/b, 1/2b)$ on the curve of Fig. 8-10. Since values of x are shown as multiples of a/b, we note that

$$\frac{a}{b} = \frac{C_{b'e} + C_{b'c}}{C_{b'c}} \quad \text{or} \quad \frac{119 + 3}{3} = 40.6$$

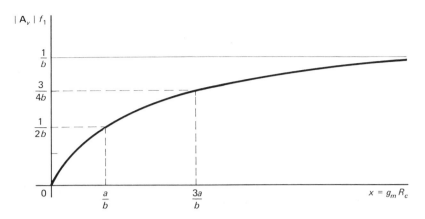

Fig. 8-10 Plot of $A_{vo}f_1$ versus x.

Thus for $g_m R_c = x = a/b$, we have

$$R_c = \frac{1}{g_m} \cdot \frac{a}{b} = \frac{40.6}{0.077} = 528 \ \Omega$$

The corresponding bandwidth must be

$$\frac{1}{2b} = \frac{1}{2 \times 2\pi \times (2050)3 \times 10^{-12}} = \frac{10^9}{2 \times 2\pi \times 2.05 \times 3} = 1.30 \times 10^7 \ \text{Hz}$$

or 13 MHz.

For the case that we analyzed previously, R_c was equal to 2 KΩ. This corresponds to a value of x equal to $0.077 \times 2000 = 154$. Expressed as a multiple of a/b, this is equivalent to $154/40.6 = 3.8a/b$, which is somewhat beyond the point shown on the curve at $x = 3a/b$. The corresponding bandwidth has to be worked out from the equation $x/(a + bx)$. This yields $(3.8a/b)/[a + (3.8a/b)b]$, or $3.8/4.8b$, which is

$$\frac{3.8}{4.8 \times 2\pi \times 2050 \times 3 \times 10^{-12}} = \frac{3.8 \times 10^9}{4.8 \times 2\pi \times 2.05 \times 3} = 20.5 \ \text{MHz}$$

Since we previously calculated the gain with R_c equal to 2 KΩ to be about 75, this figure corresponds to a bandwidth of $20.5/75 = 274$ KHz, which agrees well with our previous calculation.

A more general calculation of gain-bandwidth products can be accomplished by using Prog. 8-1.

```
10   PRINT "GAIN-BANDWIDTH PRODUCT VARIATION FOR SINGLE-TRANSISTOR COMMON-E
     MITTER AMPLIFIER"
20   PRINT "MODEL USES HYBRID-PI PARAMETERS."
30   PRINT "VARIATION IN AVO*F3DB IS CALCULATED AS A FUNCTION OF GM*RC,"
40   PRINT "WHERE RC IS THE VALUE OF THE EXTERNAL COLLECTOR RESISTOR IN KIL
     OHMS."
50   INPUT "ENTER THEVENIN RESISTANCE OF INPUT VOLTAGE SOURCE, RS.";RS
```

Prog. 8.1.

```
60   INPUT "ENTER COLLECTOR RESISTOR VALUE IN KILOHMS, RC.";RC
70   INPUT "ENTER TRANSCONDUCTANCE GM IN MILLIMHOS.";GM
75   INPUT "ENTER BASE SPREADING RESISTANCE RBB IN OHMS.";RBB
80   INPUT "ENTER CAPACITANCE CBE IN PICOFARADS.";CBE
90   INPUT "ENTER CAPACITANCE CBC IN PICOFARADS.";CBC
100  X = GM * RC
110  P = 2 * 3.14159 * (RS * 1E3 + RBB)
120  A = P * (CBE + CBC) * 1E - 12
130  B = P * CBC * 1E - 12
140  GBW = (X / (A + B * X)) * 1E - 6
150  PRINT "GAIN-BANDWIDTH PRODUCT EQUALS ";GBW;" MEGAHERTZ"
160  END
```

Prog. 8.1 (cont.)

The point of all this manipulation is to show the nature of the gain-bandwidth limitation. We see that while we have some room to maneuver, there is an ultimate limit that cannot easily be exceeded. Although some circuit modifications exist that will allow us to squeeze out a little more bandwidth, the study of such circuits is beyond the scope of this book.

8-5 CASCADED WIDE-BAND AMPLIFIERS

It might seem at first glance that to obtain any desired level of gain using amplifiers such as those we have been considering, we need only to connect enough of them in cascade. This procedure is reasonable enough insofar as it relates to gain at low frequencies. But trouble arises with regard to the high-frequency performance. Suppose, for example, that we have two wide-band amplifiers whose low-frequency voltage gains are A_{10} and A_{20} and whose high-frequency 3-db points occur at frequencies of f_1 and f_2, respectively. The response of each amplifier is described in the following expressions

$$\mathbf{A}_1(f) = \frac{A_{10}}{1 + jf/f_1}$$

$$\mathbf{A}_2(f) = \frac{A_{20}}{1 + jf/f_2}$$

Since the magnitudes of the responses are of interest we write

$$|\mathbf{A}_1(f)| = \frac{A_{10}}{\sqrt{1 + (f/f_1)^2}}$$

and

$$|\mathbf{A}_2(f)| = \frac{A_{20}}{\sqrt{1 + (f/f_2)^2}}$$

Because the two amplifier stages are connected in cascade, the second one amplifies the output of the first. Therefore, the overall gain is equal to the product of the individual gains or

$$|\mathbf{A}_{\text{overall}}| = |\mathbf{A}_1(f)| \cdot |\mathbf{A}_2(f)|$$

On a logarithmic basis, this result takes the form

$$\log|A_{\text{overall}}| = \log|A_1(f)| + \log|A_2(f)|$$

Consequently, if the gains are expressed in decibels, the overall gain in decibels is the sum of the individual decibel gains. The situation can be represented graphically as shown in Fig. 8-11. Notice that when the 3-db frequencies are different, the effective bandwidth of the cascade connection is the same as the smaller bandwidth of the two individual amplifiers.

We can readily see that when identical amplifiers are connected in cascade, the 3-db frequency for a single amplifier will be the frequency at which the gain is down by 6 db for the cascaded pair. Where, then, is the 3-db frequency for the pair? To answer this question, we recall that at the 3-db frequency the magnitude of the gain of an amplifying system is reduced by a factor of $1/\sqrt{2}$.

For two identical amplifiers having low-frequency gains of A_{10} and 3-db frequencies of f_1, the overall gain (magnitude) is equal to

$$|A_{\text{overall}}| = \frac{A_{10}}{\sqrt{1 + (f/f_1)^2}} \cdot \frac{A_{10}}{\sqrt{1 + (f/f_1)^2}}$$

When $f \ll f_1$, $|A_{\text{overall}}| = (A_{10})^2$. Therefore, at the 3-db frequency for the cascade connection, $|A_{\text{overall}}| = (A_{10})^2/\sqrt{2}$. If we call this frequency f_o, we can write

$$\frac{(A_{10})^2}{\sqrt{2}} = \frac{(A_{10})^2}{[\sqrt{1 + (f_o/f_1)^2}]^2}$$

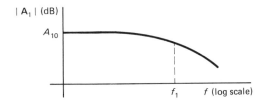

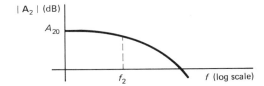

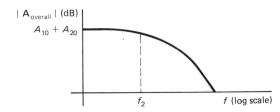

Fig. 8-11   Effect of cascading amplifiers.

It follows that

$$\sqrt{2} = 1 + \left(\frac{f_o}{f_1}\right)^2$$

and that

$$f_o/f_1 = \sqrt{\sqrt{2} - 1} \qquad (8\text{-}7)$$

The quantity on the right-hand side of Eq. (8-7) is called the *bandwidth reduction factor*. Its numerical value is 0.644.

To account for changes in bandwidth when the number of stages is other than 2, we can work out a general formula. To do this, let us suppose that n identical stages are cascaded. Then the overall gain at the 3-db frequency f_o is given by

$$\frac{(A_{10})^n}{\sqrt{2}} = \frac{(A_{10})^n}{[\sqrt{1 + (f_o/f_1)^2}]^n}$$

It is easy to show that

$$(\sqrt{2})^{1/n} = \sqrt{1 + \left(\frac{f_o}{f_1}\right)^2}$$

or

$$2^{1/n} = 1 + \left(\frac{f_o}{f_1}\right)^2$$

Finally, we obtain the general equation

$$\frac{f_o}{f_1} = \sqrt{2^{1/n} - 1} \qquad (8\text{-}8)$$

The results given by Eq. (8-8) are given in Table 8-1.

Let us now see how the idea of the bandwidth reduction factor applies to actual examples. Suppose that we wish to use CA3001s cascaded to produce a minimum low-frequency gain of 50-db. Since each unit provides 18.6 db, as we have seen, we require three units in cascade. What happens to the bandwidth?

TABLE 8-1. HIGH-FREQUENCY BANDWIDTH REDUCTION FACTORS FOR *N* IDENTICAL STAGES.

Number of stages n	f_o/f_1
1	1
2	0.644
3	0.510
4	0.435
5	0.386

According to Table 8-1, for $n = 3$, $f_o/f_1 = 0.510$. Since the 3-db bandwidth for the CA3001 is 29 MHz, we see that the bandwidth for the cascade connection will be 0.51 × 29, or 14.8 MHz.

In the case of RC-coupled, discrete amplifiers, the amplitude response does not extend without change to zero frequency (DC) but instead falls off as the frequency becomes very low. Therefore, an effect similar to the reduction of the high-frequency 3-db point occurs in such amplifiers when they are cascaded. In this case, however, the effect is to move the low-frequency 3-db point to a higher frequency. We can get a quantitative picture of the effect on this low-frequency 3-db point of the cascading of identical stages from the following analysis.

The low-frequency behavior of the RC-coupled amplifier is described by the equation

$$\frac{\mathbf{A}_{lo}(f)}{\mathbf{A}_o} = \frac{jf/f_{lo}}{1 + jf/f_{lo}}$$

where $\mathbf{A}_{lo}(f)$ is the amplitude of the output at low frequencies and $\mathbf{A}_o$ is the amplitude of the output at midband frequencies.

The magnitude of $\mathbf{A}_{lo}(f)/\mathbf{A}_o$ is equal to

$$\left| \frac{\mathbf{A}_{lo}(f)}{\mathbf{A}_o} \right| = \frac{f/f_{lo}}{\sqrt{1 + (f/f_{lo})^2}}$$

When $f \gg f_{lo}$, the term under the radical is approximately equal simply to f/f_{lo}. Therefore, under this condition $|\mathbf{A}_{lo}(f)/\mathbf{A}_o| = 1$. Consequently, at the 3-db frequency $|\mathbf{A}_{lo}(f)/\mathbf{A}_o|$ must equal $1/\sqrt{2}$. We can therefore write

$$\frac{f/f_{lo}}{\sqrt{1 + (f/f_{lo})^2}} = \frac{1}{\sqrt{2}}$$

This leads to

$$2(f/f_{lo})^2 = 1 + (f/f_{lo})^2$$

from which we conclude that the 3-db frequency is f_{lo}.

Now suppose that we cascade n identical stages and that we wish to find the new 3-db frequency. The response of n stages must be given by

$$\left| \frac{\mathbf{A}_{lo}(f)}{\mathbf{A}_o} \right|^n = \frac{(f/f_{lo})^n}{[\sqrt{1 + (f/f_{lo})^2}]^n}$$

At the 3-db frequency, this quantity must equal $1/\sqrt{2}$. Therefore, we have

$$\frac{(f/f_{lo})^n}{[\sqrt{1 + (f/f_{lo})^2}]^n} = \frac{1}{\sqrt{2}}$$

This equation leads to

$$(2^{1/n})\left(\frac{f}{f_{lo}}\right)^2 = 1 + \left(\frac{f}{f_{lo}}\right)^2$$

TABLE 8-2. LOW-FREQUENCY BANDWIDTH REDUCTION FACTORS FOR N IDENTICAL STAGES.

Number of stages n	$f_{3\,db}/f_{lo}$
1	1
2	1.554
3	1.961
4	2.299
5	2.593

A little manipulation yields

$$\frac{f_{3\,db}}{f_{lo}} = \frac{1}{\sqrt{2^{1/n} - 1}} \tag{8-9}$$

Equation (8-9) involves the same factor as Eq. (8-8), $\sqrt{2^{1/n} - 1}$. This time, however, the factor appears in the denominator of the expression. Thus the low-frequency 3-db points are moved higher in contrast to the high-frequency 3-db points. The effect is summarized in Table 8-2.

Either high-frequency or low-frequency bandwidth reduction factors may readily be calculated using Prog. 8-2.

```
10   PRINT "COMPUTATION OF BANDWIDTH REDUCTION FACTORS"
20   PRINT "FOR N IDENTICAL STAGES"
30   INPUT "ENTER NUMBER OF STAGES.";N
40   INPUT "DO YOU WANT HIGH- OR LOW-FREQUENCY 3-DB POINT? (H/L)";A$
50   IF A$ = "L" THEN 90
60   X =  SQR (2 ^ (1 / N) - 1)
70   PRINT "F1/FO = ";X
80   END
90   X = 1 / ( SQR (2 ^ (1 / N) - 1))
100  GOTO 70
```

Prog. 8-2.

In most circumstances there is only a small effect upon the overall bandwidth due to the increase in frequency at which the low-frequency 3-db point occurs. For example, suppose that we have three identical stages, each having a low-frequency 3-db point at 100 Hz and a high-frequency 3-db point at 5 MHz. When the three stages are cascaded, the new 3-db points become 196 Hz and 2.55 MHz. The original bandwidth was 5 MHz $-$ 100 Hz $\approx$ 5 MHz, whereas the new bandwidth is 2.55 MHz $-$ 196 Hz $\approx$ 2.55 MHz.

REFERENCES

1. M. E. Van Valkenburg, *Network Analysis*, 3rd ed. (Englewood Cliffs, N.J.: Prentice-Hall, Inc., 1974).
2. Harry E. Stewart, *Engineering Electronics* (Boston: Allyn and Bacon, Inc., 1969).
3. George E. Valley, Jr., and Henry Wallman, eds., *Vacuum Tube Amplifiers* (New York: Dover, 1965).

PROBLEMS

8-1. Alphanumeric characters are sometimes displayed on a CRT using a vertical-stroke scanning scheme that differs from the standard television raster-scanning method. One such scanning arrangement is shown.

    (a) If the wave form of a series of dots is as shown, calculate the time duration of one dot.

    (b) If the rise time required to produce a clearly defined dot is half the dot duration, find the required video-amplifier bandwidth.

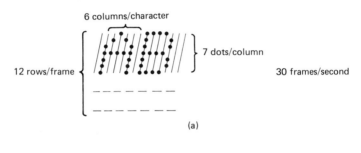

(a)

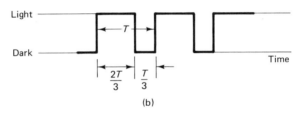

(b)

8-2. When the DC value of I_E is 1.3 mA, a junction transistor has the following parameters:

$$h_{ie} = 1100 \ \Omega$$

$$h_{fe} = 50$$

    (a) Find r_{bb}'.
    (b) Find h_{ie} when $I_E = 2$ mA.
    (c) Find g_m.

8-3. Transistor measurements made with I_E (DC) equal to 5 mA yield these data:

$$h_{fe} = 100 \qquad\qquad h_{ie} = 600 \Omega$$

$$\left| \frac{I_c}{I_b} \right| = 10 \text{ for } f = 10 \text{ MHz} \qquad C_{ob} = 3 \text{ pF}$$

Find f_β, f_T, $C_{b'e}'$, $r_{b'e}$, and r_{bb}'.

8-4.

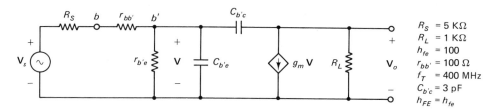

R_S = 5 KΩ
R_L = 1 KΩ
h_{fe} = 100
$r_{bb'}$ = 100 Ω
f_T = 400 MHz
$C_{b'c}$ = 3 pF
$h_{FE} = h_{fe}$

(a) Find the gain and bandwidth of the amplifier.

(b) At what frequency is the gain equal to 1?

8-5. A single-stage BJT amplifier is to be designed to provide A_v of at least 85 when connected between a 50-Ω source and a 300-Ω collector resistor. The high-frequency 3-db point is to be at least 4.5 MHz. For the transistor being considered for this application, h_{fe} = 100, C_{ob} = 4 pF, $r_{bb'}$ = 50 Ω, and f_T = 400 MHz. The quiescent point needed to provide the required signal amplitude requires I_c = 10 mA (DC). Determine whether or not this transistor can meet the gain and bandwidth requirements.

8-6. A BJT has the following parameters: $r_{bb'}$ = 50 Ω, $r_{b'e}$ = 1000 Ω, g_m = 0.03 mho, $C_{b'e}$ = 500 pF, $C_{b'c}$ = 5 pF, and r_{ce} = ∞. Find the value of collector resistance R_c that will make the 3-db frequency for *current gain* A_i equal to $f_β/2$.

8-7. Find the midband current gain A_i and the upper 3-db frequency. Assume C_1, C_2, and C_E are short circuits.

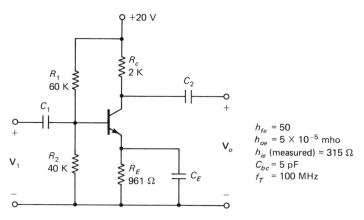

h_{fe} = 50
h_{oe} = 5 × 10^{-5} mho
h_{ie} (measured) = 315 Ω
C_{bc} = 5 pF
f_T = 100 MHz

8-8.

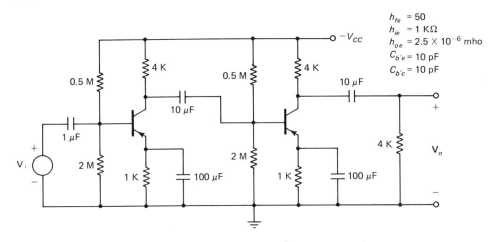

h_{fe} = 50
h_{ie} = 1 KΩ
h_{oe} = 2.5 × 10⁻⁶ mho
$C_{b'e}$ = 10 pF
$C_{b'c}$ = 10 pF

(a) Compute midband voltage gain using simplifying assumptions.

(b) Find the upper 3-db frequency.

8-9. (a) Determine the quiescent operating conditions for the circuit shown in Fig. 8-7 with + 12 V applied to terminal 2, terminals 5 and 11 grounded, terminal 9 connected to terminal 7, and terminals 8, 3, and 1 left unconnected. *Neglect the current through the 4.5-K resistor.*

(b) Estimate h_{ie} for Q_3 and Q_4 and the voltage gain with v_i applied to terminal 4 (terminal 6 at signal ground) and v_o measured at terminal 12.

8-10. An approximate small-signal model for the differential cascode amplifier used in the CA3040 video amplifier of Fig. 8-7 is shown.

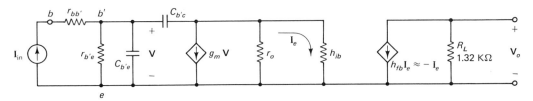

(a) If h_{fe} = 50, h_{ie} = 590 Ω, and r_{bb}' = 30 Ω, find the value of g_m that applies to this *differential* amplifier.

(b) Find the Miller capacitance seen looking into $b'e$ if $C_{b'e}$ = 2 pF and $C_{b'c}$ = 2 pF.

(c) Find the high-frequency 3-db breakpoint for this circuit.

8-11. Repeat Problem 8-9 with terminal 9 connected to terminal 3 instead of terminal 7. All other conditions remain the same.

8-12. The schematic diagram of the μA 733 integrated-circuit, differential, video amplifier is shown.

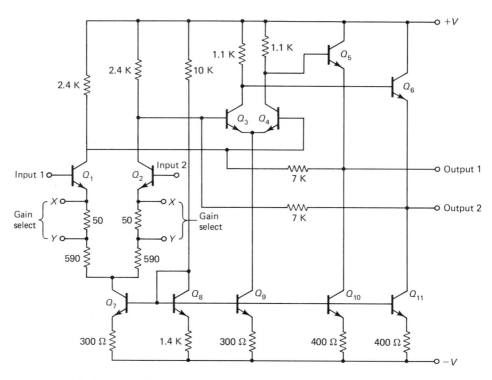

(a) With the $+V$ and $-V$ terminals connected to $+8$ V and -8 V, respectively, and $h_{fe} \approx 200$, find the input resistance to the bases of Q_3 and Q_4.

(b) Develop a small-signal equivalent circuit that incorporates the collector circuit of Q_2, Q_3 from the differential amplifier and the emitter-follower Q_6. Use this circuit to find **T** and **K** and thus the gain of the second differential-amplifier stage. Assume Q_6 has infinite input resistance, zero output resistance, and a voltage gain equal to 1.

(c) Compute the gain of the first differential stage (Q_1 and Q_2) assuming the bases of Q_3 and Q_4 are not connected and gain-select terminals XX and YY are left open.

(d) Repeat (c) with YY connected and XX open. Repeat (c) with XX connected.

(e) Calculate overall gain for the three cases of (c) and (d).

8-13. (a) For the 2N2616 transistor discussed in this chapter, determine the gain-bandwidth product if the transistor is operated at I_E equal to 50 mA and $R_s = 2$ KΩ but with $R_c = 200$ Ω.

(b) Estimate the supply voltage required with this value of R_c.

(c) What is the low-frequency value of A_v for this case?

8-14. For the transistor of Prob. 8-3, choose R_c to make $x = a/b$ and determine the corresponding bandwidth. Assume $R_S = 0$.

8-15. Verify that the bandwidth determined in Prob. 8-4 agrees with the value obtained using Eq. (8-6).

8-16. For the CA3040 video amplifier of Fig. 8-7 connected as shown in Fig. 8-8, calculate the upper 3-db frequency for three units connected in cascade.

8-17. A single-stage amplifier has a gain of 24 db and upper and lower 3-db cutoff frequencies of 40 Hz and 28 KHz, respectively. We want to use several such units in cascade to provide at least 60 db gain. Find the number of stages required and the resultant upper and lower cutoff frequencies.

8-18. Three identical amplifier stages are cascaded. Parameters for each stage are gain = 15 (absolute, not decibel) and 3-db frequencies of 50 Hz and 1.4 MHz.

   (a) Determine overall gain in decibels and the resultant 3-db frequencies.

   (b) Plot the overall frequency response on semilog paper. Show also the response for a single stage.

CHAPTER 9

Tuned Amplifiers

9-1 THE NEED FOR TUNED AMPLIFIERS

When we speak of tuned amplifiers, we mean amplifiers that have narrow band-widths, in contrast to the wide-band amplifiers that we considered in the last chapter. Why such amplifiers are useful can be easily appreciated when we consider the problem of getting an FM receiver to respond to a single FM channel out of the dozens that occupy the 88-MHz–108-MHz band. Clearly, some sort of fre-quency-selective circuit is needed that will allow us to amplify the narrow frequency band that belongs to the signal we want to hear and to reject all the other unwanted signals.

9-2 PROPERTIES OF PARALLEL-RESONANT CIRCUITS

To gain an understanding of tuned amplifiers, we must first review some of the basic principles that apply to parallel-resonant RLC circuits. From these principles the definitions of certain terms will follow. These definitions will be important even when extended to apply to tuned circuits that use resistors, capacitors, and oper-ational amplifiers rather than the original RLC networks.

We begin by defining a quantity denoted by the letter Q. The symbol is derived from the idea that we are defining a *quality factor*. That is, the quality of a coil is judged by how well it serves as an inductor relative to its undesired role as a resistor in the circuit in which it is placed. Rather than compare the inductance of a coil directly to its resistance, we take the ratio of the inductive reactance to the resistance. In that way we divide reactance in ohms by resistance in ohms and obtain the dimensionless quantity Q, which is defined as

$$Q = \frac{\omega L}{R} = \frac{2\pi f L}{R} \tag{9-1}$$

Note that Q is not a constant but a linear function of f.

Let us turn now to the parallel RLC circuit shown in Fig. 9-1. The network is driven by the phasor current generator $\mathbf{I}$. The quantity of interest is the phasor voltage $\mathbf{V}$. This voltage is equal to the current $\mathbf{I}$ times the parallel impedance of the entire RLC network. Another way to state this is that $\mathbf{V}$ is equal to $\mathbf{I}$ divided by the admittance of the network. This admittance is given by

$$\mathbf{Y} = j\omega C + \frac{1}{R + j\omega L}$$

Then $\mathbf{V}$ must equal

$$\mathbf{V} = \frac{\mathbf{I}}{\mathbf{Y}} = \frac{\mathbf{I}}{j\omega C + 1/(R + j\omega L)} \tag{9-2}$$

Equation (9-2) can be written in the form

$$\mathbf{V} = \frac{R + j\omega L}{1 - \omega^2 LC + j\omega RC}\mathbf{I}$$

At this point we make use of the following definitions. The *series* resonant frequency (in radians per second) of the coil if it were resonated with capacitor C is ω_o, which is

$$\omega_o = \frac{1}{\sqrt{LC}}$$

The value of Q for the coil *at the frequency* ω_o is called Q_o and—in view of Eq. (9-1)—is

$$Q_o = \frac{\omega_o L}{R}$$

Using these quantities we can, after a bit of manipulation, rewrite the expression for $\mathbf{V}$ in the form

$$\mathbf{V} = \frac{R[1 + j(\omega L/R)]}{1 - \omega^2/\omega_o^2 + j\omega/\omega_o Q_o}\mathbf{I}$$

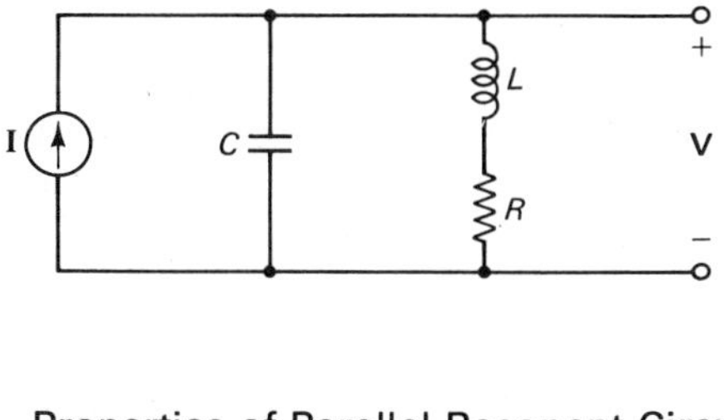

Fig. 9-1 *RLC* Circuit.

Sec. 9-2 Properties of Parallel-Resonant Circuits

If we multiply numerator and denominator by $-j$ and by Q_o, we can rearrange terms and arrive at

$$\mathbf{V} = \frac{Q_o R(-j1 + \omega L/R)\mathbf{I}}{\omega/\omega_o - jQ_o[(\omega_o^2 - \omega^2)/\omega_o^2]}$$

At this point we must make some approximations in order to get this expression into manageable form. First of all, we note that since this is a tuned circuit intended to provide a sizeable voltage $\mathbf{V}$ for only a narrow range of frequencies, we can assume that ω does not change much from the value ω_o. Accordingly, we conclude that

$$\frac{\omega L}{R} \approx \frac{\omega_o L}{R} = Q_o$$

Furthermore, for the approximate results to be useful at all, we require that $Q_o \gtrsim 10$. Therefore, we shall consider that $\omega L/R \approx Q_o \gg 1$. Consequently, we shall neglect the $-j1$ term in the numerator. Again, since ω does not vary much from ω_o, we shall assume that $\omega/\omega_o \approx 1$.

With these approximations, the result obtained for $\mathbf{V}$ becomes

$$\mathbf{V} \approx \frac{Q_o^2 R\mathbf{I}}{1 - jQ_o[(\omega_o - \omega)(\omega_o + \omega)/\omega_o^2]}$$

Notice that we have written $\omega_o^2 - \omega^2$ as $(\omega_o - \omega)(\omega_o + \omega)$. Since $\omega \approx \omega_o$, the second factor can be approximated by $2\omega_o$. If we define the fractional frequency deviation by

$$\delta = \frac{\omega - \omega_o}{\omega_o}$$

the expression for $\mathbf{V}$ becomes

$$\mathbf{V} \approx \frac{Q_o^2 R\mathbf{I}}{1 + j2Q_o\delta} \tag{9-3}$$

When δ equals zero, which occurs when $\omega = \omega_o$, $\mathbf{V} \approx Q_o^2 R\mathbf{I}$. Or, looking at the situation another way, the impedance of the network has its maximum value, which equals $Q_o^2 R$, at the frequency of resonance, which is $\omega = \omega_o$ radians per second, or $f = f_o$ Hertz.

At first glance we might be led to suppose that the narrower the bandwidth of our tuned circuit, the better. This is true to some extent, since we do want the circuit to present maximum impedance to the signal in which we are interested and (ideally) zero impedance to any unwanted signal. However, the signals of interest consist not just of a single frequency but of a range of frequencies, and it is important that our circuit pass all of these. Thus, for example, a typical signal in the FM band may be centered at 100 MHz but may contain components as high as 100 MHz + 37.5 KHz and as low as 100 MHz − 37.5 KHz. Therefore, to preserve all these components, our resonant network must have a bandwidth of 75 KHz.

If we consider the magnitude of **V** as given by Eq. (9-3), we can write

$$|\mathbf{V}| \approx \frac{Q_o^2 R |\mathbf{I}|}{\sqrt{1 + 4Q_o^2\delta^2}}$$

The value of $|\mathbf{V}|$ is reduced by $1/\sqrt{2}$ relative to its value for $\delta = 0$ when the denominator of this expression equals $\sqrt{2}$. Therefore, we have

$$2 = 1 + 4Q_o^2\delta_{3db}^2$$

from which we can see that

$$\delta_{3db} = \pm \frac{1}{2Q_o}$$

The difference between these two values of δ_{3db} can be written

$$\delta_{3db\ (hi)} - \delta_{3db\ (lo)} = \frac{1}{Q_o}$$

In terms of frequencies in Hertz, we can write this as

$$\frac{f_{3db\ (hi)} - f_{3db\ (lo)}}{f_o} = \frac{1}{Q_o}$$

The bandwidth (BW) in Hertz is given by

$$BW = \frac{f_o}{Q_o} \qquad (9\text{-}4)$$

Equation (9-4) brings out the importance of the value of Q_o in determining bandwidth.

Example 9-1

For the circuit of Fig. 9-1, an inductor of 10 μH is to be used. For a center frequency of 100 MHz and a bandwidth of 75 KHz, we wish to find Q_o, R, and C.

From Eq. (9-4)

$$Q_o = \frac{100\ \text{MHz}}{75\ \text{KHz}} = \frac{100 \times 10^6}{75 \times 10^3} = 1330$$

For an inductor of 1 μH, the associated resistance would have to be

$$R = \frac{\omega_o L}{Q_o} = \frac{2\pi \times 10^8 \times 10^{-6}}{1330} = 0.47\ \Omega$$

The capacitor required would be equal to

$$C = \frac{1}{4\pi^2 f_o^2 L} = \frac{1}{4\pi^2 \times 10^{16} \times 10^{-6}} = 2.53\ \text{pF}$$

To construct a circuit with the parameter values determined in Example 9-1 strains our capabilities. The value of Q required is extremely high, and—as we shall see later—the effect of other circuit components upon the parallel-resonant circuit will require us to construct a circuit with yet a higher value of Q_o. The capacitance is exceedingly small, comparable in fact to stray circuit capacitances. Altogether, then, these are not practical values.

Fortunately, the 50-year-old *superheterodyne principle* provides a way out of this dilemma. To see how this principle works, we refer to the diagram of Fig. 9-2.

The RF (radio-frequency) amplifier is tuned by means of a variable capacitor to the frequency of the signal that we wish to receive. This frequency would be 100 MHz. The local oscillator generates a frequency that is 10.7 MHz less than 100 MHz, or 89.3 MHz. The tuning capacitor of the local oscillator is mechanically coupled (ganged) to the RF-amplifier capacitor. In this way the frequencies of the two units are made to track, always maintaining a constant difference between them of 10.7 MHz.

The mixer is a nonlinear device (it may be a transistor or merely a diode) that generates a variety of frequencies, including 189.3 MHz (the sum frequency) and 10.7 MHz (the difference frequency). At the input to the IF (intermediate-frequency) amplifier, there is another parallel-resonant circuit. In contrast to the other tuned circuits, this one is tuned to a fixed frequency, 10.7 MHz. The lower frequency to which the IF stage is tuned makes it easier to achieve a reasonable value of Q_o, which is needed to get the required bandwidth.

For a bandwidth of 75 KHz at a center frequency of 10.7 MHz, we need a value of Q_o equal to $10.7 \times 10^6/75 \times 10^3 = 143$. If we assume a coil inductance of 10 μH, the required values of C and R turn out to be 22.2 pF and 4.7 Ω, respectively.

9-2.1 The effect of loading upon bandwidth and Q

In general, a parallel-resonant network will be shunted by external resistors. This is true for two reasons. First, such networks are driven not by ideal current sources

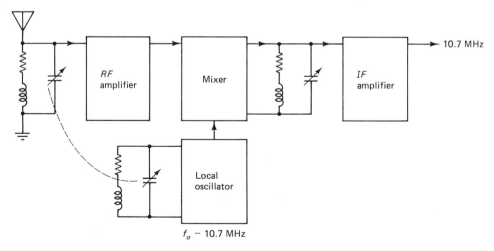

Fig. 9-2 Partial diagram of an FM receiver.

but by practical ones that always have associated with them some value of shunt resistance. Second, these networks are coupled to some other circuit or device so that the parallel-resonant network sees an external load resistance shunted across it. These shunt resistances affect the bandwidth of the circuit. We must therefore study this effect quantitatively.

First of all, Eq. (9-3) tells us that at the resonant frequency where $\delta = 0$, the impedance of the parallel-resonant network is equal to $Q_o^2 R$. If the resistance associated with the inductor approaches zero, the value of Q_o approaches infinity. At the same time, the impedance also becomes infinite, as can be seen from the following equation.

$$\lim_{R \to 0} \frac{\omega_o^2 L^2}{R^2} \cdot R = \lim_{R \to 0} \frac{\omega_o^2 L^2}{R} = \infty$$

From Eq. (9-4) we see that when R approaches zero, the bandwidth becomes zero.

These facts suggest that there might be an alternative way of representing a parallel-resonant circuit, which is by placing in parallel a perfect inductor (one that has no resistance and hence has infinite Q), a capacitor, and a resistor whose value is $Q_o^2 R$ as derived from the original network as shown in Fig. 9-3. Clearly, the infinite impedance of the LC parallel-resonant branches, when combined with the resistance value $Q_o^2 R$, will yield the correct value for the total network resistance. However, the question that remains is whether the bandwidth of the new network will be the same as that of the original one.

To find the bandwidth of the network of Fig. 9-3 we begin by writing the expression for the admittance, which is

$$\mathbf{Y} = \frac{1}{j\omega L} + j\omega C + \frac{1}{Q_o^2 R}$$

We make use of the fact that $\omega_o^2 = 1/LC$ and rearrange the expression to obtain

$$\mathbf{Y} = j\frac{C(\omega^2 - \omega_o^2)}{\omega} + \frac{1}{Q_o^2 R}$$

The impedance $\mathbf{Z}$ is equal to $1/\mathbf{Y}$. With a little manipulation, we can write the expression for $\mathbf{Z}$ in the form

$$\mathbf{Z} = \frac{\omega Q_o^2 R}{jC(\omega^2 - \omega_o^2) Q_o^2 R + \omega}$$

Further manipulation produces

$$\mathbf{Z} = \frac{Q_o^2 R}{1 + jQ_o^2[RC(\omega + \omega_o)(\omega - \omega_o)/\omega]}$$

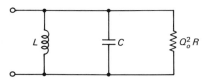

Fig. 9-3 Equivalent parallel-resonant network.

Using the same approximations that were used to obtain Eq. (9-3) and again recalling that $\omega_o^2 = 1/LC$, we obtain

$$\mathbf{Z} = \frac{Q_o^2 R}{1 + j2Q_o\delta} \tag{9-5}$$

Since this equation is identical to Eq. (9-3), all the arguments previously used apply here as well. We are therefore justified in concluding that the equivalent network of Fig. 9-3 has the same bandwidth as the original network of Fig. 10-1.

It is reasonable to suppose that since the actual parallel-resonant network can be replaced by an ideal one (that is, a network having a lossless inductor) in parallel with a resistance of $Q_o^2 R$, then a parallel-resonant network loaded by an external resistance can be treated simply by combining the external, shunt resistance with the resistance $Q_o^2 R$. Suppose we call this combined resistance R_p. That is, if the external resistance is R_L, then

$$R_p = R_L \| Q_o^2 R \tag{9-6}$$

We must be careful to note that the symbol Q_o refers to the coil alone. That is, if the actual coil resistance is R, then $Q_o = \omega_o L/R$, as previously defined. However, the presence of an external resistance loading the network suggests that a new, *equivalent* coil resistance might be found, which would, in turn, imply that a new, *equivalent* value of Q could be determined. This new value of Q would then account for the resistance R_p as well as for the resultant, loaded, bandwidth of the network. Let us designate this new value of Q by the symbol Q_e. Then Q_e must be equal to

$$Q_e = \frac{\omega_o L}{R_e}$$

where R_e is the new, *equivalent* coil resistance. Note that ω_o and the inductance L remain unchanged.

The total parallel resistance R_p, in terms of Q_e and R_e, must be

$$R_p = Q_e^2 R_e$$

$$= \frac{(\omega_o L)^2}{R_e} = \omega_o L Q_e$$

We therefore obtain the relationship

$$Q_e = \frac{R_p}{\omega_o L} \tag{9-7}$$

This new value of Q_e replaces Q_o in Eq. (9-3) and the relationships derived from it.

A further extension of Eq. (9-7) can be made that will more clearly show the connection between Q_o and Q_e. We can find this relationship by combining Eq. (9-6) and Eq. (9-7) to obtain

$$Q_e = \frac{(Q_o^2 R) \| R_L}{\omega_o L} = \frac{Q_o^2 R R_L/(Q_o^2 R + R_L)}{\omega_o L}$$

This leads to the equation

$$Q_e = \frac{Q_o}{Q_o(\omega_o L/R_L) + 1} \tag{9-8}$$

The advantage offered by Eq. (9-8) is that it relates Q_o of the coil alone to Q_e, the effective Q of the loaded circuit, without requiring that we know the coil resistance R.

Calculations on parallel-resonant circuits may be facilitated by the use of Prog. 9-1.

```
10    REM   CALCULATION OF PARALLEL-RESONANT FREQUENCY, EFFECTIVE Q, AND BAND
      WIDTH
20    REM   FOR RLC NETWORK WITH LOAD RESISTANCE.
25    PRINT "PARALLEL-RESONANT RLC NETWORK CALCULATIONS"
26    PRINT
30    PRINT "SELECT ONE OF THE FOLLOWING CALCULATION OPTIONS."
31    PRINT
35    PRINT "(A) GIVEN COIL INDUCTANCE AND RESISTANCE, CAPACITANCE AND LOAD
      RESISTANCE,"
40    PRINT "CALCULATE Fo AND BANDWIDTH."
45    PRINT "(B) GIVEN Fo, BANDWIDTH, CAPACITANCE, AND LOAD RESISTANCE, CALC
      ULATE"
46    PRINT "COIL INDUCTANCE AND RESISTANCE, AND TOTAL PARALLEL RESISTANCE."
55    PRINT "(C) GIVEN Qo, Fo, COIL INDUCTANCE AND RESISTANCE, CALCULATE Qe
      "
56    PRINT "CAPACITANCE, AND BANDWIDTH."
60    INPUT "SELECT ONE OF THE OPTIONS ABOVE. (A,B,C).";X$
70    IF X$ = "A" THEN 85
75    IF X$ = "B" THEN 190
80    IF X$ = "C" THEN 320
85    INPUT "ENTER INDUCTANCE IN MICROHENRYS";L
86    L = L * 1E - 6
90    INPUT "ENTER COIL RESISTANCE IN OHMS.";R
95    INPUT "ENTER LOAD RESISTANCE IN OHMS.";RL
100   INPUT "ENTER CAPACITANCE IN PICOFARADS";C
105   C = C * 1E - 12
110   FO = 1 / (2 * 3.14159 *  SQR (L * C))
120   QO =  SQR (L / C) * (1 / R)
130   QE = QO / (QO * (1 /  SQR (L * C)) * L / RL + 1)
140   BW = FO / QE
150    PRINT : PRINT "RESONANT FREQUENCY= ";FO;" HZ"
155    PRINT "Q OF COIL, Qo = ";QO
160    PRINT : PRINT "EFFECTIVE Q= ";QE
170    PRINT : PRINT "BANDWIDTH= ";BW;" HZ"
171   RP = ((QO ^ 2 * R) * RL / ((QO ^ 2 * R) + RL)
175    PRINT : PRINT "TOTAL PARALLEL RESISTANCE RP= ";RP
180    GOTO 420
190    INPUT "ENTER RESONANT FREQUENCY IN HZ.";FO
200    INPUT "ENTER BANDWIDTH IN HZ.";BW
210    INPUT "ENTER LOAD RESISTANCE RL IN OHMS.";RL
220    INPUT "ENTER CAPACITANCE IN PICOFARADS";C
225   C = C * 1E - 12
230   QE = FO / BW
240   L = 1 / ((2 * 3.14159 * FO) ^ 2 * C)
250   QO = 1 / ((1 / QE) - 2 * 3.14159 * FO * L / RL)
251    PRINT "QO= ";QO
260   R = 2 * 3.14159 * FO * L / QO
270   RP = 2 * 3.14159 * QE * FO * L
275   L = L * 1E6
280    PRINT "COIL INDUCTANCE = ";L;" MICROHENRY"
290    PRINT "COIL RESISTANCE= ";R;" OHMS": PRINT
300    PRINT "PARALLEL RESISTANCE= ";RP;" OHMS": PRINT
301    PRINT : PRINT "EFFECTIVE CIRCUIT Q, Qe = ";QE
```

Prog. 9-1

```
310   GOTO 420
320   INPUT "ENTER COIL Qo.";Q0
330   INPUT "ENTER RESONANT FREQUENCY IN HZ.";FO
340   INPUT "ENTER LOAD RESISTANCE IN OHMS.";RL
350   INPUT "ENTER COIL INDUCTANCE IN MICROHENRYS";L
355   L = L * 1E - 6
360   QE = Q0 / (Q0 * 2 * 3.14159 * FO * L / RL + 1)
370   C = 1 / ((2 * 3.14159 * FO) ^ 2 * L)
375   C = C * 1E12
380   BW = FO / QE
390   PRINT "EFFECTIVE Q, Qe = ";QE
400   PRINT "CAPACITANCE C = ";C;" PICOFARADS"
410   PRINT "BANDWIDTH = ";BW;" HZ"
420   INPUT "DO YOU WANT ANOTHER CALCULATION? (Y/N)";A$
430   IF A$ = "Y" THEN 60
440   END
```

Prog. 9-1 (cont.)

Example 9-2

Let us apply the ideas developed thus far to extend the results of Example 9-1. For a 75-KHz bandwidth and a center frequency of 10.7 MHz, we decided to use a coil whose inductance was 10 μH and whose value of Q_o was 143. The tuning capacitor had a value of 22.2 pF. Now suppose that the circuit is loaded by a resistance of 10 KΩ, and let us find the effect on the bandwidth.

$$Q_e = \frac{143}{143 \times (2\pi \times 10.7 \times 10^6 \times 10^{-5}/10^4) + 1}$$

$$= 13.4$$

For this value of Q, the bandwidth becomes

$$BW = \frac{10.7 \times 10^6}{13.4} = 796 \text{ KHz}$$

This result shows very clearly how loading the parallel-resonant circuit broadens its bandwidth. In this case, the effect is disastrous, since the bandwidth now turns out to be far greater than the 75-KHz value that we require.

If the effective Q_e for the *circuit* were made equal to 143, then the 75-KHz bandwidth requirement could be met. However, if we try to maintain all other parameters at the values used above, we find that the value of Q_o needed for the *coil* cannot be physically realized.

Suppose that we modify the design so that the coil inductance is 3 μH. The capacitance value then must be $1/4\pi^2f^2L$, or $1/[4\pi^2 \times (10.7)^2 \times 10^{12} \times 3 \times 10^{-6}]$ = 73.7 pF. For the bandwidth of 75 KHz, we still require that Q_e = 143. If R_L is equal to 40 KΩ, then Q_o must equal 510 for this coil. While this is quite a high value, the result does show that our modifications are in the right direction and that a little more modification should result in a suitable design.

9-2.2 The Q meter

At this point it is useful to review the basic ideas underlying the operation of a very useful laboratory instrument—the Q meter. The principal use of this instrument is in making a direct measurement of the value of Q of a given coil at the frequency at which it is to operate. The circuit in simplified form is shown in Fig. 9-4.

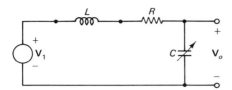

Fig. 9-4 Simplified circuit of Q meter.

The internal source $\mathbf{V}_1$ is a variable-frequency oscillator. After setting it to the desired frequency, f_o, the magnitude of $\mathbf{V}_1$ is adjusted to a calibrated value; after further suitable adjustment, the reading of $\mathbf{V}_o$ will give the value of Q_o directly. The adjustment consists of tuning the capacitor C to series resonance causing $\mathbf{V}_o$ to reach its maximum value. Since at series resonance the capacitive and inductive reactances are equal and of opposite sign, the current in the circuit is simply equal to $\mathbf{V}_1/R$. The voltage $\mathbf{V}_o$ will then equal $(\mathbf{V}_1/R)(1/j\omega_o C) = -j\mathbf{V}_1/\omega_o RC$. At resonance, $\omega_o L = 1/\omega_o C$, so $\mathbf{V}_o$ then equals $-j\mathbf{V}_1\omega_o L/R$. Thus the ratio of the magnitudes of $\mathbf{V}_o$ and $\mathbf{V}_1$ is simply $|\mathbf{V}_o/\mathbf{V}_1| = Q_o$.

The circuit shown in Fig. 9-4 is a simplified version of the actual circuit; you are advised to consult the instruction manual for the instrument for a detailed explanation of the actual circuit.

9-3 CASCADED, TUNED AMPLIFIERS

When several tuned amplifiers are cascaded, bandwidth narrowing takes place in a manner somewhat similar to that of the wideband amplifiers of Chap. 8. The change in amplitude of response as a function of frequency for a single amplifier stage is given by Eq. (9-3). We must view the current $\mathbf{I}$ in that equation as representing the current generated, for example, by the collector-emitter circuit of a transistor. Furthermore, we note that instead of the actual frequency in Hertz or radians per second, we are dealing with the fractional frequency deviation δ. For constant phasor current $\mathbf{I}$, the ratio of the voltage $\mathbf{V}$ at some frequency deviation δ relative to the voltage when δ is zero is equal to

$$\frac{\mathbf{V}(\delta)}{\mathbf{V}(o)} = \frac{1}{1 + j2Q_e\delta}$$

The 3-db points occur when $\delta = \pm\, 1/2Q_e$, as we saw earlier. In terms of the magnitude $|\mathbf{V}(\delta)/\mathbf{V}(o)|$, we can write the expression above in the form

$$\left|\frac{V(\delta)}{V(o)}\right| = \frac{1}{\sqrt{1 + (2Q_e\delta)^2}}$$

For n cascaded, identical stages, the ratio becomes

$$\left|\frac{\mathbf{V}(\delta)}{\mathbf{V}(o)}\right| = \frac{1}{\left[\sqrt{1 + (2Q_e\delta)^2}\right]^n}$$

The 3-db points occur when this expression becomes equal to $1/\sqrt{2}$. Hence we can write

$$(1 + [2Q_e\delta]^2)^n = 2$$

The quantity $2Q_e\delta$ must therefore equal

$$2Q_e\delta = \pm \sqrt{2^{1/n} - 1}$$

so that the value of δ for this situation is given by

$$\delta = \pm \frac{\sqrt{2^{1/n} - 1}}{2Q_e}$$

Since the bandwidth is the difference between the frequencies that correspond to the positive and negative δ-values, we have, as the bandwidth for the n stages,

$$BW_n = 2|\delta|f_o = \frac{f_o}{Q_e} \sqrt{2^{1/n} - 1}$$

As we saw earlier, the bandwidth for a single tuned-amplifier stage is BW $= f_o/Q_e$. Consequently we can apply the bandwidth reduction factors of Table 8-1 to the case of cascaded, identical, tuned amplifiers.

When each of the amplifiers in a cascade arrangement is tuned to the same frequency, we call the arrangement *synchronous tuning*. We might guess that if each amplifier is detuned just a little bit, some to frequencies above ω_o and others to frequencies below ω_o, we might be able to broaden the overall bandwidth. Such an arrangement is called *stagger tuning*. The problem in stagger tuning is to determine just how much to detune the individual amplifiers; too much detuning will make the overall response uneven, while too little will make it too narrow. Somewhere between these extremes lies the maximally flat tuning arrangement that will result in just the right overall bandwidth with the flattest possible response in the pass band.

To see how maximally flat stagger tuning is achieved, we refer once again to the network shown in Fig. 9-1 and write the impedance of this network, this time in terms of the Laplace frequency variable S. The result is

$$\mathbf{Z} = \frac{1}{SC + 1/(R + LS)}$$

A little rearranging leads to

$$\mathbf{Z} = \frac{S + R/L}{C(S^2 + SR/L + 1/LC)}$$

In terms of Q_o and ω_o, which were defined previously, the expression for $\mathbf{Z}$ can be written

$$\mathbf{Z} = \frac{S + \omega_o/Q_o}{C[S^2 + S(\omega_o/Q_o) + \omega_o^2]}$$

The roots of the quadratic denominator are equal to

$$S_p = -\frac{\omega_o}{2Q_o} \pm \frac{\omega_o}{2} \sqrt{\frac{1}{Q_o^2} - 4}$$

Since we are interested only in those cases where the roots are complex conjugates, we require that $Q_o > \frac{1}{2}$. This is not at all a stringent requirement, since, as we have seen, the values of Q that are of interest are far greater than $\frac{1}{2}$.

In terms of the complex conjugate roots of the denominator, $\mathbf{Z}$ can be written

$$\mathbf{Z} = \frac{S + \omega_o/Q_o}{C[S + \omega_o/2Q_o - j\omega_o \sqrt{1 - (1/2Q_o)^2}][S + \omega_o/2Q_o + j\omega_o \sqrt{1 - (1/2Q_o)^2}]} \tag{10-9}$$

Following the procedure that we used in Chap. 7, we plot the poles and the zero of Eq. (9-9). These are shown in Fig. 9-5.

As we have seen, the bandwidth of this circuit, expressed in radians per second, is ω_o/Q_o. That fact suggests the geometrical representation shown in Fig. 9-6.

Although Fig. 9-6 seems complicated and cluttered with semicircles, arrows, and algebraic expressions, it is essentially simple. First of all, the diagram shows that the poles are displaced by $\omega_o/2Q_o$ to the left of the imaginary axis and by $\omega_o\sqrt{1 - (1/2Q_o)^2}$ upward and downward from the real axis. The length of the vector from the origin to either pole turns out to be simply ω_o. Near the upper pole, if we choose as a center the point $+j\omega_o\sqrt{1 - (1/2Q_o)^2}$ and construct a semicircle of radius $\omega_o/2Q_o$, this semicircle will intersect the imaginary axis to produce a diameter of ω_o/Q_o, which is precisely the bandwidth.

There is some similarity between the semicircular constructions in Fig. 9-6 and those of Fig. 7-40. We might thus be led to suppose that if we could position more poles in the same sort of pattern as in Fig. 9-6, somehow making them fall in a symmetrical array on the semicircles, a maximally flat response would be produced. This is indeed the case, but the proof of that fact is beyond the scope of this book. For convenience, however, Prog. 9-2 is included so that you can readily calculate pole locations for band-pass amplifiers of this type if you are interested.

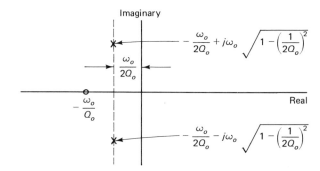

Fig. 9-5   S-plane plot of poles and zero.

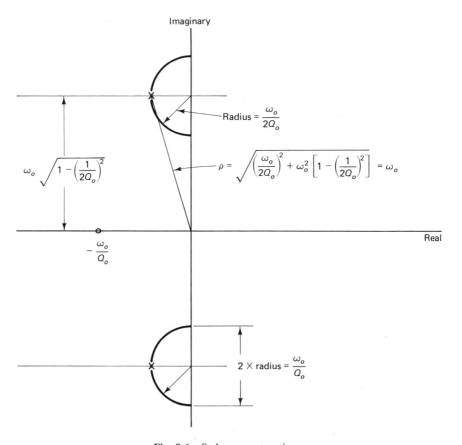

Fig. 9-6 *S*-plane construction.

```
10   PRINT "CALCULATION OF POLE LOCATIONS FOR BUTTERWORTH BAND-PASS AMPLIFI
     ERS"
30   INPUT "ENTER ORDER OF FILTER";N
35   PRINT
40   INPUT "ENTER BAND-CENTER FREQUENCY IN KILORAD/SEC";FO
45   INPUT "ENTER 3-DB BANDWIDTH IN KILORAD/SEC";BW
46   PRINT
47 THETA = 3.14159 / 2
50   PRINT "S-PLANE COORDINATES IN KILORADIANS/SEC"
51   PRINT : PRINT
52   PRINT "POLE NO."; SPC( 5);"X"; SPC( 5);" +/- Y "; SPC( 5);"EFF Q"
53   PRINT
55 PHI = 3.14159 / (2 * N)
60   FOR K = 1 TO N
65 THETA = (2 * K - 1 + N) * PHI
80 X = (BW / 2) *  COS (THETA)
90 Y = FO + (BW / 2) *  SIN (THETA)
94   IF X = 0 THEN Q = 9999999: GOTO 100
95 Q = Y / ( ABS (2 * X))
100  PRINT  TAB( 1);K; TAB( 4);X; TAB( 17);Y; TAB( 29);Q
110  PRINT
130  NEXT K
140  END
```

Prog. 9-2

In setting up tuned amplifier stages so that the poles will produce a maximally flat response with the correct bandwidth, we must be aware that the results we are going to obtain involve certain approximations. To begin with, the center frequency as used to define δ in Eq. (9-3) (itself an approximation) is taken to be ω_o. In the diagram of Fig. 9-7, the center frequency is $\omega_o\sqrt{1 - (1/2Q_o)^2}$. Since Q_o is large in all cases of interest to us, the quantity under the radical can be approximated by $1 - 1/8Q_o^2$. If Q_o has the modest value of 20, the difference in center-frequency values will amount only to about 0.03%.

For two cascaded, stagger-tuned amplifiers, let us use the pole-zero pattern shown in Fig. 9-7 and then calculate the required parameter values.

The two principal parameters are the center frequency ω_o and the bandwidth BW. As we see in the diagram of Fig. 9-6, the 45° pole pattern requires the poles to be displaced to the left of the imaginary axis by (BW/2) cos 45°. They are also displaced upward and downward from the center frequency by (BW/2)sin 45°. Based on this information, we can write the following expressions for the pole locations:

$$\mathbf{S}_1 = -\frac{BW}{2}\cos 45° + j\left(\omega_o - \frac{BW}{2}\sin 45°\right)$$

$$\mathbf{S}_2 = -\frac{BW}{2}\cos 45° + j\left(\omega_o + \frac{BW}{2}\sin 45°\right)$$

To place these poles at these locations, we require tuned circuits whose center frequencies equal the magnitudes of $\mathbf{S}_1$ and $\mathbf{S}_2$. That is,

$$\omega_1 = |\mathbf{S}_1| = \sqrt{\left(\frac{BW}{2\sqrt{2}}\right)^2 + \left(\omega_o - \frac{BW}{2\sqrt{2}}\right)^2}$$

and

$$\omega_2 = |\mathbf{S}_1| = \sqrt{\left(\frac{BW}{2\sqrt{2}}\right)^2 + \left(\omega_o + \frac{BW}{2\sqrt{2}}\right)^2}$$

The next few manipulations are aimed at getting these equations into a form suitable for easy calculation.

We begin by expanding the expressions under the radicals to obtain

$$\omega_1 = \sqrt{\frac{(BW)^2}{8} + \omega_o^2 - \omega_o\frac{BW}{\sqrt{2}} + \frac{(BW)^2}{8}} = \sqrt{\omega_o^2 + \frac{(BW)^2}{4} - \omega_o\frac{BW}{\sqrt{2}}}$$

and

$$\omega_2 = \sqrt{\frac{(BW)^2}{8} + \omega_o^2 + \omega_o\frac{BW}{\sqrt{2}} + \frac{(BW)^2}{8}} = \sqrt{\omega_o^2 + \frac{(BW)^2}{4} + \omega_o\frac{BW}{\sqrt{2}}}$$

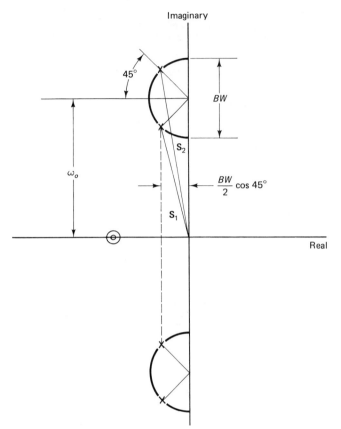

Fig. 9-7 Pole-zero pattern for two-stage, maximally flat amplifier.

We can factor out ω_o^2 from the terms under the radicals and write them:

$$\omega_o \sqrt{1 + \left(\frac{\text{BW}}{2\omega_o}\right)^2 \mp \sqrt{2}\left(\frac{\text{BW}}{2\omega_o}\right)}$$

Since $(\text{BW}/2\omega_o)^2 << 1$, we can neglect it and arrive at the approximation

$$\omega_o \sqrt{1 \mp \sqrt{2}\left(\frac{\text{BW}}{2\omega_o}\right)}$$

Finally, since $\sqrt{2}\,\text{BW}/2\omega_o$ is small compared to 1, we can approximate ω_1 and ω_2 by

$$\omega_1 \approx \omega_o \left[1 - \frac{1}{\sqrt{2}}\left(\frac{\text{BW}}{2\omega_o}\right)\right] \quad \text{and} \quad \omega_2 \approx \omega_o \left[1 + \frac{1}{\sqrt{2}}\left(\frac{\text{BW}}{2\omega_o}\right)\right] \qquad (9\text{-}10)$$

From Eq. (9-10) we find the fractional amounts by which the two tuned-amplifier stages must be detuned from the center frequency ω_o. These are simply

$$\pm \frac{1}{\sqrt{2}}\left(\frac{\text{BW}}{2\omega_o}\right)$$

To place the poles correctly according to the diagram of Fig. 9-7, a lateral displacement of (BW/2) cos 45° must exist. It is important to remember that each of the two pairs of complex conjugate poles follows the geometry given in Fig. 9-6. However, the radius ω_o of Fig. 9-6 must now be replaced by ω_1 or ω_2, whose values are given by Eq. (9-10). Accordingly, we now see that

$$\frac{BW}{2} \cos 45° = \frac{\omega_1}{2Q_1}$$

and

$$\frac{BW}{2} \cos 45° = \frac{\omega_2}{2Q_2}$$

where Q_1 is the value of Q_o for amplifier 1 and Q_2 is the value of Q_o for amplifier 2. The values of Q are thus given by

$$Q_1 = \frac{\sqrt{2}\,\omega_1}{BW} \quad \text{and} \quad Q_2 = \frac{\sqrt{2}\,\omega_2}{BW} \tag{9-11}$$

We can now give a simple procedure for designing a two-stage, maximally flat, stagger-tuned amplifer, given the center frequency and the bandwidth:

1. Calculate the individual frequencies to which each stage is tuned using Eq. (9-10).
2. Find the required Q for each circuit from Eq. (9-11).

Example 9-3

Let us apply the procedure to the design of a two-stage, stagger-tuned amplifier to operate at 10.7 MHz with a bandwidth of 75 KHz. The fractional detuning factor is BW/$(2\sqrt{2}\omega_o)$, which equals $75 \times 10^3/2\sqrt{2} \times 10.7 \times 10^6 = 2.48 \times 10^{-3}$. This yields $f_1 = 10.7$ MHZ $- 26.6$ KHZ and $f_2 = 10.7$ MHZ $+ 26.6$ KHz. The Q values are

$$Q_1 = \frac{2(10.7 \text{ MHz} - 26.6 \text{ KHz})}{75 \text{ KHz}} = 202 - 0.50 \approx 202$$

and

$$Q_2 = \frac{2(10.7 \text{ MHz} + 26.6 \text{ KHz})}{75 \text{ KHz}} = 202 + 0.50 \approx 202$$

Clearly, in this narrow-band example the Q values are affected very little by the detuning. This suggests that we can simplify the calculations by modifying Eq. (9-11) to read

$$Q_o \approx \frac{\sqrt{2}\,\omega_o}{BW} \tag{9-12}$$

Throughout the development of these formulas, we have used Q_o (or Q_1 or Q_2) to represent the Q of the coil in the particular parallel-resonant circuit of interest at the resonant frequency. You should understand that, just as we saw earlier,

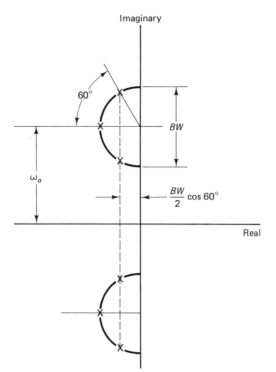

Imaginary

$60°$

BW

$\dfrac{BW}{2}\cos 60°$

ω_o

Real

Fig. 9-8 Pole arrays for three-stage, maximally flat, stagger-tuned amplifier.

circuit loading must be taken into account and the effective (or loaded) Q, which we have called Q_e, must be used wherever applicable.

Again following up on the ideas suggested by Fig. 7-40, we can work out the required values of Q_o and the detuning needed to produce a three-stage, maximally flat, stagger-tuned amplifier. The required pole array is shown in Fig. 9-8.

It should be clear that one amplifier stage is tuned to frequency ω_o and that its parallel-resonant circuit must have a value of Q_o equal to $\omega_o/$BW. The other two amplifiers are detuned by frequencies of approximately (BW/2) sin 60° or BW$\sqrt{3}/4$ rad/s. The Q values for these stages must be approximately $\omega_o/(2$BW/2) cos 60° or $2\omega_o/$BW. Once again we use the 10.7-MHz IF amplifier as a numerical example.

One stage, as we have seen, is tuned to ω_o, or 10.7 MHz. This stage must have $Q_o = 10.7 \times 10^6/75 \times 10^3$, or 143. The other two stages must be detuned by ± 32.5 KHz. Each of these stages has a Q value equal to 2×143, or 286.

To appreciate the value of using stagger tuning, we can study Fig. 9-9, in which response curves are plotted for two and three stages of maximally flat, stagger-tuned amplifiers. Also shown for comparison is the response of a single tuned amplifier. For purposes of comparison, all curves are normalized so that they have a maximum value of $|\mathbf{V}_o/\mathbf{V}_1|$ equal to 1. The ratio of center frequency to bandwidth is chosen equal to 100. Coordinates along the horizontal axis are given in values of δ, the fractional frequency deviation measured with respect to the center frequency.

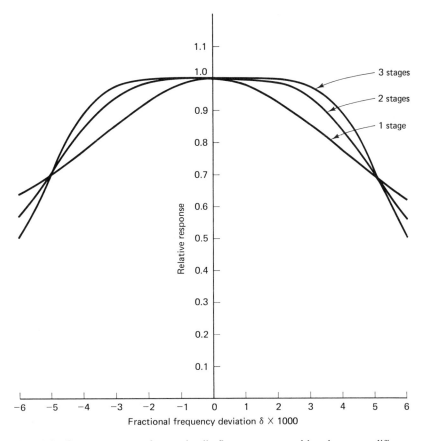

Fig. 9-9 Response curves for maximally flat, stagger-tuned band-pass amplifiers.

The bandwidth for all three cases plotted in Fig. 9-9 is the same. Hence, for $\delta = +0.005$ or $\delta = -0.005$, the amplitude is down to 0.707, corresponding to the -3-db point. The improvement in flatness of response as well as in the steepness of roll-off with increasing numbers of stages, is very clear.

We might wonder what would happen if we maintained the Q values for the several stages of a stagger-tuned amplifier but changed the tuning of the individual stages from the frequencies that yield maximal flatness. The effect of such detuning upon a two-stage amplifier is shown in Fig. 9-10. The figure clearly shows that detuning may result either in an inadequate bandwidth (curve I) or in an uneven response (curve III). Curve II is the maximally flat curve that also is shown in Fig. 9-9.

Piezo-electric, band-pass, interstage networks make use of the extremely high value of Q_o and the precise setting of resonant frequencies that are possible with piezo-electric crystals. Each prefabricated, interstage network may contain several interconnected crystals so that, for example, a high-order, maximally flat network results.

Sec. 9-3   Cascaded, Tuned Amplifiers   **305**

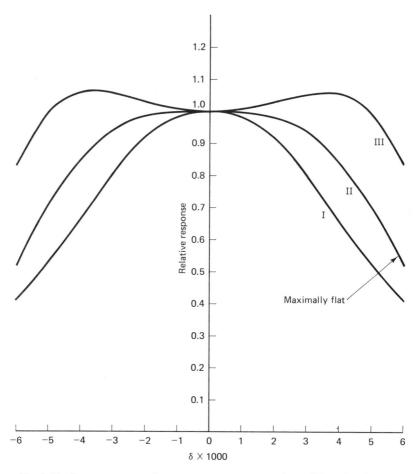

Fig. 9-10 Response curves for two-stage, stagger-tuned amplifiers showing maximally flat and mistuned responses.

9-4 OPERATIONAL-AMPLIFIER BAND-PASS FILTERS

We have seen how operational amplifiers may be incorporated in networks to yield adjustable low-pass filters and oscillators. We should not be surprised to learn that operational-amplifier circuits can also be constructed to function as tunable, narrow-band amplifiers. To see how this may be done, let us consider the circuit shown in Fig. 9-11.

As usual, what we wish to find for this circuit is the transfer function $\mathbf{V}_o/\mathbf{V}_1$. The voltage $\mathbf{V}$ shown in the diagram is simply an auxiliary quantity to aid us in our calculations. Assuming that the operational amplifier is ideal (infinite input impedance, infinite gain, zero output impedance, and unlimited bandwidth) then the negative terminal is at ground (zero) potential. Therefore, adding currents at this node, we get

$$\mathbf{V}SC_2 + \mathbf{V}_o \left(\frac{1}{R_2}\right) = 0$$

where S is the Laplace frequency variable. Clearly, the voltage $\mathbf{V}$ must equal

$$\mathbf{V} = -\frac{1}{SC_2} \cdot \frac{\mathbf{V}_o}{R_2}$$

Now, adding currents at the $+\mathbf{V}$ node, we get

$$\frac{\mathbf{V}_1 - \mathbf{V}}{R_1} + (\mathbf{V}_o - \mathbf{V})SC_1 - \mathbf{V}SC_2 = 0$$

After some rearranging and upon substituting the expression above for $\mathbf{V}$, we obtain

$$\frac{\mathbf{V}_o}{\mathbf{V}_1} = -\frac{1}{R_1 C_1} \cdot \frac{S}{S^2 + S\left[(C_1 + C_2)/R_2 C_1 C_2\right] + 1/R_1 R_2 C_1 C_2}$$

This expression can be simplified:

$$\frac{\mathbf{V}_o}{\mathbf{V}_1} = -\frac{1}{R_1 C_1} \cdot \frac{S}{S^2 + S\left[(1/R_2)(1/C_1 + 1/C_2)\right] + 1/R_1 R_2 C_1 C_2} \qquad (9\text{-}13)$$

For the network of Fig. 9-11 to have a resonant frequency, the poles of Eq. (9-13) must be complex conjugates. If we apply the quadratic formula to the denominator of Eq. (9-13), we find the condition that

$$\frac{1}{R_1 R_2 C_1 C_2} > \left[\frac{1}{2R_2}\left(\frac{C_2 + C_1}{C_1 C_2}\right)\right]^2$$

in order for the poles to be complex conjugates. One case of interest occurs when $C_1 = C_2 = C$. When this is the case, this last inequality becomes

$$\frac{1}{R_1 R_2 C^2} > \left[\frac{1}{2R_2 C/2}\right]^2 = \frac{1}{R_2^2 C^2}$$

or

$$\frac{1}{R_1} > \frac{1}{R_2}$$

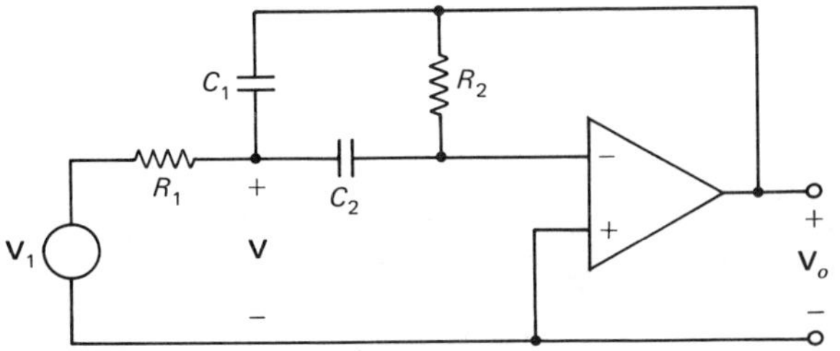

Fig. 9-11 Operational-amplifier band-pass circuit.

which reduces to $R_2 > R_1$, a condition that is easy to satisfy.

To appreciate the significance of Eq. (9-13), we revert to the circuit of Fig. 9-1. Using the Laplace variable S rather than $j\omega$, we rewrite Eq. (9-2). This yields

$$V = \frac{I}{SC + 1/(R + SL)}$$

which can be written in the form

$$V = I \frac{R + SL}{S^2LC + RCS + 1} = I \frac{R + SL}{LC[S^2 + S(R/L) + 1/LC]}$$

If we now make use of the definitions previously given for ω_o, Q_o, and the bandwidth BW, we can write

$$V = I \frac{R + SL}{LC[S^2 + S(\omega_o/Q_o) + \omega_o^2]} \tag{9-14}$$

Comparing the denominators of Eq. (9-13) and Eq. (9-14), we see that we can make the following identifications with respect to the circuit of Fig. 9-10:

$$\frac{\omega_o}{Q_o} = \frac{1}{R_2}\left(\frac{1}{C_1} + \frac{1}{C_2}\right) \tag{9-15a}$$

and

$$\omega_o^2 = \frac{1}{R_1R_2C_1C_2} \tag{9-15b}$$

By combining Eq. (9-15a) and Eq. (9-15b), we can find the expression for Q_o, which is

$$Q_o = \frac{\sqrt{R_2/R_1}}{\sqrt{C_1/C_2} + \sqrt{C_2/C_1}}$$

For simplicity, let us again take $C_1 = C_2 = C$. Then the expressions for ω_o and Q_o become

$$\omega_o = \frac{1}{C\sqrt{R_1R_2}}$$

and

$$Q_o = \frac{1}{2}\sqrt{\frac{R_2}{R_1}}$$

The bandwidth is given by

$$BW = \frac{\omega_o}{Q_o} = \frac{2}{R_2C}$$

We can now make use of these newly defined quantities to modify Eq. (9-13). This procedure yields

$$\frac{\mathbf{V}_o}{\mathbf{V}_1} = \frac{-2Q_o^2 \, \text{BW} \cdot S}{S^2 + \text{BW} \cdot S + \omega_o^2} \tag{9-16}$$

For design purposes, we take R_1 to be given. Then, knowing the desired center frequency and bandwidth, we solve for R_2 and C. For this purpose it is convenient to solve the equations above to obtain

$$R_2 = 4Q_o^2 R_1 \quad \text{and} \quad C = \frac{2}{R_2 \cdot \text{BW}}$$

The reason for considering R_1 to be given is that, as we see in Fig. 9-11, it appears in series with the input voltage source. In other words, R_1 is the source resistance. It is thus less convenient to adjust than the other parameters.

A more flexible procedure for calculating circuit parameters for the band-pass amplifier of Fig. 9-11 is given in Prog. 9-3. Here the restriction that $C_1 = C_2$ has been removed. You are cautioned that the price of this added flexibility is that the basic inequalities needed to produce complex conjugate poles may not be satisfied. When that situation occurs, the program generates a message: SELECTED VALUES OF R1 AND C2 LEAD TO UNREALIZABLE VALUES FOR R2 AND C1.

```
10   PRINT "CALCULATION OF R2 AND C1 FOR OP-AMP BAND-PASS CIRCUIT OF FIG.
     9-11."
20   REM  SINCE FOUR PARAMETERS ARE NEEDED, BUT ONLY TWO CONDITIONS APPLY,
30   REM  TWO PARAMETERS MUST BE CHOSEN ARBITRARILY.
40   INPUT "SPECIFY CENTER FREQUENCY IN HERTZ, FO";FO: PRINT
50   INPUT "SPECIFY CIRCUIT Qo VALUE";Q: PRINT
60   INPUT "SELECT A VALUE FOR R1 IN OHMS";R1: PRINT
70   INPUT "SELECT A VALUE FOR C2 IN PICOFARADS";C2: PRINT
80   W = 2 * 3.14159 * FO
81   C2 = C2 * 1E - 12
85   P = 1 - Q * W * R1 * C2
86   IF P < 0 THEN 140
90   R2 = Q / (W * C2 * P)
100  C1 = 1 / (R1 * C2 * W ^ 2 * R2)
110  C1 = C1 * 1E12
115  PRINT "R2= ";R2;" OHMS": PRINT
120  PRINT "C1= ";C1;" PICOFARADS": PRINT
130  GOTO 170
140  PRINT "SELECTED VALUES OF R1 AND C2 LEAD TO UNREALIZABLE VALUES FOR R
     2 AND C1."
150  PRINT "SELECT NEW VALUES."
160  GOTO 60
170  INPUT "DO YOU WANT TO SOLVE ANOTHER CASE? (Y/N)";A$
180  IF A$ = "Y" THEN 40
190  END
```

Prog. 9-3.

Example 9-4

Suppose that, for a source resistance R_1 equal to 100 Ω, we wish to design a band-pass circuit having a center frequency of 2000 rad/s and a bandwidth of 40 rad/s. The corresponding value of Q_o must be 2000/40 = 50. Thus we find that $R_2 = 4 \times (50)^2 \times 100 = 1$ MΩ. Then $C = 2/(10^6 \times 40) = 0.05$ μF.

9-5 TUNABLE BAND-PASS AMPLIFIERS

A useful variation of the circuit of Fig. 9-11 is the one shown in Fig. 9-12. The modification consists of the replacement of R_1 by R_a and the addition of R_b. The result of this modification is that the center frequency ω_o is now adjustable. Let us see how this result is achieved.

The network consisting of V_1, R_a, and R_b can be replaced by a Thevenin equivalent. It should be easy to see that this replacement is equivalent to changing V_1 in Fig. 9-11 to the value $V_1 R_b/(R_a + R_b)$ and to replacing R_1 by $R_a \| R_b$. With these modifications, Eq. (9-13) retains its form except for the scale factor $R_b/(R_a + R_b)$ and the fact that R_1 now equals $R_a \| R_b$, as we have noted.

Since, as before, $\omega_o = 1/C\sqrt{R_1 R_2}$, we see that adjusting R_b will change R_1 and also ω_o. Furthermore, since the bandwidth equals $2/R_2 C$, it is not affected by the change in center frequency resulting from the adjustment of R_b.

With sinusoidal input to this band-pass filter, which represents practical applications, the variable S is replaced by $j\omega$. Then at the center frequency when $\omega = \omega_o$, Eq. (9-16) yields the following expression for the magnitude of the transfer ratio:

$$\left|\frac{V_o}{V_1}\right| = \left|\frac{-2Q_o^2 \, \text{BW} \, j\omega_o}{-\omega_o^2 + j\omega_o \text{BW} + \omega_o^2}\right| = 2Q_o^2$$

With the multiplier $R_b/(R_a + R_b)$ applied, this last equation becomes

$$\left|\frac{V_o}{V_1}\right| = \frac{R_b}{R_a + R_b} \cdot 2Q_o^2$$

Since $Q_o^2 = R_2/4R_1 = R_2/4(R_a\|R_b)$, we can write

$$\left|\frac{V_o}{V_1}\right| = \frac{R_b}{R_a + R_b} \cdot \frac{2R_2(R_a + R_b)}{4R_a R_b} = \frac{R_2}{2R_a}$$

This last result means that the magnitude of the peak value of the voltage transfer ratio is independent of the value to which R_b is adjusted. Thus the circuit of Fig. 9-12 has a number of advantageous features, including constant bandwidth, adjustable center frequency, and constant voltage-transfer ratio at the center frequency.

We can extend the calculations performed earlier to make them apply to this

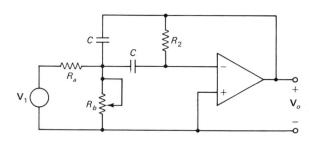

Fig. 9-12 Tunable band-pass filter.

Tuned Amplifiers   Chap. 9

modified circuit. If we make $R_a = 10$ KΩ, then the maximum gain at the center frequency will be $R_2/2R_a$, or $10^6/2 \times 10^4 = 50$. It is important to keep this figure fairly low; higher gain values could run afoul of the gain-bandwidth product of the operational amplifier and require us to take into account its high-frequency gain roll-off. In all the discussion in this section, we have made the tacit assumption that the gain-frequency characteristic of the operational amplifier is flat.

For $R_a = 10$ KΩ, a value of R_b equal to 101 Ω yields an equivalent value of R_1 of 100 Ω, which is the value we used in the original example. If R_b is variable and if its high value is, for instance, 1000 Ω, then the resulting value of R_1 will be 909 Ω. The center frequency will now become equal to $1/(0.05 \times 10^{-6} \sqrt{909 \times 10^6})$ or 663.36 rad/s. Thus, using a tuning potentiometer having a range of from 100 Ω to 1000 Ω, we can construct a band-pass filter having a gain at the center frequency equal to 50, a constant bandwidth of 251 Hz, and a center frequency ranging from 4168 Hz to 12,566 Hz.

9-6 SENSITIVITY

An important concept pertaining to operational-amplifier filter circuits that we have not considered up to this point is that of sensitivity. *Sensitivity* is a measure of how much an operating parameter of a circuit varies when a circuit parameter changes. We can distinguish between an operating parameter and a circuit parameter by saying that an operating parameter is a measured quantity such as a current, a voltage, or a frequency, whereas a circuit parameter is a quantity such as a resistance or a capacitance. Thus the quiescent operating point—for example, current—of a transistor may vary when h_{FE} changes, and the relationship between these two parameters may be expressed as a sensitivity. In general, if the variation of one parameter, for example, x, results in a change in a second parameter, which we shall call y, the sensitivity is written as S_x^y and is given by

$$S_x^y = \frac{\Delta y/y}{\Delta x/x} \tag{9-17}$$

Equation (9-17) expresses the sensitivity as a ratio of the fractional change in the effect divided by the fractional change in the cause. It should be clear that the smaller is the numerator of Eq. (9-17) with respect to the denominator, the better is the circuit performance. We can rearrange the equation to obtain

$$S_x^y = \frac{\Delta y}{\Delta x} \cdot \frac{x}{y}$$

For purposes of calculation, it is convenient to use the derivative of y with respect to x and write

$$S_x^y \approx \frac{dy}{dx} \cdot \frac{x}{y} \tag{9-18}$$

Let us now use Eq. (9-18) to examine the sensitivities of the circuit of Fig.

9-10. To begin with, we find the sensitivity of the center frequency to changes in R_1, that is, $S_{R_1}^{\omega_o}$.

$$S_{R_1}^{\omega_o} = \frac{d\omega_o}{dR_1} \cdot \frac{R_1}{\omega_o} = \frac{d}{dR_1}\left(\frac{1}{C\sqrt{R_1 R_2}}\right) \cdot R_1 C \sqrt{R_1 R_2}$$

or

$$S_{R_1}^{\omega_o} = \frac{1}{C\sqrt{R_2}} \cdot \left[-\frac{1}{2} R_1^{-3/2}\right] R_1 C \sqrt{R_1 R_2} = -\frac{1}{2}$$

This result indicates that a 1% increase in R_1 will result in a $\frac{1}{2}$% downward shift in the resonant frequency ω_o. As it turns out, the values of $S_{R_2}^{\omega_o}$ and $S_C^{\omega_o}$ are also $-\frac{1}{2}$.

If the variation of Q_o is of interest, the sensitivity $S_{R_1}^{Q_o}$ is equal to

$$S_{R_1}^{Q_o} = \frac{dQ_o}{dR_1} \cdot \frac{R_1}{Q_o} = \frac{d}{dR_1}\left(\frac{1}{2}\sqrt{\frac{R_2}{R_1}}\right) \cdot \frac{R_1}{\frac{1}{2}\sqrt{R_2/R_1}}$$

or

$$S_{R_1}^{Q_o} = \frac{\sqrt{R_2}}{2}\left(-\frac{1}{2} R_1^{-3/2}\right) \frac{2R_1^{3/2}}{\sqrt{R_2}} = -\frac{1}{2}$$

It is easy also to compute $S_C^{Q_o}$. Since Q_o is equal to $\frac{1}{2}\sqrt{R_2/R_1}$, it is not a function of C, and therefore dQ_o/dC is zero. Therefore, $S_C^{Q_o}$ is zero.

REFERENCES

1. Harry E. Stewart, *Engineering Electronics* (Boston: Allyn and Bacon, Inc., 1969).
2. George E. Valley, Jr., and Henry Wallman, eds., *Vacuum Tube Amplifiers* (New York: Dover, 1965).

PROBLEMS

9-1. The oscillator frequency is varied, but $|V_1|$ is maintained constant at 0.1 V. At a frequency of 1 MHz, $|V_o|$ reaches a maximum value of 14 V.
(a) Calculate Q_o, R, and L.
(b) What is the Q of this coil at a frequency of 2 MHz?

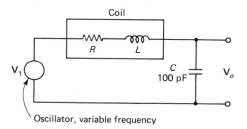

9-2. Find (a) the resonant frequency, (b) Q_o, and (c) the magnitude of **Z** at a frequency 1% above resonance.

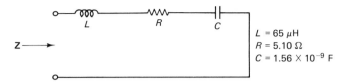

$L = 65\ \mu H$
$R = 5.10\ \Omega$
$C = 1.56 \times 10^{-9}\ F$

9-3. For the circuit of Prob. 2, find the values of f and the corresponding values of $|\mathbf{Z}|$ that occur when $Q_o\delta$ has values of $-\frac{1}{2}, -1$, and -2. What is the bandwidth for this circuit?

9-4. The circuit of Prob. 2 is rearranged so that the capacitor is connected across the series RL circuit. Find the total impedance (magnitude only) of the entire parallel circuit at resonance and at a frequency 1% above resonance.

9-5. The magnitude of the impedance of the circuit comprising *only* L, r, and C is 50,000 Ω at resonance. Find the value of the shunt resistance R so that $|\mathbf{Z}|$ is kept within 10% of the maximum value over the 100-KHz pass band.

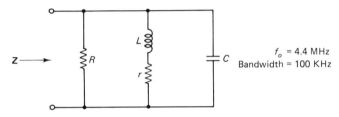

$f_o = 4.4$ MHz
Bandwidth = 100 KHz

9-6. Find the maximum value of $|\mathbf{V}_o|/|\mathbf{V}_1|$ if the parallel resonant circuit is arranged so that $C = 20$ pF and the bandwidth is 50 KHz.

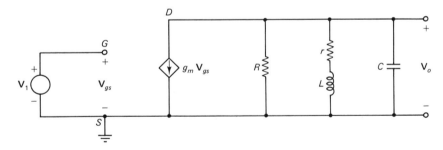

9-7. **(a)** Find the center frequency.
   (b) Find $|V_o/V_1|_{max}$.
   (c) What is the 3-db bandwidth?

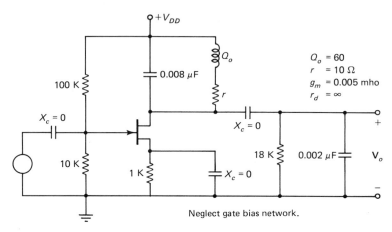

$Q_o = 60$
$r = 10\ \Omega$
$g_m = 0.005$ mho
$r_d = \infty$

Neglect gate bias network.

9-8. **(a)** Find the Q_o of the coil if the circuit is to resonate at 10.7 MHz.
   (b) Find the bandwidth and $|V_o/V_1|$ at resonance.

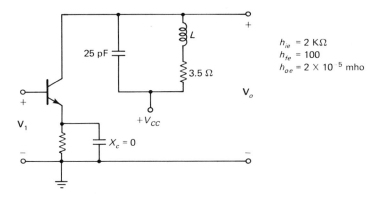

$h_{ie} = 2\ K\Omega$
$h_{fe} = 100$
$h_{oe} = 2 \times 10^{-5}$ mho

9-9. The circuit shown is to be tuned so that $f_o = 1$ MHz and the bandwidth is 20 KHz.

    (a) Find L and C

    (b) Estimate $|V_o/V_1|_{max}$.

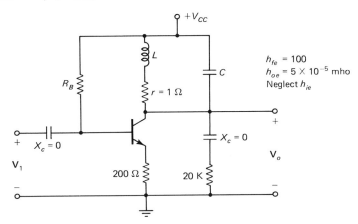

$h_{fe} = 100$
$h_{oe} = 5 \times 10^{-5}$ mho
Neglect h_{ie}

9-10. (a) If $R = 80$ KΩ, find $|V_o/V_1|$ at f_o and the bandwidth.

    (b) What value should R have to make the bandwidth 10 KHz?

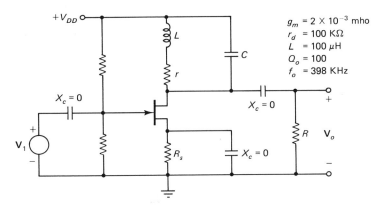

$g_m = 2 \times 10^{-3}$ mho
$r_d = 100$ KΩ
$L = 100$ μH
$Q_o = 100$
$f_o = 398$ KHz

9-11. A two-stage synchronous-tuned amplifier has $f_o = 10$ MHz and *overall* bandwidth equal to 200 KHz.

    (a) What must be the bandwidth of each stage?

    (b) How many decibels below the output at 10 MHz will the output be at a frequency of 9.5 MHz?

9-12. Two identical amplifiers are connected in cascade. Each has $A_v = 28$ db, $f_o = 10$ MHz, and a bandwidth of 50 KHz.

    (a) Find the overall gain and bandwidth.

    (b) Each amplifier is slightly detuned to provide stagger tuning with maximal flatness. *Overall* f_o is maintained at 100 MHz. Find the resultant bandwidth and the individual center frequencies.

9-13. Given $R_1 = 1\ \text{K}\Omega$, use the circuit above to design a band-pass amplifier with center frequency of 8 KHz and a bandwidth of 300 Hz.
  (a) Assuming that $C_1 = C_2 = C$, calculate values for C and R_2. Is the solution practical?
  (b) Assume that $C_2 = 5C_1$ and $C_1 = 10^{-9}\ \text{F}$. Find R_1 and R_2 required to meet the specifications above.

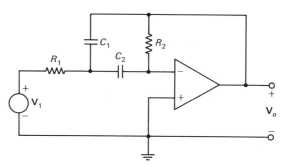

9-14. Using the circuit configuration of Prob. 13 with $C_1 = C_2 = 0.1\ \mu\text{F}$, find R_1 and R_2 to yield $f_o = 440$ Hz and a bandwidth of 60 Hz.

9-15. If the value of R_2 is limited to a maximum of 270 KΩ and the value of R_1 is limited to a minimum of 3 KΩ, find the minimum bandwidth that the circuit of Prob. 14 can provide at a center frequency of 100 Hz. Use $C_1 = C_2$. Find the required value of C.

9-16. A band-pass amplifier using the circuit of Prob. 13 has the following parameters:

$$R_1 = 1\ \text{K}\Omega \qquad R_2 = 1.6\ \text{M}\Omega \qquad C = 1989\ \text{pF}.$$

  (a) Find the center frequency and bandwidth.
  (b) If R_1 changes value to 950Ω, R_2 to 1.7 MΩ, and C to 2200 pF, by what percentage will the center frequency change?

9-17. Determine the resistance range required for R_b, values of C, and R_2 for the circuit of Fig. 9-11 that will provide a range of center frequencies for 400 Hz to 4 KHz with a constant center-frequency gain of 2 and a bandwidth of 80 Hz. Assume that $R_a = 500$ KΩ.

9-18. If two identical circuits with the parameters determined in Prob. 14 are connected in cascade, determine the resultant bandwidth.

9-19. Determine the values of R_1 and R_2 for each of two amplifiers of the configuration of Prob. 14 to be connected in cascade to produce $f_o = 440$ Hz and BW $= 60$ Hz with maximal flatness.

9-20. The given circuit is proposed as a means of realizing a band-pass amplifier.

(a) Considering only the operational amplifier labeled A, show that

$$\mathbf{V}_2 = \left(\frac{R_2}{R_1 + R_2}\right)\left(\frac{R + R_3}{R_3}\right)\mathbf{V}_1 + \left(\frac{R_1}{R_1 + R_2}\right)\left(\frac{R + R_3}{R_3}\right)\mathbf{V}_3 - \left(\frac{R}{R_3}\right)\mathbf{V}_4$$

[*Hint:* Assume voltages $\mathbf{V}_x$ and $\mathbf{V}_y$, as shown. Then use Kirchoff's current law at the negative and positive nodes, respectively. Use the fact that $\mathbf{V}_2 = A(\mathbf{V}_x - \mathbf{V}_y)$ and take the limit as $A \to \infty$.]

(b) Find the transfer function

$$\frac{\mathbf{V}_3}{\mathbf{V}_1}(s)$$

(c) If $R = 4.5$ MΩ, find f_o and the bandwidth of the function $|\mathbf{V}_3/\mathbf{V}_1|$.

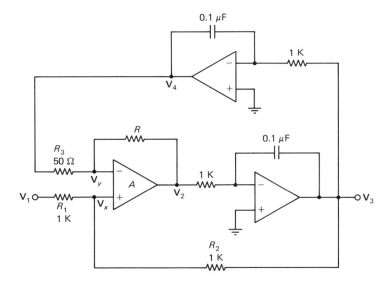

9-21. Determine f_o and Q_e for the circuit shown.

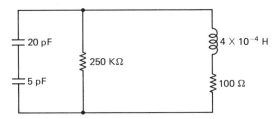

CHAPTER 10

POWER AMPLIFIERS

10-1 POWER AMPLIFIERS COMPARED TO VOLTAGE AMPLIFIERS

Up to this point we have considered amplifiers whose primary function was to magnify a small signal (specified in terms either of its voltage or of its current amplitude) in order to yield an appropriate output signal amplitude. We turn now to the study of those amplifiers whose purpose is to deliver a specified level of power to a load that typically has fairly low resistance. A classic example of such an amplifier is the audio output amplifier used to drive a low-impedance loudspeaker.

A good way to gain some appreciation of the special requirements assigned to a power amplifier is to consider an example. Suppose that we wish to deliver 10 W to a 4-Ω load. (These are typical figures for an audio power amplifier.) To dissipate 10 W in a 4-Ω load, we require an rms voltage equal to $\sqrt{10 \times 4} = 6.32$ V. If the voltage wave form is sinusoidal, the corresponding peak-to-peak voltage value is 17.89 V. The rms load current must then be $10/6.32 = 1.58$ A, and the corresponding peak-to-peak value for the current wave must be 4.97 A. These figures mean that voltage at the output of our amplifier must swing over a range of 17.89 V, whereas the current that the amplifier delivers to the load must swing over a range of 4.47 A.

For maximum power transfer, the load resistance connected across the collector-emitter terminals of a transistor ought to equal the internal resistance of $1/h_{oe}$ ohms. However, values of $1/h_{oe}$ yield resistances much larger than 4 Ω. What this means is that for such small load resistances, the transistor approximates an ideal current source, and considerations of maximum power transfer do not apply. Furthermore, h_{oe} is a small-signal parameter and is therefore of little significance in dealing with large-signal operation of transistors.

The analysis and design of large-signal (that is, power) transistor amplifiers is best done by making use of the collector characteristic curves. Just as we did in Chap. 1, we can plot the load line and, from the circuit parameters, calculate the quiescent operating point. Then, from the movement along the load line of the instantaneous operating point, we can evaluate the output voltage and current and the internal power dissipation. Before considering these matters in detail, we must first take note of the fact that the location of the quiescent operating point determines the major categories by which power amplifiers are classified. The capabilities of these categories of amplifier and the techniques used to analyze them are somewhat different. To see why this is so, let us consider the hypothetical collector characteristics and load line shown in Fig. 10-1.

If the quiescent operating point is at Q_1, the no-signal value of i_C is about 2.4 A, and the corresponding value of v_{CE} is about 10 V. When variations in base current are applied, Q_1 can swing up to allow i_C to become approximately 4.4 A, with v_{CE} equal to about 1 V. On downward swings, i_C can approach 0 while v_{CE} approaches 20 V. When the quiescent operating point is at Q_2, which is at the point of cutoff, i_C can swing only in the upward direction to a maximum value of about 4.4 A.

In the first case, if i_C is to vary about Q_1 in a sinusoidal fashion, the amplitude of the sine wave of current would have to be $4.4 - 2.4 = 2.0$ A. In the second case, with the quiescent point at Q_2, a sinusoidal current wave form cannot be obtained. These two dissimilar operating modes lead us to the following definitions.

A *Class A amplifier* is one in which, with sinusoidal input, the output current flows during the entire cycle. A *Class B amplifier* is one in which, with sinusoidal input, the output current flows only over one-half cycle.

The category of Class A amplifiers includes all the small-signal amplifiers we have studied up to this point as well as power amplifiers having large voltage and current variations.

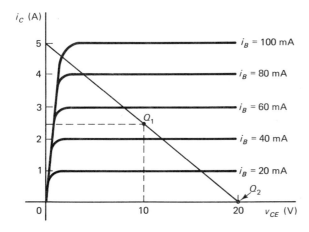

Fig. 10-1 Collector characteristics and load line.

10-2 CLASS A POWER AMPLIFIERS

The simplest form of Class A power amplifier is the circuit shown in Fig. 10-2. Note that this circuit is identical to the elementary ones considered as small-signal amplifiers in Chap. 1. The difference between those circuits and this one is that here we are using a transistor having a higher power capability and that we are driving the transistor with a large enough base-current signal to produce the required collector current swing.

In the circuit of Fig. 10-2, resistor R_L serves both as load resistor and as the means by which collector current is supplied to transistor Q. This is not an advantageous arrangement, since R_L must dissipate the power due to the DC component of its current as well as that due to the signal component. For that reason, it is common to arrange the circuit in such a way as to separate these two components in as much as this is possible. This may be done either by using a coupling capacitor, as in the circuit of Fig. 10-3, or by using transformer coupling, as in the circuit of Fig. 10-5. Capacitor coupling is cheaper than transformer coupling, but in the capacitor-coupled circuit, the collector resistor still must carry some signal current component in addition to DC. The presence of the collector resistor means that, in any case, there is signal power dissipated in this resistor that does not reach R_L. Consequently, the capacitor-coupled circuit is inherently less efficient than the transformer-coupled one.

To see how this comes about, let us find the circuit parameters for a capacitor-coupled Class A amplifier that will deliver 10 W to a 4-Ω load. The important feature here is the capacitor, which we shall assume to be very large so that its reactance is negligible compared to R_L and R_C. The presence of C and R_L means that, with respect to the AC component of voltage appearing at the collector-emitter terminals of Q, R_C and R_L appear to be in parallel. With regard to DC, only R_C is of interest. This means that, in contrast to our usual methods of plotting load lines on the transistor collector characteristics, we now must consider not one but two different load lines. One of these, the DC load line, represents only the

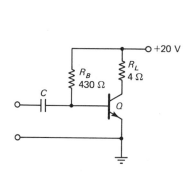

Fig. 10-2 Class A power amplifier.

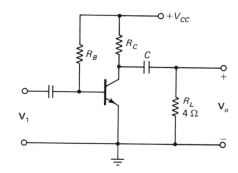

Fig. 10-3 Capacitor-coupled Class A power amplifier.

Power Amplifiers   Chap. 10

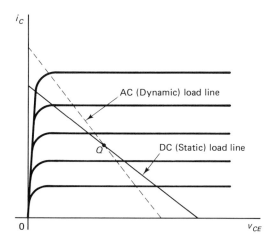

Fig. 10-4 Static and dynamic load lines.

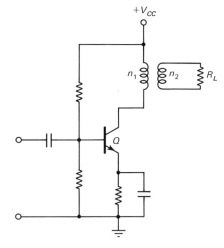

Fig. 10-5 Transformer-coupled amplifier.

effect of R_C and so has a slope equal to $-1/R_C$. The other is called the *dynamic load line* and has a slope equal to $-1/(R_C \| R_L)$. The two load lines are shown in Fig. 10-4.

Although these load lines have different slopes, they coincide at Q, the quiescent operating point. This must be true, since, when the AC output signal passes through 0, the instantaneous operating point of the composite load line must be at its neutral position, which is the point Q of the DC load line.

Figure 10-5 shows how a transformer may be used to couple the load resistor to the transistor. The principle upon which this circuit rests is that the transformer reflects the load resistance into the collector circuit so that the collector sees only a dynamic (AC) load. If we assume that the transformer is ideal, then we realize that its windings have zero resistance. This means that, so far as the static (DC) load line is concerned, this load line is simply a vertical line passing through the voltage V_{CC}. On the other hand, the AC load line has a slope determined by the reflected value of R_L. From transformer theory we know that this reflected resistance R_L' is given by

$$R_L' = \left(\frac{n_1}{n_2}\right)^2 R_L$$

where n_1 and n_2 are the number of turns on the primary and secondary windings, respectively.

The situation is represented in the diagram of Fig. 10-6, which shows the load lines drawn on the idealized collector characteristics. Note that the value of I_Q is determined as usual by the base bias network and h_{FE}.

When signal power is delivered to R_L, the instantaneous operating point moves along the dynamic load line so that, for symmetrical excursions of i_C, i_C

Sec. 10-2   Class A Power Amplifiers

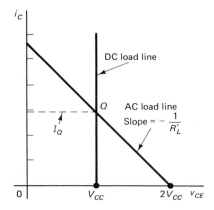

DC load line

Q AC load line
Slope = $-\dfrac{1}{R'_L}$

I_Q

0   V_{CC} $2V_{CC}$ v_{CE}

Fig. 10-6 Load lines for transformer-coupled amplifier.

varies from I_Q up to $2I_Q$ and down to zero. This means that the AC load power is equal to

$$\left(\frac{I_Q}{\sqrt{2}}\right)^2 R'_L = \frac{I_Q^2 R'_L}{2}$$

The input power from the battery is $I_Q V_{CC}$. Note that because of the symmetrical operation along the AC load line, the highest instantaneous voltage is $2V_{CC}$ so that $I_Q = V_{CC}/R'_L$.

We thus have

$$P_{\text{out}} = \frac{V_{CC}^2 R'_L}{2(R'_L)^2} = \frac{V_{CC}^2}{2R'_L}$$

and

$$P_{\text{batt}} = I_Q V_{CC} = \frac{V_{CC}^2}{R'_L}$$

The efficiency is therefore equal to $\frac{1}{2}$, or 50%.

Example 10-1

To illustrate these ideas, we once again consider the design of a Class A power amplifier intended to deliver 10 W to a 4-Ω load, but this time with transformer coupling. The rms load current must, of course, equal $\sqrt{2.5}$, or 1.58 A. Hence, the peak-to-peak load current (for a sinusoidal wave form) must be 4.47 A. Suppose that we wish to operate the transistor so that the peak-to-peak signal component of the collector current shall be 2 A. To deliver 10 W to the load with a signal current of $2/2\sqrt{2}$ = 0.707 A requires a signal voltage of $10/0.707 = 14.14 V_{\text{rms}}$. This corresponds to a peak-to-peak voltage of 40 V. The situation is illustrated by the construction shown in Fig. 10-7. Note that the slope of the resulting load line corresponds to an effective load resistance of 40 V/2 A $= 20\ \Omega$. Since the load is actually 4 Ω, a transformer having the turns ratio of $\sqrt{\frac{20}{4}} = 2.236$ is required. The supply voltage is seen to be 20 V and the transistor collector-power rating must be at least $20 \times 1 = 20$ W.

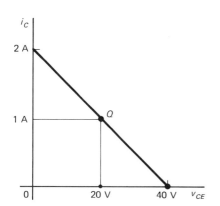

Fig. 10-7 Construction to determine the AC load line.

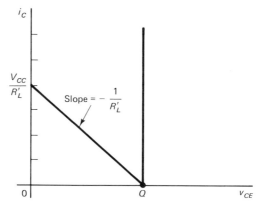

Fig. 10-8 Effect of moving Q to the cutoff point to obtain Class B operation.

Although V_{CC} is 20 V, the transistor must be capable of handling a maximum collector-emitter voltage of 40 V.

10-3 CLASS B AMPLIFIERS

The circuit of Fig. 10-5 will operate as a Class B amplifier simply by changing the bias resistors to bring the current level of I_Q down to 0 A. The situation is illustrated in Fig. 10-8. It is clear from Fig. 10-8 that i_C can have only positive excursions from the point Q; negative excursions cannot occur, since reverse collector current is inconsistent with normal transistor operation. This means that sinusoidal operation is impossible; only the positive halves of sine waves of current can be obtained. This is consistent with the definition of Class B operation given earlier.

To provide Class B operation and to allow for full-cycle output wave forms, a two-transistor network is needed. One transistor conducts over positive half-cycles, whereas the other conducts during negative half-cycles. Such an arrangement, called a *push-pull amplifier*, is shown in Fig. 10-9.

Actually, by adjusting R_1 and R_2 so that the quiescent operating point for Q_1 and Q_2 falls at the midpoint of the operating range, this amplifier could be operated as a Class A type. In that case, both transistors are always conducting. The signal voltage $\mathbf{V}_s$ (assumed sinusoidal) is applied to transformer T_1 and equal sinusoidal voltages, phased as shown, are developed in the base circuit. As a result, at the instant when the base of Q_1 is driven positive with respect to ground, the base of Q_2 is driven negative. Hence i_1 increases, whereas i_2 decreases. In effect, this means that i_1 and i_2 are 180° out of phase. But because of their reference directions, these currents are actually aiding and producing the load current i_L through transformer action.

With sinusoidal input, as we have already seen, each transistor of the push-pull pair conducts only over one half-cycle in Class B operation. This means that the movement of the instantaneous operating point of, for example, transistor Q_1 of Fig. 10-9 is along the dogleg load line shown in Fig. 10-10.

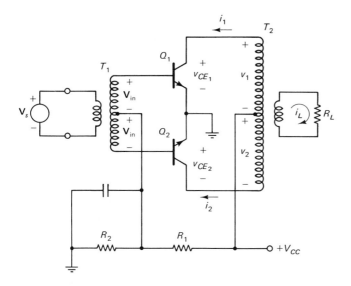

Fig. 10-9 Push-pull amplifier.

To see why this kind of performance occurs, consider the KVL equation for the loop involving V_{CC}, the upper half of the primary of T_1, and Q_1. This equation is

$$v_{CE_1} - V_{CC} - v_1 = 0$$

which leads to

$$v_{CE_1} = V_{CC} + v_1 \qquad (10\text{-}1)$$

The KVL equation for the lower loop yields

$$v_{CE_2} - V_{CC} + v_2 = 0$$

from which we get

$$v_{CE_2} = V_{CC} - v_2 \qquad (10\text{-}2)$$

At the instant when Q_2 is fully turned on, Q_1 is cut off. Under this condition, v_{CE_2} must be 0; therefore, from Eq. (10-2), $v_2 = V_{CC}$. Because of the coupling between the two sections of the primary of T_2, the induced voltage in the upper

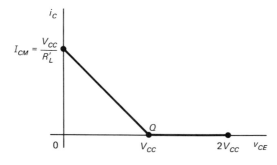

Fig. 10-10 Trajectory of instantaneous operating point.

Power Amplifiers   Chap. 10

section, v_1, equals the voltage across the lower section, v_2. Therefore,

$$v_1 = v_2 = V_{CC}$$

From Eq. (10-1) we must therefore conclude that

$$v_{CE_1} = V_{CC} + V_{CC} = 2V_{CC}$$

This explains the dogleg extension of the load line in Fig. 10-10 to the point where $v_{CE} = 2V_{CC}$. We also see that the transistors have to be able to withstand voltages of $2V_{CC}$.

With one transistor always cut off at any particular instant, coupling to the load occurs only from the conducting transistor and therefore through only half of the primary of T_2. Consequently, the slope of the load line is determined by the reflected resistance R'_L, given by

$$R'_L = \left(\frac{n_1/2}{n_2}\right)^2 R_L$$

Notice that at the quiescent point (no signal), the collector current for both transistors is, ideally, 0. Already, we see one advantage in Class B operation— namely, that less power is dissipated at the collector-base junctions of the transistors than in Class A operation.

If we assume that a sinusoidal signal is applied to the input, we can sketch the wave forms of v_{CE_1} and i_1 and then calculate the efficiency of the amplifier. Since, over the positive half cycle, i_1 varies from 0 to I_{CM} and back again, its wave form yields the loop shown in Fig. 10-11. From Fig. 10-10 we see that when $i_1 = I_{CM}$, $v_{CE_1} = 0$. When i_1 just reaches 0, $v_{CE_1} = V_{CC}$; a quarter-cycle later it becomes $2V_{CC}$. These ideas lead to the wave forms shown in Fig. 10-11.

The maximum power delivered to the load is $I_{rms}^2 R'_L$. Here, I_{rms} represents the rms value of the wave forms of both i_1 and i_2, since both transistors actually drive the load. This is the same as the rms value of a sinusoidal wave, or

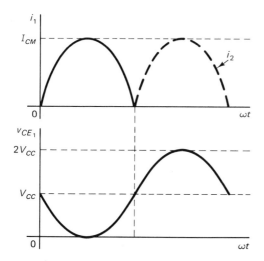

Fig. 10-11   Wave forms in ideal class B operation.

$I_{CM}/\sqrt{2}$. Hence

$$P_L = \left(\frac{I_{CM}}{\sqrt{2}}\right)^2 R_L' = \left(\frac{V_{CC}}{\sqrt{2}R_L'}\right)^2 R_L' = \frac{V_{CC}^2}{2R_L'}$$

The battery power is equal to the average of $(i_1 V_{CC} + i_2 V_{CC})$ or simply to the average of $2i_1 V_{CC}$. This yields

$$P_{batt} = 2V_{CC}\frac{1}{2\pi}\int_0^\pi I_{CM}\sin\omega t\, d(\omega t)$$

or

$$P_{batt} = \frac{V_{CC}I_{CM}}{\pi}(-\cos\omega t\big|_0^\pi) = \frac{2V_{CC}I_{CM}}{\pi}$$

$$= \frac{2V_{CC}^2}{\pi R_L'}$$

The efficiency is therefore equal to

$$\eta = \left(\frac{V_{CC}^2}{2R_L'}\right)\bigg/\left(\frac{2V_{CC}^2}{\pi R_L'}\right) = \frac{\pi}{4} = 78.5\%$$

This figure is considerably better than the maximum that can be obtained in Class A operation.

10-4 CROSSOVER DISTORTION

As we saw in Chap. 2 in connection with logic circuits, the base-emitter junction of a transistor behaves much as a diode. Consequently, there is a turn-on voltage V_γ that must be reached by v_{BE} before any appreciable forward base current begins to flow. The result of this fact is that the collector current of transistors biased to cutoff does not begin to flow in a significant amount immediately upon the application of input signal voltage. As a result, the ideal wave forms that we have envisioned up to this point are in fact somewhat distorted. What we wish to examine now is the source of this distortion and its cure.

In Fig. 10-12 the curve at the upper left is a "double-headed" plot of the input characteristics of transistors Q_1 and Q_2. That is, there are no negative axes in the plot; the first quadrant represents the base-emitter characteristic of Q_1, whereas the third quadrant applies to Q_2. As we now understand, when v_{in} is positive, only Q_1 is active, and with v_{in} negative, only Q_2 is active. For each transistor, *active* means that it operates in its own forward-conducting region. Because of the circuit connection, when Q_1 conducts, i_{C_1} flows in a direction opposite to that taken by i_{C_2} when Q_2 conducts. Hence we can choose, for convenience, to plot the double-headed characteristic as shown.

If we set up a time axis running vertically and plot on it the sinusoidal input voltage v_{in}, we obtain the curve at the bottom of Fig. 10-12. Next, we choose convenient points on the wave form of v_{in} and project these onto the $i_B - v_{BE}$

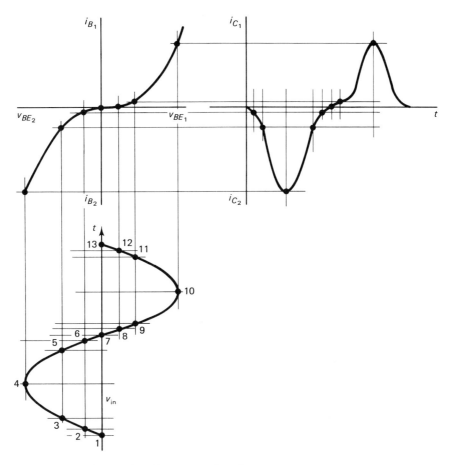

Fig. 10-12 Generation of crossover distortion.

plot. Finally, we project the resulting i_B points to the right and plot them on the time axis set up there using time points corresponding to those chosen initially on the v_{in} wave form. (It is assumed that the i_B values are multiplied by a constant value of h_{FE} in forming the i_C-vs.-t curve. We shall consider this and related matters more carefully later.)

The "dead zones" on the t axis in the i_C-vs.-t curve constitute the phenomenon called *crossover distortion*. In some applications such distortion may not be important; certainly it must be minimized for an audio amplifier. Since the problem arises essentially from the transition of v_{BE} through the turn-on voltage V_γ, a cure would seem to be possible if Q_1 and Q_2 were biased to a point a little above V_γ. This is easily done by adjusting R_1 and R_2 in the circuit of Fig. 10-9 to a point such that with $v_{in} = 0$, Q_1 and Q_2 are both conducting a little. Operation in this mode is referred to as *Class AB*. Figure 10-13 helps explain how it works.

Writing KVL equations for the base circuits of Q_1 and Q_2, we get for Q_1

$$-V_\gamma - v_{in} + v_{be_1} = 0$$

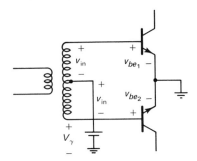

Fig. 10-13 Base circuit of push-pull amplifier.

or (10-3)

$$v_{be_1} = V_\gamma + v_{in}$$

For Q_2 we have

$$-V_\gamma + v_{in} + v_{be_2} = 0$$

or (10-4)

$$v_{be_2} = V_\gamma - v_{in}$$

We now see from Eq. (10-3) that when v_{in} becomes positive, v_{be_1} exceeds V_γ and i_{C_1} will reproduce the positive loop of the input sine wave. Equation (10-4) shows that at the same time, v_{be_2} is either less than V_γ or negative so that Q_2 is not conducting. When v_{in} reverses sign, Eq. (10-3) shows v_{be_1} to be less than V_γ or negative so that Q_1 now becomes nonconducting. Again, v_{be_2} now exceeds V_γ, so that the negative loop of the input sine wave is reproduced by i_{C_2}. There is now no longer a dead zone.

In practice, the mode of operation just described is achieved by simply applying an input signal from a sinusoidal oscillator, observing the output wave form on an oscilloscope, and adjusting the base bias voltage so that the dead zone visible on the output wave is minimized.

10-5 PUSH-PULL CIRCUITS WITHOUT TRANSFORMERS

A disadvantage of push-pull amplifiers of the type shown in Fig. 10-9 is that two transformers are required. Transformers are among the most expensive of components and, as a consequence, we avoid using them whenever possible. To do this in the present case requires that we consider first the essential function performed by each transformer.

It is easy to see that the function of T_1 is to deliver equal, out-of-phase voltages to the bases of Q_1 and Q_2. This function can be performed by a phase-inverter circuit, which consists simply of a transistor amplifier whose voltage gain is adjusted to provide equal in-phase and out-of-phase signals. Two possible phase-inverter circuits are shown in Fig. 10-14.

In network (a), if the output were taken from the collector, the voltage gain would be approximately $-h_{fe}(R_1 + R_2)/h_{ie}$. However, with the output taken at

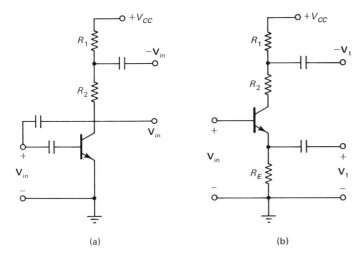

(a)   (b)

Fig. 10-14 Two phase-inverter circuits (base bias networks omitted).

the tap between R_1 and R_2, the gain is

$$\frac{-h_{fe}(R_1 + R_2)}{h_{ie}} \cdot \frac{R_1}{R_1 + R_2} = -h_{fe}\frac{R_1}{h_{ie}}$$

By making $R_1 = h_{ie}/h_{fe}$, the gain will be -1, so that an output signal equal to $-\mathbf{V}_{in}$ is available. Using the input line and this output, both $\mathbf{V}_{in}$ and $-\mathbf{V}_{in}$ are available to drive the push-pull pair.

The network of Fig. 10-14(b) works in a similar manner. This time, however, one output is derived from the emitter-resistor R_E. The gain between input and collector in this circuit is about $-(R_1 + R_2)/R_E$, while the gain between input and emitter is some value less than 1. By making $R_1 + R_2 = R_E$, we obtain two output channels providing equal signals of opposite phase. You should note that these results are only approximate and that, in practice, adjustment of circuit values will be necessary to obtain closely balanced outputs.

The output transformer in Fig. 10-9, T_2, couples the output of the push-pull transistors to the loads. This function can be performed quite well by the complementary-symmetry circuit shown in Fig. 10-15.

Note that Q_1 is an *npn* transistor, whereas Q_2 is a *pnp* type. If the two bases are driven in phase with each other, then when the base potentials rise, Q_1 is made to conduct more and Q_2 conducts less. On downward excursions of the base potentials, Q_1 conducts less and Q_2, more. If both transistors are biased close to cutoff, Class AB operation will result. By using two power sources of equal but oppositely polarized voltages and by careful adjustment of the quiescent operating points of Q_1 and Q_2, the output terminal can be made to rest at ground potential. In this way direct coupling to the load R_L is possible. This is the basis of operation of the output circuits of many operational amplifiers.

Because Q_1 and Q_2 are complementary types, no phase inverter is required; only a single driver transistor is necessary, as should be clear from the discussion above. Unfortunately, Q_3, operating as a Class A driver, must dissipate relatively large amounts of heat. There are two ways to offset this requirement. One is to

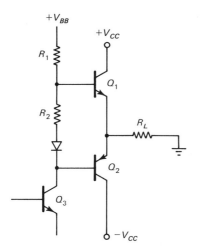

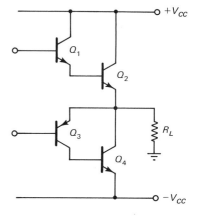

Fig. 10-15 Complementary-symmetry circuit.

Fig. 10-16 Basic quasi-complementary circuit.

substitute complementary types of Darlington-connected transistors for Q_1 and Q_2. The other way is to use what is known as the quasi-complementary circuit. The basic features of this circuit are shown in Fig. 10-16.

In this circuit, Q_1 and Q_2 form a Darlington-connected *npn* array. The new feature of this circuit is the arrangement of the *pnp* transistor, Q_3, and the *npn* type, Q_4.

It is not difficult to see that when the lower transistors, Q_3 and Q_4, are cut off, Q_1 and Q_2 act as an emitter-follower driving R_L. When Q_3 and Q_4 are active, Q_3 provides the necessary polarity inversion to drive Q_4. Moreover, because the emitter of Q_3 is connected to the "hot" side of R_L, the Q_3–Q_4 combination acts as an effective emitter-follower because of the feedback inherent in this connection. As a result, even though Q_2 and Q_4 are not complementary types, the behavior of the circuit is the same as that of a true complementary-symmetry amplifier. The true complementary circuit is used mainly for audio amplifiers rated from 3 to 20 W. For higher power levels, the quasi-complementary circuit is usually found.

A practical example of the quasi-complementary amplifier is shown in Fig. 10-17. Note that the output transistors Q_8 and Q_9 (RCA design uses two type 1C09 transistors) are of the same type. Both are lightly turned on in the quiescent state, so that the output sits at 0 V. DC feedback to Q_2 controls this aspect of operation. The driver transistors Q_4 and Q_5 are complementary types. However, their lower power rating makes them less expensive than comparable, high-power, output transistors would be. Offset voltage between the complementary driven transistors is provided by the three series diodes *D1*, *D2*, and *D3*. Note that Q_4 drives Q_8 as an emitter load while Q_5 drives Q_9 as a collector load. This arrangement provides the large voltage offset required by the quasi-complementary connection of Q_8 and Q_9. Except for these features, the operation of the circuits of Figs. 10-16 and 10-17 is similar.

It is interesting to recall the output circuits used in operational amplifiers.

Power Amplifiers   Chap. 10

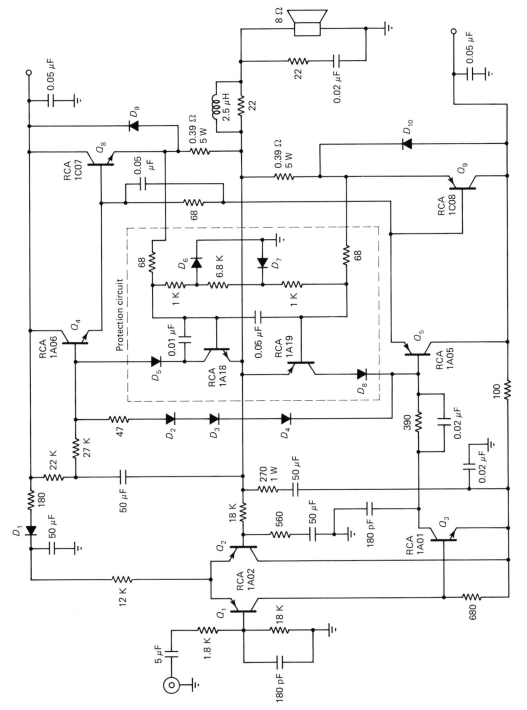

Fig. 10-17 Quasi-complementary-symmetry amplifier. (Courtesy RCA)

That of the 741 is repeated here in Fig. 10-18. (The full circuit diagram is given in Fig. 3-31.)

Details of the operation of the 741 unit were discussed adequately in Chap. 3. At this point it is necessary only to observe that the output transistors Q_{14} and Q_{20} are a complementary pair. Since the operational amplifier is intended to be used in circuits where external feedback is provided by the user, no DC feedback is provided internally to stabilize the operating point. The only internal feedback is that provided by the 30-pF capacitor discussed in Chap. 3.

10-6 DISTORTION ANALYSIS

Distortion is a measure of the extent to which the output wave form of an amplifier fails to be an exact replica of the input. When two or more different frequency components are fed to an amplifier, any nonlinearity in the amplifier may cause the generation of spurious frequency components, which is the source of one form of distortion. This process is called *intermodulation*. Another simpler view of distortion is had simply by feeding a single, pure, sinusoidal input to the amplifier and observing by how much the output differs from a pure sinusoid. It is this latter approach with which we are concerned here. What we wish to develop is a method for determining, from either a predicted or an observed wave form, the harmonic content of the signal.

The standard tool for harmonic analysis is the Fourier series. As you should recall, any periodic wave form that occurs in practical situations can be represented

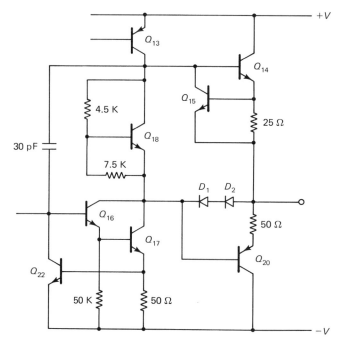

Fig. 10-18 Complementary-symmetry output stage of the 741 operational amplifier.

by a series of the form

$$f(t) = \frac{a_0}{2} + \sum_{n=1}^{\infty} a_n \cos n\omega t + \sum_{n=1}^{\infty} b_n \sin n\omega t$$

where $\omega = 2\pi/T$ and T is the period of the wave. The quantity $a_0/2$ is the average value of the function $f(t)$. Thus if $f(t)$ represents a voltage, $a_0/2$ would be the average, or DC, value of that voltage.

Evaluation of the coefficients a_n and b_n is readily done using well-known integral formulas. However, this is a tedious procedure. Fortunately, a_n and b_n can be approximated fairly well by using Prog. 10-1. To use this program, the first step is to develop an accurate sketch of the variation of $f(t)$ with time, such as that shown for i_C in Fig. 10-20.

```
10   PRINT "FOURIER ANALYSIS OF ARBITRARY WAVEFORMS."
20   PRINT "NOTE: TO OBTAIN N HARMONICS, AT LEAST 2*N EQUALLY SPACED"
25   PRINT "SAMPLES OF F(T) ARE NEEDED. MAXIMUM N = 18."
30   PRINT "USE OF MORE THAN THE MINIMUM REQUIRED NUMBER OF SAMPLE "
35   PRINT "POINTS WILL YIELD MORE PRECISE RESULTS."
40   DIM F(36),A(18),B(18)
60   INPUT "ENTER NUMBER OF SAMPLE POINTS.";M
65   INPUT "ENTER NUMBER OF HARMONICS DESIRED.";N
70   PRINT
75   PRINT "ENTER SEQUENTIAL VALUES OF F(K)."
80   FOR K = 0 TO M
85   PRINT "ENTER F(";K;")": INPUT F(K)
90   NEXT K
100  REM   CALCULATION OF AVERAGE (DC) VALUE
110  FOR K = 0 TO M
120  SUM = SUM + F(K)
130  NEXT K
140  AO = SUM / M
150  PRINT "AVERAGE VALUE EQUALS ";AO
160  FOR I = 1 TO N
165  SUM = F(0)
170  FOR K = 1 TO M
175  PHI = 2 * 3.14159 * I * K / M
176  PRINT "PHI= ";PHI
180  SUM = SUM + F(K) *  COS (PHI)
190  NEXT K
200  A(I) = SUM
210  NEXT I
220  FOR I = 1 TO N
230  SUM = F(0)
240  FOR K = 1 TO M
245  PHI = 2 * 3.14159 * I * K / M
250  SUM = SUM + F(K) *  SIN (PHI)
260  NEXT K
270  B(I) = SUM
275  NEXT I
280  PRINT "DC COMPONENT EQUALS ";AO
285  PRINT : PRINT
290  PRINT "HARMONIC NUMBER"; SPC( 10);"AMPLITUDE"
300  PRINT
310  FOR K = 1 TO N
320  C(K) = (2 / M) *  SQR (A(K) ^ 2 + B(K) ^ 2)
330  PRINT   TAB( 8);K; SPC( 18);C(K)
340  NEXT K
345  PRINT : PRINT
350  PRINT "PERCENT RELATIVE HARMONICS": PRINT
370  PRINT "HARMONIC NUMBER"; SPC( 10);"PERCENT"
375  PRINT
```

Prog. 10-1.

```
380 PRINT    TAB( 8);"1"; SPC( 18);"100"
390 FOR I = 2 TO N
400 P = (C(I) / C(1)) * 100
410 PRINT    TAB( 8);I; SPC( 18);P
420 NEXT I
430 END
```
Prog. 10-1 (cont.)

We assume that the nonlinearity in the wave forms we encounter results only from the nonlinearity of transistor characteristics. This means that as the instantaneous operating point moves along a load line, it will trace through exactly the same points on the way up as it will on the way down. This will result in a symmetrical wave form, however much that wave form fails to be sinusoidal. To see how this works, refer to Fig. 10-19.

If we assume that the transistor whose characteristics are shown operates in the Class B mode with $i_B = 40 \sin \omega t$ milliamperes, we can plot the wave form of i_C, as shown in Fig. 10-20. This is done by assuming values of ωt (in degrees), finding the corresponding values of i_B, entering Fig. 10-19 to find values of i_C, and plotting these values. Table 10-1 aids the process.

Of course, because of the transistor paired with this one, there must be a negative loop exactly similar to the positive one that we have obtained.

A different type of shape occurs if the paired transistors are not identical. In this case, the effect will be to cause the positive peak value to differ from the negative one and the entire 360-degree cycle must be plotted.

Once the wave form of i_C has been determined, our next task is to choose equal intervals along the ωt axis and find the corresponding values of i_C at those points. These values provide the data points that are entered successively in Prog. 10-1. The program produces as output the harmonic amplitudes which, for a harmonic number n, are equal to $\sqrt{a_n^2 + b_n^2}$. In addition, the program also produces values of percent relative harmonics. These quantities are defined as follows: Suppose that the amplitude of the fundamental (first harmonic) term is I_1. Then, if the second-harmonic amplitude is I_2, the percent relative harmonic for the second harmonic is $100 \times I_2/I_1$. This number is denoted by the symbol D_2. Similarly, for harmonic number n, the percent relative harmonic D_n is given by $100 \times I_n/I_1$. The

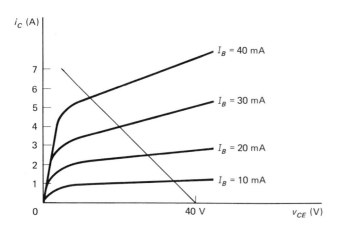

Fig. 10-19 Transistor collector characteristics.

Power Amplifiers   Chap. 10

TABLE 10-1

ωt	i_B (mA)	i_C (A)
0°	0	0
14.5°	10	1
30°	20	2
48.6°	30	4
90°	40	5.3

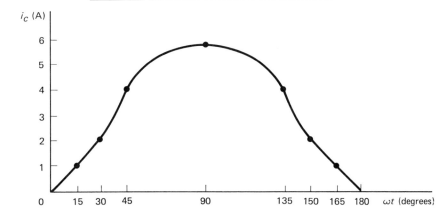

Fig. 10-20   Wave form of i_c.

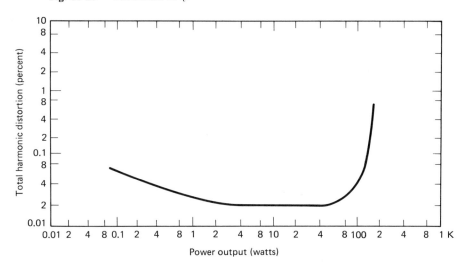

Fig. 10-21   Distortion versus output.

total harmonic distortion is given by the expression

$$D = \sqrt{D_2^2 + D_3^2 + \cdots}$$

Ideally, we would like to have high output power with low distortion. In general, these two requirements are contradictory; increasing the output power usually causes increased distortion. A typical distortion-versus-power curve is shown in Fig. 10-21.

10-7 HEAT SINKS AND TRANSISTOR POWER RATINGS

As we know, a transistor is rated by, among other things, the maximum collector current and the maximum collector-base voltage that it can withstand and by the maximum power that can be developed at the collector-base junction. It is this last parameter upon which we now focus our attention.

The amount of power at the collector-base junction that a transistor can tolerate is determined by the temperature developed at the junction. For a silicon transistor, this temperature limit is about 200°C. We can see intuitively that a transistor having a larger junction area and thus more material in its structure will develop a lower temperature for a given amount of power than one with a smaller junction area. It is also evident intuitively that if some means is provided to carry away the heat produced at the junction, the transistor can tolerate the generation of a larger amount of heat power than would otherwise be the case.

What we must deal with is a heat-flow problem in which heat power developed at the junction gives rise to a high junction temperature. Heat then flows out of the junction from high-temperature regions to the region of lowest temperature. The lowest temperature in the vicinity is that of the surrounding air. This may be either the room temperature or the temperature of the air in an equipment cabinet in which the apparatus using the transistor in question is housed. Either temperature is spoken of as the *ambient temperature*.

The fact that heat flows through material objects from high-temperature regions to low-temperature regions reminds us of the flow of current from high-potential to low-potential through a resistor. This analogy leads naturally to the concept of thermal resistance. It also suggests the analogy between temperature and voltage. Evidently the flow of heat power can be considered to be analogous to the flow of current.

Putting these ideas together leads us to the analog of the heat-flow problem provided by the circuit shown in Fig. 10-22.

The current source P_C represents the heat power developed at the collector-base junction of the transistor. Also, T_j, T_c, T_s, and T_a are the temperatures of the junction, the case, the heat sink, and the ambient region, respectively. It is customary to use the symbol θ for thermal resistance. Accordingly, θ_{jc}, θ_{cs}, and θ_{sa} are the thermal resistances between junction and case, case and heat sink, and heat sink and ambient, respectively. Application of Ohm's law leads directly to the equation

$$P_C(\theta_{jc} + \theta_{cs} + \theta_{sa}) = T_j - T_a \tag{10-5}$$

Since P_C is given in watts and $T_j - T_a$ is measured in degrees Celsius, it is clear from Eq. (10-5) that values of θ have the dimensions of degrees Celsius per watt.

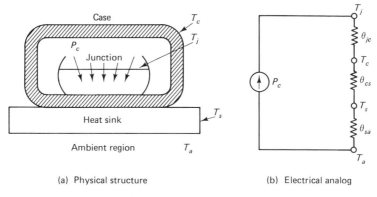

| (a) Physical structure | (b) Electrical analog |

Fig. 10-22 Heat-flow problem and its electrical analog.

Example 10-2

Let us illustrate these ideas with an example. Suppose that $\theta_{jc} = 1.5°C/W$, $\theta_{cs} = 2.2°C/W$, and $\theta_{sa} = 4.05°C/W$. If the ambient temperature is 20°C and the maximum junction temperature must not exceed 175°C, we wish to find the maximum collector power that the transistor in question can tolerate. Using Eq. (10-5), we have

$$P_C(1.5 + 2.2 + 4.05) = 175 - 20$$

This yields $P_C = 155/7.75 = 20$ W.

The maximum junction temperature and the thermal resistances associated with the transistor determine a maximum value of P_C, as we have seen. This means that for a given value of V_{CE}, there is a corresponding value of I_c that cannot be exceeded without our exceeding P_C and therefore damaging the transistor. Since P_C equals the product of V_{CE} and I_c, a locus is defined by the equation

$$I_c = P_C/V_{CE} \tag{10-6}$$

Equation (10-6), when plotted, yields a hyperbola, and when this curve is plotted on the same axes as the transistor characteristics, we obtain the diagram shown in Fig. 10-23.

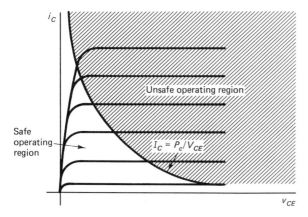

Fig. 10-23 Transistor characteristics showing locus of constant collector power.

Note the safe and unsafe operating regions shown in the figure. Strictly speaking, it is possible for the instantaneous operating point to "zip" through the unsafe zone, providing that it starts and ends in the safe zone and that the transition through the unsafe zone occurs so rapidly that the junction does not get a chance to reach a dangerous temperature. Some switching circuits do operate in this manner, but, in general, such a design tempts fate.

REFERENCES

1. Robert Boylestad and Louis Nashelsky, *Electronic Devices and Circuit Theory*, 3rd ed., (Englewood Cliffs, N.J.: Prentice-Hall, Inc., 1982).
2. RCA Corporation, *Audio Amplifier Manual APA 551*, (Somerville, N.J.: RCA Corporation, 1979).

PROBLEMS

10-1. In the circuit shown, R_B is so chosen that the DC base current I_B is 100 mA. Idealized collector characteristics are shown.
   (a) What is the maximum possible undistorted collector current? Give the peak-to-peak value, assuming i_S is sinusoidal.
   (b) Find the power furnished by the DC source under the conditions of part (a).

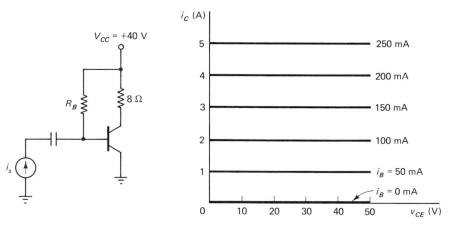

10-2. **(a)** Sketch the maximum collector-dissipation curve for $P_{max} = 50$ W on the characteristics of Prob. 1.
   (b) If, for the circuit of Prob. 1, I_B is maintained at 100 mA and V_{CC} is increased to 50 V, determine whether the limit on collector dissipation will be exceeded.

10-3. **(a)** Find I_m for maximum undistorted AC current in R_L.
  (b) Find the total power dissipated in R_L.
  (c) Find the power dissipated by the collector-base junction.

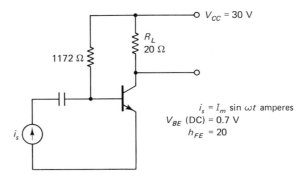

$i_s = I_m \sin \omega t$ amperes
V_{BE} (DC) $= 0.7$ V
$h_{FE} = 20$

10-4. The class-A amplifier shown is to be designed for maximum P_L with sinusoidal wave form and with the smallest possible input signal. Three watts is to be delivered to the 5-Ω load. The output transformer has an efficiency of 75%.
  (a) Choose the proper transistor for the design.

1. $I_{c\max} = 500$ mA, $V_{c\max} = 50$ V, $P_{c\max} = 4$ W, $h_{FE} = 150$
2. $I_{c\max} = 2.0$A, $V_{c\max} = 40$ V, $P_{c\max} = 8$ W, $h_{FE} = 80$
3. $I_{c\max} = 1.5$A, $V_{c\max} = 60$ V, $P_{c\max} = 6$ W, $h_{FE} = 100$

  (b) Determine V_{CC}, R_E, and n_1/n_2.

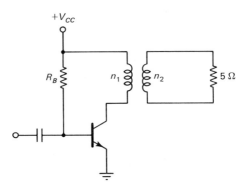

10-5. A class A, transformer-coupled, power amplifier is to deliver 4 W to a 4-Ω load. Also, $V_{CC} = 12$ V. Assume ideal collector characteristics, ideal transformer, and a quiescent operating point chosen for symmetrical swing. Find (a) transformer turns ratio, (b) operating point, (c) quiescent (standby) power requirement, and (d) peak collector current.

10-6. The circuit shown is to operate as an ideal, Class A power amplifier. Undistorted power delivered to the 8-Ω load is to be 3 W and I_{CQ} is to be the minimum permissible value.

(a) Find I_{CQ}.

(b) Find n_1/n_2.

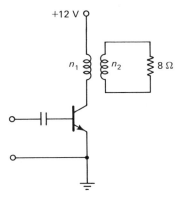

+12 V

n_1 n_2 8 Ω

10-7. A transistor dissipates 125 mW at its collector junction. The total thermal resistance comprises the resistance between junction and interior of the case and the resistance between the case interior and the part of the case that is in contact with either a heat sink or the ambient region. That is, $\theta_T = \theta_{jc} + \theta_{cs} = 0.6°C/mW$. If the ambient temperature is 40°C, what is the junction temperature?

10-8. Addition of a heat sink lowers θ_T to 0.45°C/mW. For the conditions of Prob. 7, find the junction temperature.

10-9. If the maximum allowable junction temperature for the transistor of Probs. 7 and 8 is 125°C, find the maximum allowable collector dissipation for each case.

10-10. A transistor has $\theta_{jc} = 4°C/W$ and a maximum junction temperature of 150°C. The ambient temperature is 45°C. The transistor is used with a mica washer ($\theta_{cs} = 1°C/W$) and a heat sink ($\theta_{sa} = 3°C/W$).

(a) What is the maximum allowable power dissipation in the transistor?

(b) When dissipating this amount of power, what is the case temperature?

10-11. For Prob. 4, determine the area of an aluminum heat sink if maximum junction temperature is to be 182°C, ambient temperature is 22°C, $\theta_{jc} = 20°C/W$, and $\theta_{cs} = 10°C/W$. The heat sink area in square inches is given by $A = 450/\theta_{sa}^2$.

10-12. The collector of the transistor in the circuit of Prob. 3 is coupled by means of a large capacitor ($X_c \approx 0$) to a load resistor of 20 Ω. Calculate the AC power in this added load resistor. What is the efficiency in this case?

10-13. **(a)** Using the transistor ratings, choose a quiescent operating point that will allow maximum undistorted output in class A operation.

(b) If the junction temperature can reach 175°C and the ambient temperature is 25°C, find the total thermal resistance required from junction to ambient.

(c) Find the maximum output power and the turns ratio required if $R_L = 10\ \Omega$.

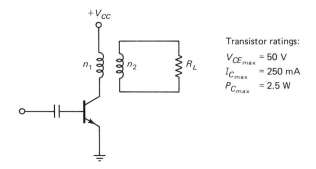

Transistor ratings:
$V_{CE_{max}} = 50$ V
$I_{C_{max}} = 250$ mA
$P_{C_{max}} = 2.5$ W

10-14. A class B amplifier is shown. Meter A is a DC ammeter that reads 0.75 A when the amplifier delivers power to the 40-Ω load. The turns ratio of each *half* of the transformer primary to the secondary is n_1/n_2. Find the power delivered to the load.

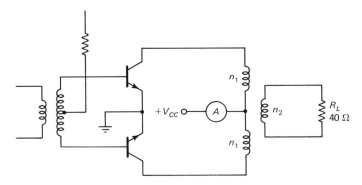

10-15. A push-pull, class B amplifier is to deliver maximum power to a 10-Ω load. The *npn* transistors to be used have the following ratings: $P_{C_{max}} = 4$ W, $V_{CE_{max}} = 40$ V, and $I_{C_{max}} = 1$ A. Find V_{CC}, peak collector current, transformer turns ratio, and power delivered to the load.

10-16. A class B power amplifier is to operate from a 12-V supply. Determine (a) the required minimum collector voltage rating for the transistors, (b) the average current taken from the power supply for a 20-W sinusoidal output, (c) the load resistance seen by each half of the transformer, and (d) the minimum rated collector dissipation for each transistor for an unidentified signal wave shape of any amplitude up to the maximum ability of the amplifier.

10-17. The circuit shown is to function as a class B amplifier. The transistors have the following maximum ratings: $I_{C\text{max}} = 2$ A, $V_{CE\text{max}} = 80$ V, $P_{C\text{max}} = 24$ W (instantaneous). The amplifier is to be designed for maximum power output with maximum efficiency under sinusoidal conditions.

(a) Determine the maximum power output.

(b) Determine the greatest average collector dissipation under these conditions.

(c) Determine the maximum instantaneous collector dissipation.

(d) Determine the load resistance seen from the primary.

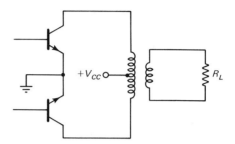

10-18. The complementary-symmetry amplifier is to be analyzed; $V_{CC} = 24$ V and $R_L = 8$ Ω. The value of V_o is such that the transistors are biased to minimize crossover distortion.

(a) When the sinusoidal signal V_s is applied, the DC current drawn from each supply is 0.5 A. Find the power in the load and the efficiency under these conditions.

(b) What is the maximum undistorted power that this circuit can deliver to the load?

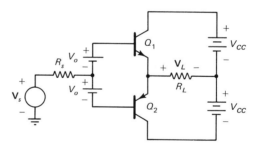

10-19. A complementary-symmetry power amplifier operating in class B mode has the circuit shown in Prob. 18.

(a) Find an expression for the average load power in terms of I_{dc} and R_L.

(b) What is the maximum theoretical efficiency obtainable from a complementary-symmetry amplifier of this type? Assume ideal transistor characteristics.

CHAPTER 11

Digital Memory Devices

In this chapter we consider certain applications of the basic digital logic devices that we considered in Chap. 2. We shall see that by suitably interconnecting some of these devices, we can create new structures whose outputs depend not only upon the inputs presently applied to them but also upon the outputs that existed prior to the application of these inputs. In other words, these new structures appear to remember the last output. Such structures fall within the general category of *memory devices*.

We shall first examine the connection of basic NOR gates in a feedback circuit to form the simplest type of digital memory device—the *RS* flip-flop. Next we shall see how the *RS* flip-flop is modified to overcome certain inherent limitations, thus yielding the *JK* flip-flop. Further modifications of the *JK* flip-flop then yield special-purpose flip-flops—the *T* and *D* types.

For many applications, when an array of flip-flops is interconnected to form a larger system, it is important that switching occur only at specified instants. We shall consider how this is accomplished through the use of synchronizing clock pulses.

Having a variety of basic flip-flops at our disposal, we are in a position to connect them in various arrays to create counters and registers, which are basic building blocks in forming computers and microprocessors.

Up to this point, the integrated circuits we have considered, whether logic gates as in Chap. 2 or operational amplifiers as in Chap. 3, are of the small-scale integrated type. When large numbers of gates are fabricated and interconnected on a single chip to form registers, the resulting type of integrated circuit is referred to as having *medium-scale integration* (MSI). Arrays of registers on a single chip, forming memories and microprocessors, are said to have *large-scale integration* (LSI).

11-1 THE *RS* FLIP-FLOP

Figure 11-1 shows an interconnection of NOR gates that form an *RS* flip-flop. Before analyzing this device, we must take note of an essential feature of all flip-flops—not just the *RS* flip-flop: After inputs have been applied, the two outputs *must* come to rest in such a way that one is at logic 1 and the other is at logic 0. In other words, one output is said to be the *complement* of the other. For that reason, one output terminal is marked Q, whereas the other one is marked $\overline{Q}$. It follows that when another set of inputs is applied, the next output configuration (or state) must proceed from one in which the outputs were *initially complementary*. For that reason, we must always analyze the flip-flop by assuming that the Q and $\overline{Q}$ outputs are in a complementary condition before inputs are applied.

Let us trace the operation of the *RS* flip-flop of Fig. 12-1 by assuming that Q is at logic 0 and $\overline{Q}$ is at logic 1. Suppose also that both R and S are at logic 0. The inputs to the upper NOR gate are 0 and 1. Hence its output is 0. This causes the inputs to the lower gate to be 0 and 0, yielding an output of 1. We see then that Q will be 0 and $\overline{Q}$ is 1, which is the condition assumed initially. From the symmetry of *RS* flip-flop, we can see at once that if initially Q had been at logic 1, whereas $\overline{Q}$ was at logic 0, the application of logic 0 to both R and S would leave Q and $\overline{Q}$ unchanged.

Once again, suppose Q is at 0 and Q is at 1. Further, suppose R is at logic 0 and S is at logic 1. The inputs to the upper gate will be 0 and 1, yielding an output of 0. The inputs to the lower gate are 0 and 1, also yielding an output of 0. This means that $\overline{Q}$ must change from 1 to 0. In turn, this changes the inputs to the upper gate to 0 and 0. Therefore, Q must change from 0 to 1. We must check the lower gate again, since its inputs have now changed. We see that these inputs are now 1 and 1, so that $\overline{Q}$ remains at 0. From the symmetry of the flip-flop, we see that if Q had been initially at 1, whereas $\overline{Q}$ was at 0, then with R at 1 and S at 0, Q would switch to 0 and $\overline{Q}$ would switch to 1.

In this last analysis we had to trace the pattern of switching through the upper gate, the lower gate, and back again through the upper gate until we found a set of conditions that produced no further changes. Something of the sort actually happens, so that the change from one output state to the next requires a definite amount of time.

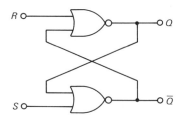

Fig. 11-1 Basic *RS* flip-flop.

The final condition that we have to consider is that in which both inputs, R and S, are set to 1. If Q is initially at 1 and $\overline{Q}$ is at 0, we see that the inputs to the upper gate are 1 and 0, so that Q must switch to 0. The lower gate will now have inputs of 1 and 0, so that its output must be 0. No further change occurs at the upper gate. Thus we have arrived at a state where *both* Q and $\overline{Q}$ outputs are 0. But this contradicts our fundamental requirement that Q and $\overline{Q}$ be complementary. Such a condition cannot be accepted, so we must stipulate that *inputs R and S must never have simultaneous values of logic* 1. It is up to the user to so arrange the circuit operation that this forbidden condition never occurs.

The behavior of the *RS* flip-flop is summarized in Table 11-1. Note that only the Q outputs are listed; it is understood that the $\overline{Q}$ outputs are complementary. The X entries indicate that forbidden inputs have been applied.

It is easy to remember the information presented in Table 11-1 by recognizing that the symbols S and R stand for *set* and *reset*. The term *set* implies that the Q output is switched to the logic 1 state. *Reset* implies that Q is switched to logic 0. Thus Q is *set* when $S = 1$ and $R = 0$; Q is *reset* when $S = 0$ and $R = 1$. With $S = 0$ and $R = 0$, the flip-flop does not switch at all but merely maintains its already existing output condition.

11-2 THE *JK* FLIP-FLOP

Figure 11-2 shows a modified *RS* flip-flop. Its operation can be understood if we assume that inputs are applied to the *JK* terminals while the outputs Q and $\overline{Q}$ are in either of their two possible initial states. We then determine the resulting R and S inputs. From what we now know about the *RS* flip-flop, we can predict the new output state that will then arise. The procedure is summarized in Table 11-2. You can verify the table entries.

The important feature revealed by Table 11-2 is that regardless of the inputs applied to the J and K terminals, there is no condition in which the R and S inputs are both at logic 1. Thus the forbidden condition on the *RS* flip-flop is avoided. Note that with J and K both at logic 1, the output Q becomes 1 if it is initially 0. If Q is initially 1, it becomes 0.

TABLE 11-1 STATE TRANSITIONS OF THE *RS* FLIP-FLOP

Inputs R	S	Present State Q	Next State Q
0	0	0	0
0	0	1	1
0	1	0	1
0	1	1	1
1	0	0	0
1	0	1	0
1	1	0	X
1	1	1	X

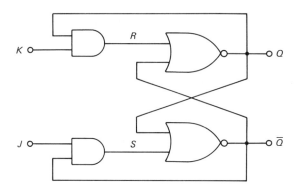

Fig. 11-2 The *RS* flip-flop modified to form a *JK* flip-flop.

The action of the *JK* flip-flop when both inputs are at logic 1 is called *toggling*. That is, whatever the initial state of *Q*, it will switch to the complementary state.

We might wonder what would happen if the *J* and *K* inputs were maintained at logic 1 for a long time. If we take the trouble to trace the operation through the four gates of Fig. 11-2, we find that, in fact, rapid alternation of the output between the 0 and 1 levels will occur. The speed of this alternation will depend on the propagation time of the gates.

In general, such behavior is undesirable. After all, the reason we are interested in flip-flops at all is because they can be set to some output condition that can be retained for some extended period of time. A flip-flop that switches its output uncontrollably is not useful as a memory element.

11-3 CLOCKED FLIP-FLOPS

The oscillation of the *JK* flip-flop in its toggle mode is not the only problem encountered in using flip-flops of the types we have examined up to this point. We have only to recall the way the *RS* flip-flop passed through an intermediate state before settling into its final state when *Q* was initially 0 and inputs of *R* = 0 and *S* = 1 were applied. In a system where a number of flip-flops and gates are interconnected, the situation becomes even worse. This is because, as the system executes multiple transitions from state to state, the inputs to the various flip-flops—and, therefore, their outputs—depend on which signals arrive at what

TABLE 11-2 STATE TRANSITIONS OF THE *JK* FLIP-FLOP

J	*K*	*Present Q*	*R*	*S*	*Next Q*
0	0	0	0	0	0
0	0	1	0	0	1
0	1	0	0	0	0
0	1	1	1	0	0
1	0	0	0	1	1
1	0	1	0	0	1
1	1	0	0	1	1
1	1	1	1	0	0

times. Thus, the intermediate states as well as the final state at which the system will arrive, are uncertain. Such a condition is referred to as a *race*.

To avoid these anomalies, it is common to make the flip-flops in a digital system switch only when commanded by a clock pulse. Between clock pulses, the flip-flops remain inactive. During these inactive periods, signals can propagate through whatever gates may be in the system external to the flip-flops.

Figure 11-3 shows a form of clocked RS flip-flop that differs somewhat from that shown in Fig. 11-2. The difference lies not only in the provision that is made for the introduction of a clock pulse but also in the fact that more AND gates are used. The reason for this seemingly unnecessary complexity is that we are building up the component elements of a type of integrated-circuit flip-flop that is currently manufactured; it will be easier to understand that device if we break it up into simpler subsections like this one.

We begin our analysis of the circuit of Fig. 11-3 by noting that when the terminal marked X is at logic 1, the outputs of NOR gates 5 and 6 will not change, regardless of the inputs applied to the terminals marked R and S. This is easy to verify by assuming that Q is initially at logic 1, whereas $\overline{Q}$ must be at logic 0. We then observe that the output of AND gate 3 must necessarily be logic 1, whereas the outputs of AND gates 1 and 2 must be logic 0. These conditions occur regardless of the logic levels of R and S. The result must be that Q remains at logic 1 and $\overline{Q}$ stays at logic 0. From the symmetry of the structure, we see that if Q had been initially at logic 0 while $\overline{Q}$ had been at logic 1, these initial states would have remained unchanged. Thus when X is at logic 1, the flip-flop is in a hold condition.

If X is set to logic 0, the output of AND gates 2 and 3 will always be at logic 0. Suppose that Q is initially at logic 0, whereas $\overline{Q}$ is at logic 1. Suppose further that R is at 0 and S is at 1. It is clear that under these conditions, all four AND gates deliver outputs of logic 0. This implies that both Q and $\overline{Q}$ produce outputs equal to logic 1, a condition that is forbidden. If we agree, however, that such a condition is what happens and pursue the process further, we see that AND gate 4 will deliver a logic 1 output to NOR gate 6. Hence $\overline{Q}$ switches to logic 0. This

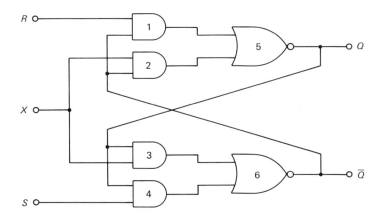

Fig. 11-3 Modified RS flip-flop.

is the final, stable state of the flip-flop. Again, arguing on the basis of symmetry, it is easy to see that if R were set to 1 and S to 0, the flip-flop would arrive at the state where Q is at logic 0 and $\overline{Q}$ is at logic 1.

It is worthwhile to note that, starting from the time when X switches from logic 1 to logic 0, the time required for the flip-flop to arrive at its final state must be the sum of the times required for the signal to propagate through (1) AND gates, (2) one NOR gate, (3) AND gate again, and (4) one NOR gate. The total time may be in the order of 20 to 40 ns.

If both R and S inputs are at logic 1 with X still at logic 0 and if Q is at 1 and $\overline{Q}$ is at 0, AND gates 1, 2, and 3 will produce outputs of logic 0. Only AND gate 4 produces a 1. Hence $\overline{Q}$ stays at 0, whereas Q maintains an output of 1. Once again, the symmetry of the structure indicates that initial values of $Q = 0$ and $\overline{Q} = 1$ would also be maintained if R and S both have logic levels of 1. Evidently $R = 1$, $S = 1$ is the hold input for this flip-flop.

If both R and S are set to logic levels of 0, with $X = 0$, we see that both Q and $\overline{Q}$ would have to switch to 1. This is a forbidden condition, although it is the opposite of the one imposed upon the circuit of Fig. 11-1, where the simultaneous occurrence $R = 1$, $S = 1$ was the problem.

Now consider the circuit shown in Fig. 11-4. The purpose of this circuit, among other things, is to serve as a front end to our RS circuit of Fig. 11-3 to convert that circuit to JK operation. Notice that when X is at the logic-0 level, the output of NAND gates 7 and 8 is logic 1. With X at the logic-1 level, we shall assume logic corresponding R and S outputs.

It is clear that $R = 0$ only when $\overline{Q}$, J, and X are all at the logic-1 level. When this occurs, it is clear that Q has to be at logic 0, and therefore, regardless of the logic level of K, S must be at logic 1. Similarly, we conclude that $S = 0$ only when Q, K, and X are all at the logic-1 level and that $\overline{Q}$ must then be 0. Hence, $R = 1$. We note also that when both J and K are 0, both R and S become 1. We summarize these conclusions in Table 11-3.

If we now connect the front-end circuit of Fig. 11-4 to the RS flip-flop of Fig. 11-3, we have the 7473 master-slave JK flip-flop shown in Fig. 11-5.

The input labeled X that appeared in both Fig. 11-3 and Fig. 11-4 has been relabeled $\overline{CP}$ in Fig. 11-5. This significance of this label will be discussed later. The

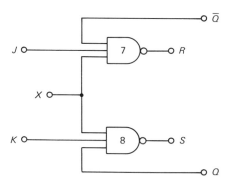

Fig. 11-4 Front-end circuit for JK flip-flop.

TABLE 11-3 RS OUTPUTS DERIVED FROM JK OUTPUTS

Q	$\overline{Q}$	J	K	R	S
1	0	0	0	1	1
1	0	0	1	1	0
1	0	1	0	1	1
1	0	1	1	1	0
0	1	0	0	1	1
0	1	0	1	1	1
0	1	1	0	0	1
0	1	1	1	0	1

purpose of the $\overline{RD}$ input, which did not appear in the other figures, will also be discussed later. For the present, we shall assume that $\overline{RD}$ is fixed at logic 1.

When the $\overline{CP}$ input is at logic 1, R and S are determined by the existing logic levels of Q, $\overline{Q}$, J, and K as specified in Table 11-2. But, as we saw in connection with the circuit of Fig. 11-3, the outputs of NOR gates 5 and 6 cannot change. When $\overline{CP}$ switches from 1 to 0, the existing logic states of the R and S inputs pass through AND gates 1 and 4. During this transition R and S are switched to logic 1, the hold signal for the NOR-gate flip-flop. However, before the hold signal arrives, the flip-flop first switches to the state called for by the inputs just received from AND gates 1, 2, 3, and 4.

We thus see that NAND gates 7 and 8 act to process the information received on the J and K lines while the $\overline{CP}$ line is at logic 1 and to deliver it to the RS inputs of the flip-flop. The flip-flop, in turn, switches to its next state only on the transition of $\overline{CP}$ signal from 1 to 0. Since this switching operation is already in progress, the $R = 1$, $S = 1$ hold signal generated by NAND gates 7 and 8 arrives too late to stop the transition. However, it does inhibit any further transitions. Details of the process are left to you.

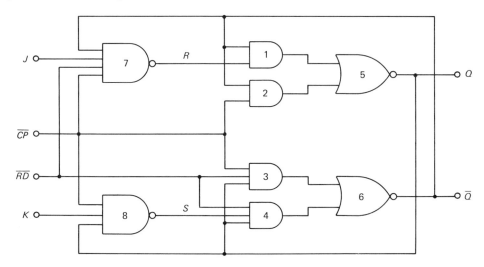

Fig. 11-5 The 7473 master-slave JK flip-flop.

NAND gates 7 and 8 are referred to as the *master* section of the system. That is, while $\overline{CP}$ is at logic 1, they prepare the input data for the flip-flop. On the transition of $\overline{CP}$ from 1 to 0, the data are processed by the *slave* section to yield the next output state. The entire system is called a *clocked, master-slave, JK flip-flop*.

To complete our analysis of the system of Fig. 11-5, we consider the function of the $\overline{RD}$ input. It is clear that with $\overline{RD}$ at logic 0, R and S are at logic 1; the outputs of AND gates 3 and 4 are 0, so that $\overline{Q}$ is forced to the logic-1 level. In turn, this forces at least one of AND gates 1 and 2 to deliver a 1 to NOR gate 5, thus forcing Q to the 0 level. Thus the $\overline{RD}$ input, when at the 0 level, overrides the $\overline{CP}$ and RS inputs and forces the flip-flop to the reset, or $Q = 0$, $\overline{Q} = 1$, condition. When $\overline{RD}$ is at the logic-1 level, operation proceeds as described previously. The $\overline{RD}$ input is referred to as an *asynchronous reset*.

11-4 GLITCHES

In our discussion of the circuit of Fig. 11-3, we saw that there was a brief period in which the clocked RS flip-flop, in switching from the reset to the set state, passed through a condition in which both Q and $\overline{Q}$ outputs were at the forbidden logic-1 state. This is one form of anomalous behavior that is referred to by the technical slang term of *glitch*. In general, glitches are the more-or-less unexpected and usually quite narrow pulse outputs generated by a digital device as it switches from one state to another.

If Q and $\overline{Q}$ were connected to the JK inputs of another flip-flop, the second flip-flop, if it were asynchronous, would toggle upon receiving this anomalous $J = 1$, $K = 1$ input. The propagation of signals arising from such a glitch can have all sorts of unexpected consequences in a complex system. The detection and cure of such problems can require much time and effort.

In the present case, if an identical JK flip-flop were cascaded with the clocked, JK flip-flop of Fig. 11-5, the glitch problem would be avoided. This is true because NAND gates 7 and 8 are disabled during the time that NOR gates 5 and 6 are switching into their final, stable states. This illustrates the usefulness of the master-slave, clocked structure.

11-5 *T* AND *D* FLIP-FLOPS

Both J and K inputs of a JK flip-flop may be wired together to form a single input. If this input is at the logic-1 level, then after each clock pulse, Q and $\overline{Q}$ will assume logic levels that are the complements of those that prevailed before the clock pulse came along. In this manner we have created a toggle, or T flip-flop. If this single input, denoted by the T input, is at the logic-0 level, then no changes take place and the flip-flop merely holds onto its existing state.

It is customary to indicate a flip-flop in a system diagram by drawing a box with inputs labeled according to function. Making use of this convention, we have in Fig. 11-6 a diagram showing how a JK flip-flop can be wired to form a D flip-flop.

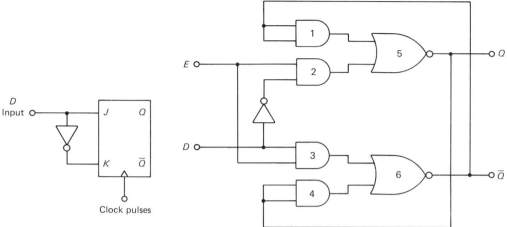

Fig. 11-6 A *JK* flip-flop
wired as a *D* flip-flop.

Fig. 11-7   Latch circuit.

The designation *D* applied to this flip-flop implies that it is a *delay* flip-flop. That is, a logic level applied to the *D* input is reproduced at the *Q* output *after the next clock pulse*. This flip-flop may also be regarded as a *data* flip-flop because a signal (1 or 0) applied to the *D* input is stored by the device after the next clock pulse has occurred.

11-6 LATCHES

We might wish to store a single piece of data, such as a logic level of 1 or 0 (such a piece of data is called a *bit*), over an extended period of time. If, during this time, the logic level on the data line is changing, then a *D* flip-flop will not be suitable for the purpose because after each clock pulse, *Q* assumes the logic level that last existed on the *D* terminal.

A solution to this problem is simply to make use of the asynchronous *RS* flip-flop of Fig. 11-1 wired in a manner similar to the *D* flip-flop of Fig. 11-6. The trouble with this approach is that while the clock pulse no longer determines when switching will occur, the device still requires some sort of additional input logic circuitry to allow us to save data even while the input is changing. Such a config-uration is called a *latch*. The circuit of one section of the 7474 quad latch is shown in Fig. 11-7.

To understand the operation of the latch, assume that the *E* line is at logic 1. AND gates 2 and 3 will then be enabled, and the logic level of data line *D* will appear at the input of NOR gate 6. Its complement will be transmitted via AND gate 2 to NOR gate 5. It is easy to see that NOR gates 5 and 6 are wired to form an *RS* flip-flop. The inverter between AND gates 2 and 3 insures that if *D* = 1, the flip-flop will set, making *Q* = 1. Conversely, if *D* = 0, the flip-flop resets, making *Q* = 0. Moreover, there is no possibility of producing the forbidden *R* 1, *S* = 1 input. When *E* is at logic 0, in effect *R* = 0, *S* = 0, and the flip-flop holds its existing state. The E line thus furnishes an enabling signal that determines when new data are to be accepted, which permits long-term storage.

At first glance AND gates 1 and 4 appear to have no function, since they simply reproduce the levels of Q and $\overline{Q}$ at the NOR gate inputs. Their purpose is to provide the same propagation time for the Q and $\overline{Q}$ signals as AND gates 2 and 3 provide for the D and E signals.

11-7. COUNTERS

A toggle, or T flip-flop, can easily be made to serve as a divide-by-two unit. To understand this idea, consider the arrangement shown in Fig. 11-8, where a JK flip-flop is shown connected as a T device. As each clock pulse switches from 1 to 0, Q begins a transition from its prior state to the complementary one. Thus we see that one positive pulse is generated at Q for every two clock pulses. Alternatively, we can say that the logic level of Q represents in binary form the number of clock pulses that have occurred since some reference time when Q was initially at the 0 level. Because Q can have only two values, it can register only zero pulses or one pulse.

Suppose now that the Q output of a T flip-flop feeds the clock terminal of a second T flip-flop, as shown in Fig. 11-9. We see that the output of flip-flop B changes at half the frequency, as does that of flip-flop A. In fact, if we take simultaneous readings of the logic levels of Q_A and Q_B in the intervals between count pulses, we can list the results, as shown in Table 11-4.

It is clear that the binary numbers represented by Q_A and Q_B, arranged as shown in the table, begin with 0 and are incremented by 1 after each clock pulse until the number 3 is reached. The next increment after 3 returns the count to 0.

The flip-flops connected as shown in Fig. 11-9 thus form a two-bit binary counter. Obviously, more bits could be added in the same manner. Each bit has a binary weight equal to 2^k, where k is an integer denoting the bit position counting from right to left, with $k = 0$ for the rightmost, or least significant, bit. We also see that the greatest number an n-bit binary counter can reach is $2^n - 1$.

Counters of the type illustrated in Fig. 11-9 are called *ripple counters*. This is because each flip-flop *begins* its switching transition only upon the downswing

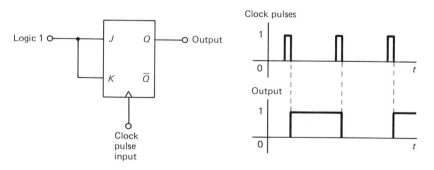

Fig. 11-8 T-connected flip-flop and associated wave forms.

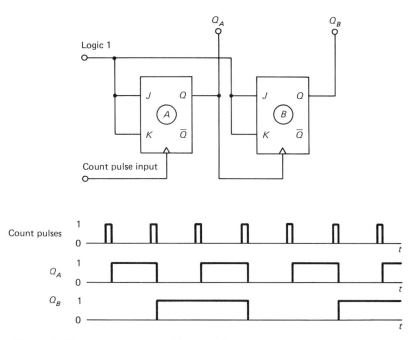

Fig. 11-9 Two-stage counter and its wave forms.

of the input to its clock terminal, which is derived from the output of the preceding flip-flop. Thus, for example, a four-bit counter switching from 1 1 1 1 to 0 0 0 0 would do so in row-of-dominoes fashion. Thus the state transition made by a given flip-flop is delayed by the sum of the propagation times of the various gates in the several flip-flops preceding the flip-flop of interest. Suppose that in each flip-flop this propagation time equals 35 ns. Then the last flip-flop of this four-bit counter would arrive at the 0 logic level $4 \times 35 = 140$ ns after the downswing of the clock pulse at the first flip-flop. In a binary counter having a greater number of bits, this "rippling through" may be undesirable.

To overcome the time-delay problems associated with ripple counters, we can make use of *synchronous counters*. A three-bit synchronous counter is shown in Fig. 11-10. We see that the input count pulses are connected to the clock terminals of all three flip-flops. Thus whatever transitions are to occur will all take place on

TABLE 11-4 SUCCESSIVE
OUTPUTS OF TWO-STAGE
COUNTER

Q_B	Q_A
0	0
0	1
1	0
1	1
0	0

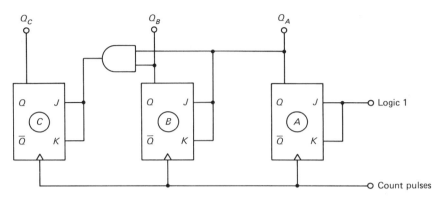

Fig. 11-10 Three-bit synchronous counter.

the downswing of the count pulse. This makes all transitions occur simultaneously. Since the J and K terminals of each flip-flop are connected together to act as a T input, we see that when these inputs are at the 1 level, the flip-flops will toggle; when the inputs are at the 0 level, the flip-flops merely hold. Flip-flop A toggles on every count pulse. It is left as a problem for you to show that, starting from 0, the three-bit counter increments to a total binary count representing 7 and then resets to 0. Finally, we note that if any anomalous internal transitions occur, they do so *between* count pulses. As a result, conditions at each pair of JK terminals are stable when the next count pulse comes along. Naturally, the time required for stable inputs to exist sets an upper limit on counting speed.

Design procedure for synchronous counters is another interesting topic that lies outside the scope of this book.

To conclude our discussion of counters, we note that many varieties of counter exist that are fabricated on a single IC chip. Some typical examples of *TTL* counter chips are listed in Table 11-5.

The divide-by-twelve unit has internal logic connections to allow this somewhat curious operation. (Note that $12 \neq 2^k$, where k is an integer.) BCD counters have internal logic connections so that they count only from 0 to 9. A *carry* feature is provided so that additional decade counters can be connected to register 10s, 100s, and so on. The counters whose structures we have studied all count *up*. That is, the indicated total count increases as more count pulses are received. Counters can be made to count *down*, or decrement. Some counters provide a control-signal input that allows either the up- or the down-count mode of operation to be selected.

TABLE 11-5 SOME *TTL* COUNTER TYPES

Device Number	Description
7492	Divide-by-twelve ripple counter
74160	Synchronous BCD decade counter
74161	Synchronous four-bit binary counter
74S169	Synchronous four-bit binary up/down counter
74192	Presettable BCD decade up/down counter

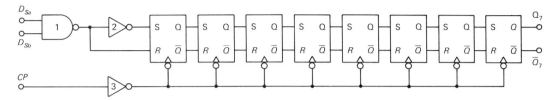

Fig. 11-11 The 7491A shift register.

A presettable counter allows the user to set up a predetermined count via appropriate input terminals. The counter then counts up or down from this point, and a *carry* or *borrow* output indicates that a desired number of count pulses has been registered.

11-8 REGISTERS

An array of flip-flops forms a register. A register is used to store and, in conjunction with other logic elements, to process digital data. Each flip-flop in the register stores one bit of data. The number of flip-flops thus determines the number of bits in the register or the size of the binary *word* that can be stored. An eight-bit word is called a *byte*.

One important use of a register is in the serial reception or transmission of data. A register used for such a purpose is called a *shift register*. Figure 11-11 shows the structure of the 7491A shift register.

The 7491A is a particularly simple sort of shift register. It provides only for data entry via NAND gate 1 and for shifted serial output data bits (delayed by eight-clock cycles) from terminal Q_7. The complementary output $\overline{Q}_7$ is provided for whatever purpose the user might have. When both D_{Sa} and D_{Sb} are at logic 1, with the aid of inverter 2, $S = 1$, $R = 0$ inputs are applied to the first flip-flop. After the first clock pulse, this flip-flop is set (that is, $Q = 1$, $\overline{Q} = 0$). As the next seven clock pulses occur, this data bit is shifted through the successive flip-flops, ultimately arriving at the last one.

No provision is made for clearing all flip-flops initially. This can be done only by shifting in eight 0s. Such an input occurs when at least one of the input lines, D_{Sa} and D_{Sb}, is at logic 0.

More elaborate shift registers are available. These provide such features as the option of shifting either to the right or to the left, a provision for external reset allowing all flip-flop Qs to be set to logic 0, and the possibility of parallel loading. In parallel loading, all eight flip-flops of an eight-bit shift register would be loaded with data bits simultaneously. Similarly, some shift registers also allow all bits to be transmitted to the outside world on a parallel basis. One can easily picture an application for such a feature in conjunction with a device such as a microprocessor that is to receive serial data. Such a string of data bits might be transmitted via telephone wires. Since a telephone line is a single-channel data source, the bits are necessarily received one at a time and shifted into the register. When all bits in the word have been received, they may then be transmitted simultaneously to the microprocessor on, for instance, eight lines if the word length is eight bits.

11-9 STATIC RAMs

It should be clear from the foregoing discussion that any register can be used to store a binary word for some period of time. We are thus led to the idea that many words could be stored in an assembly of registers. Such an assembly is called a *memory*.

The chief application of memories of this type is in the organization of computers. In such applications, memories are usually of two types—the *random-access memory*, or RAM, and the *read-only memory*, or ROM. In both types of memory, two important problems arise. One of these is the problem of selecting a single desired register from the entire array of registers. The other problem is that of obtaining all bits of the desired word simultaneously and upon command. We have already seen that by the mechanism of parallel transfer, all bits may be transmitted simultaneously. All that remains is that we have some auxiliary, external register to store these bits after they are read from the memory. Alternatively, this auxiliary register can also serve as the data source for bits to be simultaneously loaded, or written, into the memory. This external data register is called a memory *buffer* register.

The problem of selecting a single register in a memory is spoken of as selecting a single *address*. If, for example, a memory consists of 16 four-bit registers, then we could number each register successively from 0 to 15. By specifying, say, address number 11, we could read from or write into the eleventh register. At first glance it might appear that we might need to provide 16 external address lines in order to select memory locations. Only 4 address lines are required if we code the addresses into binary numbers. Clearly, if we use four bits, the largest number we can thus represent is 1 1 1 1, or 15. The smallest number is 0 0 0 0, or 0. In general, a k-bit number can specify 2^k addresses. It should be evident, however, that an address-decoding gate configuration must be provided as part of a memory structure.

Up to this point we have tacitly assumed that memories comprise an array of registers that, in turn, are made up of *cells*—one to a bit—that consist of flip-flops. In our development of flip-flop structures, we have assumed that these are formed of various interconnected logic gates. Finally, the gates are made up of interconnected transistors along the lines discussed in Chap. 2.

While some memories are structured in this way, a very large number of memories are made up of simpler bit-cell structures. Instead of the complicated flip-flop formed from gates, a simpler flip-flop structure is used. In many types of semiconductor memories, the storage element is not a flip-flop at all.

Some RAMs are built using multi-emitter bipolar transistors that resemble those used in TTL logic gates. The transistors are connected in a simple flip-flop circuit, such as that shown in Fig. 11-12.

The cell illustrated represents the most significant bit (MSB) of the twelfth four-bit word. That is, the weight assigned to this bit is $2^3 = 8$. Let us assume that

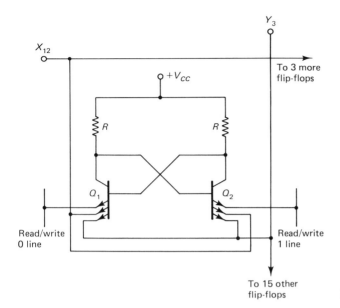

Fig. 11-12 Bipolar RAM storage cell.

the RAM consists of 16 such four-bit words. Then the X_{12} line is the word line (there must be 16) that selects the twelfth word. At the same time, by means of an enabling signal directed to this chip, all four bit-select lines are enabled (Y_3 is the bit-select line for this cell). When not enabled, the X_{12} and Y_3 lines rest at logic levels of 0. As a result, if either Q_1 or Q_2 is conducting, emitter currents flow into the X_{12} and Y_3 lines.

Suppose now that Q_1 is conducting while Q_2 is cut off. (Because of the cross coupling between collectors and bases of Q_1 and Q_2, only one transistor can conduct.) If the logic levels of the X_{12} and Y_3 lines are now raised from 0 to 1, the emitter current of Q_1 is diverted into the read/write 0 line. This line terminates in a sense amplifier, which translates the flow of current into its input into a logic-0 signal at its output. Since Q_2 was cut off, its emitters pass no current, and thus none is diverted into the read/write 1 line. Clearly, if Q_1 had been cut off while Q_2 was conducting, a logic-1 signal would have been produced at the output of the appropriate sense amplifier.

Again suppose that Q_1 is conducting with Q_2 cut off. We now wish to write a logic 1 into this flip-flop. This is accomplished by again enabling the X_{12} and Y_3 lines. If, at the same time, the read/write 1 line is switched from logic 1 to logic 0, the base-emitter junction of Q_2 will be forward biased, causing Q_2 to conduct. The resulting drop in collector potential of Q_2 pulls down the potential of the base of Q_1, cutting off that transistor. The positive feedback inherent in the cross-coupled connection enhances the speed of the switching operation. It is left for you to investigate the manner in which a logic 0 is written into the flip-flop.

Flip-flop memory cells are made using p-channel MOS transistors as well as bipolar junction transistors. A MOS-based flip-flop circuit is shown in Fig. 11-13.

Both the zero-bit and one-bit lines are maintained at voltages of $-V_{DD}$. Each line is connected to a *sense* amplifier and a *write* amplifier. Just as in the case of

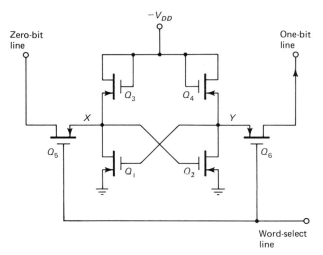

$-V_{DD}$

One-bit
line

Q_3 Q_4

X Y

Q_5 Q_6

Q_1 Q_2

Word-select
line

Fig. 11-13 A flip-flop circuit using MOS transistors.

the BJT flip-flop, the sense amplifier translates an inflow of current into the appropriate logic-level output. The *write* amplifier allows the voltage of the bit line to be raised from $-V_{DD}$ to ground.

If the word array to which this flip-flop belongs is not being selected, the word-select line rests at ground potential. In this condition, transistors Q_5 and Q_6 are turned off and the flip-flop is isolated from the rest of the system. When this word is addressed, the word-select line is pulled down to a potential of $-V_{DD}$. Then, if Q_1 is on while Q_2 is off, node X is close to ground potential, and current will flow through Q_5 into the zero-bit line. Note that Q_3 and Q_4 act simply as dynamic load resistors and that if Q_1 is on, the gate of Q_2 is close to ground potential, turning Q_2 off. This causes node Y to be at a potential equal to $-V_{DD}$. Since, as we noted earlier, the bit lines are at $-V_{DD}$, no current flows through Q_6. It is evident that in the state described, the flip-flop is storing a logic 0.

To write a logic 1 into the flip-flop, the word-select line is brought to a potential of $-V_{DD}$ and the one-bit line is driven by its *write* amplifier to ground potential. Since node Y is at $-V_{DD}$ potential, Q_6 is turned on and the gate of Q_1 is brought to a potential near ground, turning Q_1 off. Node X now moves toward $-V_{DD}$ potential, causing the turn-on of Q_2. Thus the initial state of the flip-flop has been reversed and a logic 1 has been written into it.

Memories that make use of flip-flops as bit-storage elements are called *static* memories because, as long as power is supplied to the circuits, the flip-flops hold the stored bits indefinitely.

11-10. DYNAMIC RAMs

Static RAMs using MOS devices in circuits like that of Fig. 11-13 require six transistors per bit. Since designers are constantly striving to store the maximum number of bits on a single integrated-circuit chip, it is natural that they should look to reducing the number of transistors per bit.

The number of transistors required per bit of information storage can be reduced by abandoning the flip-flop as the bit-storage cell. Instead, the logic level of the bit is stored as a voltage on a charged, gate-substrate capacitance. Unfortunately, the charge tends to leak off of this capacitance, and so it must be frequently replenished. This periodic recharging is referred to as a *refresh* operation.

One form of dynamic RAM cell is shown in Fig. 11-14. The transistors are of the p-channel, enhancement type, so that with both V_{DS} and V_{GS} equal to $-V_{DD}$, they are turned on, causing current to flow from source to drain. In the operation of this cell, the ground level corresponds to logic 0, whereas a voltage of $-V_{DD}$ (for example, -10 V) corresponds to a logic level of 1. The gate-ground capacitance of Q_2 is shown in the diagram as the capacitor C.

Suppose that the capacitor is not charged. That is, the gate of Q_2 is at ground potential. If the *read-select* line, which normally rests at ground potential, is brought to -10 V, while the *read* line, through its associated sense amplifier, is held at -10 V, then Q_3 will be turned on. However, since V_{GS} of Q_2 is at 0 V, no current flows through Q_2 and Q_3 to the *read* line. The sense amplifier therefore interprets this zero current as a logic 0.

If C had been initially charged to -10 V, then current would have passed through Q_2 and Q_3 and a logic 1 would have been exhibited at the output of the sense amplifier.

With C at logic 0 (not charged), if the *write* line is pulled down to -10 V while the *write-select* line is also at -10 V, Q_1 is turned on, causing current to flow into the *write* line and charging C to -10 V. Thus a logic 1 has been written into the cell. Notice that if we now wish to write a logic 0 into the cell, the *write* line must be held at 0 V while the *write-select* line is at -10 V. In this situation, the roles of the drain and source of Q_1 are reversed and Q_1 is made to conduct in such a direction that C is discharged to 0 V. It should be noted that Q_1, Q_2, and Q_3 are constructed symmetrically so that designations of drain and source terminals are

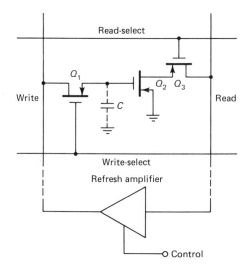

Fig. 11-14 Three-transistor dynamic RAM cell.

arbitrary. For this reason, it is common to draw diagrams like that of Fig. 11-14 with the source arrows omitted.

The *refresh* operation is performed successively on each row of cells by pulling down the potentials of both *read select* and *write select* lines and pulsing the controls of the *refresh amplifiers* of all the columns. This action momentarily causes the *read* line of each column to drive its associated *write* line. Thus if C is storing a logic 1, so that Q_2 and Q_3 turn on, the output of the *refresh* amplifier is also at logic 1 (-10 V), turning on Q_1 and recharging C in the logic-1 (-10 V) direction.

Refreshing is required at frequent intervals, typically about every 2 ms. This fact, together with certain critical timing requirements, make dynamic RAMs more difficult to use than static RAMS. It is interesting to note that in Intel's 1981 Component Data Catalog, only six dynamic RAM's are listed, whereas there are twenty-three static RAMs. Two of the dynamic RAMs are designated "pseudo-static," which means that refresh timing circuits are an integral part of the RAM chip.

The move toward ever-larger bit densities on RAM chips has led to the development of single-transistor dynamic RAM cells. These, like the circuit of Fig. 11-14, utilize a charged capacitance to store data. Reading, writing, and refreshing are more complicated for the single-transistor cell than for more complex cell structures. Furthermore, the process of reading data is *destructive*. That is, reading discharges the capacitor, so that the stored bit is thereby lost. Therefore, if the data bit is to be retained, it must be copied into a cell in a buffer register and then rewritten into the memory cell. In contrast to the single-transistor memory-cell RAMs, those that were considered earlier are said to have *nondestructive* readout.

11-11 ROMs

Read-only memories (ROMs), as the name implies, allow data to be read from them but do not have a *write* feature. Obviously, if ROMs are to be useful at all, there must be some way for data to be written into them. Thus, in a sense, the name is a misnomer. It is true, however, that unusual means must be used to write data into a ROM and that such data are stored on a more-or-less permanent basis.

Some ROMs are programmed by arranging internal metallic connections between bit cells and word lines according to a predetermined pattern. The manufacturer prescribes a format by which the user specifies the required words and their addresses. The ROM is then custom-made for the user by means of a specially prepared mask. ROMs that contain system-operating instructions, assemblers, or compilers for use with microprocessors may be of this type.

Many ROMs may be programmed by the user. One way to do this is by the use of floating-gate MOS transistors. In these transistors, a metallic gate is embedded in the SiO$_2$ above the substrate. This is in contrast with the usual construction in which the metallic gate lies on the surface of the SiO$_2$ and is available for external circuit connections. This isolated gate can be made to acquire a negative charge due to the strange phenomenon whereby the drain-substrate junction, when strongly reverse biased, injects electrons through the SiO$_2$ onto the gate. Once the gate is thus charged, the transistor is permanently biased. ROMs of this type have windows through which intense ultraviolet radiation can be projected. This radiation causes the charge on the isolated gate to leak off, erasing the bit pattern previously established. Such ROMs are called EPROMs—erasable, programmable, read-only memories.

REFERENCES

1. M. Morris Mano, *Digital Logic and Computer Design* (Englewood Cliffs, N.J.: Prentice-Hall, Inc., 1979).
2. Signetics Corporation, *TTL Logic Data Manual 1982* (Sunnyvale, Calif.: Signetics Corporation, 1982).
3. Intel Corporation, *Component Data Catalog* (Santa Clara, Calif.: Intel Corporation, 1980).

PROBLEMS

11-1. For the *JK* flip-flop of Fig. 11-2, verify the entries in Table 11-2.

11-2. If each transition through a gate requires 8 ns, calculate the time required for the Q and $\overline{Q}$ outputs of the *RS* flip-flop of Fig. 11-1 to stabilize. Assume that initially $Q = 0$ and $\overline{Q} = 1$ and that the inputs applied are $R = 0$, $S = 1$.

11-3. For the *JK* flip-flop of Fig. 11-2, if the inputs are maintained as $J = 1$, $K = 1$, calculate the frequency at which the output will oscillate, assuming that the gate propagation time is 10 ns.

11-4. For the modified *RS* flip-flop of Fig. 11-3, assume that initially $Q = 0$ and $\overline{Q} = 1$ and that the inputs are $R = 0$, $S = 1$. If the propagation time through each gate is 8 ns, calculate the time required to produce stable outputs after X changes from 1 to 0.

11-5. For the 7473 flip-flop of Fig. 11-5 with $\overline{RD} = 1$, assume that initially $Q = 0$ and $\overline{Q} = 1$.

    (a) With the *CP* input at logic 1 and $J = 1$ and $K = 0$, what are the logic levels at the R and S lines?

    (b) What are the inputs to NOR gates 5 and 6?

11-6. For the circuit and conditions of Prob. 5, complete the timing diagram, taking into account the propagation time τ through each gate.

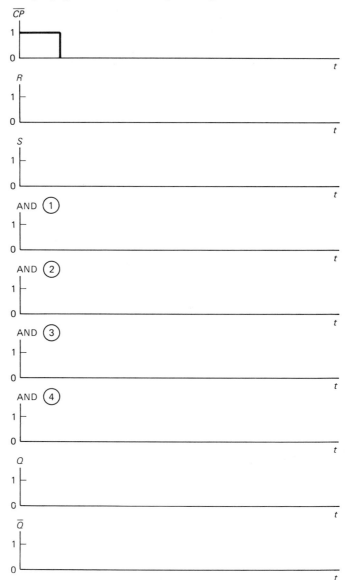

11-7. For the JK flip-flop of Fig. 11-2, assume that initially $Q = 0$, $\overline{Q} = 1$, $J = 0$, and $K = 0$. Calculate the maximum duration of a logic-1 pulse to be applied to the J and K inputs so that only a single change of state will occur. Assume gate propagation time is 10 ns.

11-8. For the circuit of Fig. 11-1, assume that the initial state is $Q = 0$, $\overline{Q} = 1$ and that both R and S are at logic 0. If, at $t = 0$, S is switched to logic 1, calculate the duration of the anomalous condition in which both Q and $\overline{Q}$ are at the logic 0 level. Assume all gates have propagation times of 12 ns.

11-9. For the four-bit binary ripple counter discussed in Sec. 11-7 in which the propagation time for each flip-flop is 35 ns, calculate the highest pulse rate that can be counted.

11-10. For the three-bit synchronous counter of Fig. 11-10, show that the count sequence, starting from 000, increments by 1 at each clock pulse until 111 is reached, after which the sequence repeats.

11-11. Assume that $Q_A = Q_B = 0$ initially. List the logic states of Q_A and Q_B for each succeeding clock pulse. After what number of clock pulses do the logic states repeat?

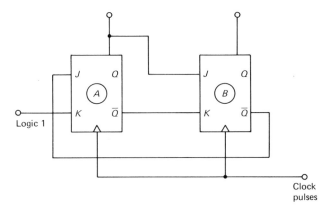

11-12. Three 7476, JK, master-slave flip-flops are connected to form a counter, as shown. The $\overline{S}$ inputs provide asynchronous *set* operations that are effective only when $\overline{S}$ is at logic 0. Similarly, the $\overline{R}$ inputs provide asynchronous *reset* operation when $\overline{R}$ goes from logic 1 to logic 0. The flip-flops, in response to the CP inputs, change state when the CP inputs go from logic 1 to logic 0. Starting with $Q_A = Q_B = Q_C = 0$, determine the sequence of states of the counter until a repeat occurs. Assume that flip-flop A registers the least significant bit.

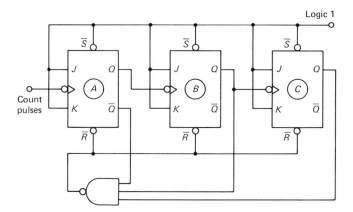

11-13. For the circuit of Prob. 12, determine whether a glitch in the counting sequence is possible. If so, at what point(s) in the sequence can it (they) occur?

11-14. Three 7476 flip-flops, whose operation is explained in Prob. 12, are connected as shown to form a counter. Starting with $Q_A = 0$, $Q_B = 0$, and $Q_C = 0$, determine the sequence of states of the counter.

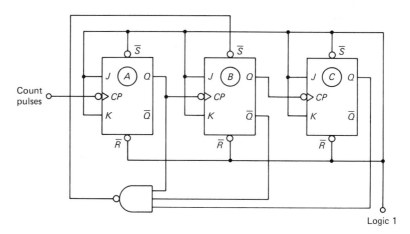

11-15. Repeat Prob. 13 for the circuit of Prob. 14.

11-16. A synchronous counter using three 7476 flip-flops is given. The $\bar{R}$ and $\bar{S}$ inputs are all wired to logic 1 and are thus made inoperable. Starting with $Q_A = 0$, $Q_B = 0$, and $Q_C = 0$, determine the sequence of states until the initial state recurs.

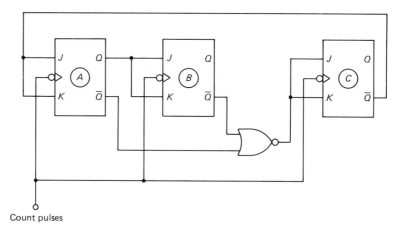

Count pulses

11-17. The 7491A shift register of Fig. 11-11 begins operation with Q_0 (the leftmost, or least significant, flip-flop) at logic 1, and all other flip-flops at logic 0. The D_{S_a} and D_{S_b} inputs are also at logic 0. List the contents of the shift register after each successive clock pulse until all flip-flops have $Q = 0$.

11-18. A static RAM has 12 address lines. How many words does the RAM contain?

11-19. A 4096 × 8-bit RAM requires, in addition to address lines and data (bit) lines, one pin for the power supply ($+V_{CC}$), one ground pin, and three control lines. Specify the minimum number of pins required to make external connections to this RAM.

11-20. Describe the process by which a logic 0 is written into the RAM storage cell of Fig. 11-12.

CHAPTER 12

Other Integrated-Circuit Devices

There are a number of other devices that do not fall within the categories we have considered up to this point. Many of these have such wide application that we can not afford to neglect them.

Several of the devices that we shall consider play important roles in interconnecting analog and digital systems. Among these are analog comparators, Schmitt triggers, and digital-analog and analog-digital converters. Other important devices will simply have to be relegated to the category of special devices. These include timers, phase-locked loops, and balanced modulators.

12-1 ANALOG COMPARATORS

An *analog comparator* accepts as an input a continuous, time-varying signal, which it compares to a given reference voltage. When the varying signal exceeds the reference value, the output of the comparator switches to some fixed value. If the varying signal falls below the reference value, the output switches to a different fixed value. Thus the output can have only two fixed values.

Suppose, for example, that a sinusoidal voltage given by $v_1 = 10 \sin \omega t$ is to be compared to a reference value of 3 V. Let us assume that we wish to use the comparator to drive TTL logic gates. Thus, when v_1 exceeds 3 V, we shall need an output of $+5$ V (logic 1); when v_1 is less than 3 V, the output must be 0. The circuit and its performance are represented in Fig. 12-1.

An important application of the analog comparator is in the construction of an analog-digital (A/D) converter. We shall consider A/D converters later in this chapter.

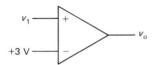

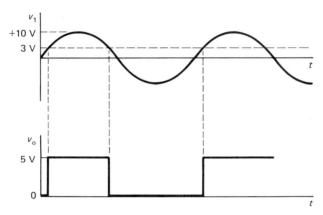

Fig. 12-1 Operation of comparator.

We might guess that any operational amplifier when operated without feed-back (open-loop) could serve as an analog comparator. This is indeed the case. However, some operational amplifiers are not designed to withstand large differences between the voltages applied to their differential input terminals. Furthermore, for some applications it is desirable to provide output circuitry that differs from the complementary-symmetry circuit typically used. An operational-amplifier circuit specifically designed for use as an analog comparator is that of the LM 111 voltage comparator shown in Fig. 12-2.

In a configuration similar to that of some operational amplifiers, the LM 111 includes three, cascaded, differential amplifiers made up of the following transistor pairs: $Q_1 - Q_2$, $Q_3 - Q_4$, and $Q_8 - Q_9$. The input transistors $Q_1 - Q_2$ operate as emitter followers. Also, Q_5 and Q_{24} operate as high-resistance dynamic loads for the emitters of Q_1 and Q_2. Therefore, it is a good approximation to consider that Q_1 and Q_2 are loaded only by the input resistances seen at the bases of Q_3 and Q_4.

The differential amplifier, $Q_3 - Q_4$, drives the third stage, $Q_8 - Q_9$, in the usual fashion. Transistor Q_8 drives the emitter-follower Q_{10}, which, in turn, drives the base of Q_{11}. The emitter of Q_{11} is driven by the collector of Q_9. Thus the two input signals driving Q_{11} are out of phase with each other. Using the superposition principle, we can show that the signal current developed at the collector of Q_{11} is approximately twice that which would result from each source taken separately.

The collector of Q_{11} is connected through R_{10}, R_{11}, R_{12}, and R_{13} to a common-return potential (which could be ground) via terminal 1. As far as signal gain is concerned, the collector resistor of Q_{11} is simply the 4-KΩ resistor of R_{10}. By means of the emitter-follower Q_{12}, this resistor is bootstrapped, so that its effective signal resistance is much greater than 4 KΩ.

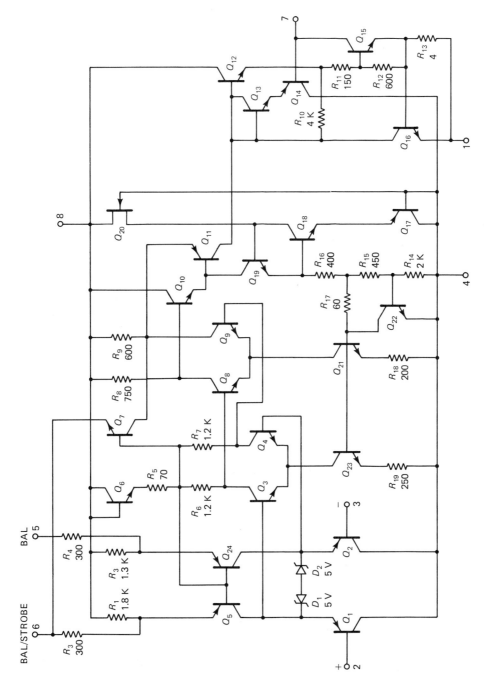

Fig. 12-2 The LM 111 voltage comparator.

The emitter follower Q_{12} drives the output transistor Q_{15}. Note that the circuit does not include a collector resistor for Q_{15}. Such an arrangement, often found in logic-gate output circuits, is called an *open-collector configuration*. It allows the user to connect a number of comparator output terminals together using a single external collector resistor. Then if any one comparator or if several comparators sink output current through the resistor, the effect is that of a logical OR. Such a connection is called a *wired OR*. (In the present case, with the collector resistor connected to $+V_{CC}$, the effect would be that of a wired NOR.)

The two Zener diodes, D_1 and D_2, are connected back-to-back between the emitters of Q_1 and Q_2. For large differences in voltage between input terminals 2 and 3, one diode will be forward biased, producing a voltage drop of 0.7 V, while the other will operate in the reverse, Zener condition to develop a voltage of 5 V. Hence the maximum voltage that can appear between the bases of Q_3 and Q_4 is 5.7 V. In this way, the comparator is protected against extreme input-voltage differences.

Terminals 5 and 6 allow the connection of an external potentiometer with its wiper connected to $+V_{CC}$. This potentiometer is used to adjust the current balance of Q_1 and Q_2 in much the same way as was done in the 741 operational amplifier. Terminal 6 is labeled BAL/STROBE because, in addition to the balance function, it also provides the means for strobing the comparator. To see what this means, consider first that terminal 6 is connected to a potential near $+V_{CC}$. Because of the voltage drop through R_5 and Q_6, the base voltage of Q_7 is lower than its emitter voltage. This transistor is therefore cut off. Hence current may be delivered through R_9 to the emitter of Q_{11}. Whether this occurs depends upon whether Q_9 is conducting or cut off. In turn, this depends upon the relative levels of the input voltages applied to terminals 2 and 3.

If terminal 6 is pulled down to a potential near ground, transistor Q_7 is turned on and deprives Q_{11} of the current that its emitter might receive. At the same time, substantial current flows through the 300-Ω resistor R_3 taking current from the emitter of Q_5. This turns on Q_{24} and Q_4 and, in turn, cuts off Q_9, whereas Q_8 conducts heavily. Less current is thus available to drive the base of Q_{10}. The emitter current of Q_{10} is thus reduced, and more current is demanded of the base of Q_{11}. This would tend to turn on Q_{11}, but the emitter current it would normally pass has been diverted through Q_7, as we have seen. The result of this action is that Q_{12} and Q_{15} are cut off, and the potential of terminal 7 (which is connected through an external resistor to $+V_{CC}$) rises to the logic-1 level.

Thus, when the strobe input is at $+V_{CC}$, the output of the comparator is at a logic level determined by its inputs. But when the strobe input is at ground, the output logic level is 1, regardless of the inputs. If a short positive pulse is applied to the strobe input, a sample of the comparator output is available during the strobe pulse interval. This action is demonstrated by the wave forms shown in Fig. 12-3.

The same strobe pulse that causes a valid signal to appear at the output of the comparator can also be used to control other digital logic elements in a system. Thus, for example, if the strobe pulse is connected to the E (Enable) input of a

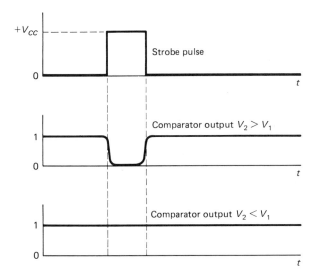

+V_{CC} ....

Strobe pulse

0 .... t

Comparator output $V_2 > V_1$

1

0 .... t

Comparator output $V_2 < V_1$

1

0 .... t

Fig. 12-3 Strobe operation.

latch, such as the 7474 shown in Fig. 11-7, whereas the output of the comparator is connected to the D (Data) input, the latch will store the comparator output until the next strobe pulse occurs.

The LM 111 comparator that we have just considered has a more complex circuit than other comparators. One simpler type is the LM 139, which is a quad voltage comparator. That is, in the LM 139, four comparators are fabricated on a single IC chip. As we might expect, some features available in the LM 111 are sacrificed to obtain the simplicity and economy of space found in the LM 139. The equivalent circuit (so called by the manufacturer because it omits internal circuit details) is shown in Fig. 12-4.

Input terminals are connected to the bases of emitter-followers Q_1 and Q_4 to provide high input resistance. These transistors drive the differential pair Q_2 and Q_3. The output of the differential pair is taken from the collector of Q_3. This collector drives the high-resistance collector load provided by Q_6 and the relatively low-resistance input to the base of the common-emitter stage Q_7. Transistor Q_7, in turn, drives the output transistor Q_8. Just as in the case of the LM 111, the output transistor has an open collector.

Diodes D_1 and D_2 protect the base-emitter junctions of Q_1 and Q_2 against large reverse voltages, such as might occur when large input voltages are applied. Diodes D_3 and D_4 protect the 3.5-μA current sources.

Both the LM 111 and the LM 139 comparators have voltage gains of 200 V/mV as stated by the manufacturer. This figure implies that if the output is used to drive a TTL logic gate requiring a voltage swing of 5 V, a differential input signal of 5/200,000, or 25 μV, will suffice. We shall consider some applications of comparators further on in this chapter.

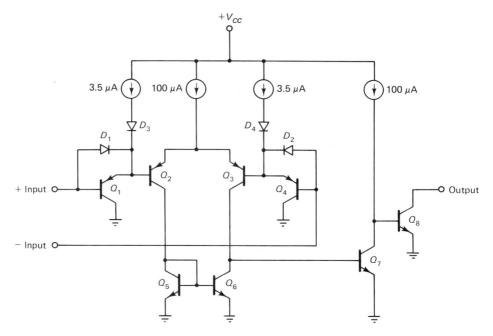

Fig. 12-4 The LM 139 comparator (one unit).

12-2 SCHMITT TRIGGERS

The Schmitt trigger is a special form of comparator that is often used to derive well-defined wave forms from noisy ones. It is also used to generate logic-signal outputs with fast rise—or fall—times when the input signals vary slowly.

Any high-gain, differential-input amplifier can be made into a Schmitt trigger, although complete Schmitt trigger circuits fabricated on a single silicon chip are available. One of these, the 74132, comprises four two-input Schmitt triggers in a NAND configuration. This means that if one of the triggers has both inputs at the logic-1 level, its output will be at the logic-0 level. For any other input combination, the output will be at the logic-1 level.

To understand the operation of typical Schmitt trigger circuits, consider Fig. 12-5. The amplifier shown in the diagram could be an operational amplifier, such as the 741, or its function could be very well performed by the LM 139 comparator with a suitable resistor added between the collector of the output transistor and $+V_{CC}$.

One important feature of Schmitt trigger circuits is immediately evident upon inspection of Fig. 12-5—the positive feedback provided by R_1 and R_2.

The relation between the output voltage v_o and the input voltage v_1 is displayed in the transfer characteristic plotted in Fig. 12-6.

Suppose that both the input voltage v_1 and the output voltage v_o are at logic 0, or ground potential. If v_1 increases, the voltage at the positive terminal will be $v_1 R_1/(R_1 + R_2)$. Because of the high gain of the amplifier, v_o will be rapidly driven

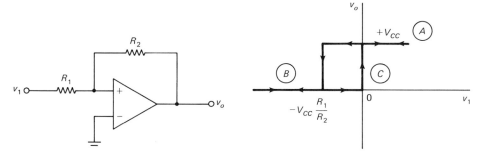

Fig. 12-5 Schmitt trigger circuit.

Fig. 12-6 Transfer characteristic of the Schmitt trigger.

to a level close to $+V_{CC}$. Any further increase in v_1 will not affect v_o. This accounts for the leg of the transfer characteristic labeled Ⓐ in Fig. 12-6.

Now suppose that v_1 is reduced. Clearly, even if v_o is brought to ground potential, there will be no effect, since at that point, the voltage at the positive terminal will be $+V_{CC}R_1/(R_1 + R_2)$. The contribution of v_1 to the voltage at the positive terminal can be determined by using the superposition principle. This leads to Eq. (12-1), which expresses the total voltage at the positive terminal due to both v_o and v_1 when $v_o = +V_{CC}$. Thus we have

$$v_+ = v_1 \frac{R_2}{R_1 + R_2} + V_{CC} \frac{R_1}{R_1 + R_2} \tag{12-1}$$

The voltage v_+ will just become zero when

$$v_1 \frac{R_2}{R_1 + R_2} = -V_{CC} \frac{R_1}{R_1 + R_2}$$

or

$$v_1 = -V_{CC} \frac{R_1}{R_2} \tag{12-2}$$

For values of v_1 more negative than that given by Eq. (12-2), the output v_o will be driven to zero. This accounts for the Ⓑ leg of the transfer characteristic.

Increasing v_1 will have no effect as long as v_1 is negative. At the point where v_1 crosses the origin, v_o will snap to the $+V_{CC}$ level, yielding the Ⓒ leg.

The direction of each transition on the transfer characteristic is indicated by an arrow. The entire characteristic reminds us of the hysteresis loop that occurs when magnetic materials are cyclically magnetized. Indeed, hysteresis is the important feature that distinguishes Schmitt triggers from other types of comparators. For this reason, the symbols used to represent Schmitt triggers include schematic representations of a hysteresis loop. An example is the 7413 dual four-input NAND Schmitt trigger shown in Fig. 12-7. For the 7413, the input voltage difference between the positive-going and negative-going branches of the hysteresis loop is about 800 mV and is fixed. This is in contrast to our hypothetical Schmitt trigger

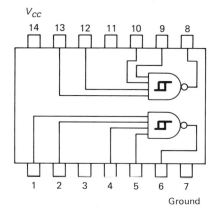

V_{CC}
14 13 12 11 10 9 8

1 2 3 4 5 6 7
Ground

Fig. 12-7 The 7413 dual four-input NAND Schmitt trigger.

of Fig. 12-5, in which the width of the hysteresis loop is determined by the resistance ratio R_1/R_2.

We might well wonder why hysteresis is a desirable feature. This question is easily answered by considering the wave forms shown in Fig. 12-8. The wave form in (a) represents an input signal free of noise. Whenever the signal crosses the axis and is going positive, the output shown in (b) is at the logic-1 level. If the input signal is noisy, as in (c), instead of being free of noise, the zero crossings are not well defined. This causes several anomalous pulses to be generated at the beginning and at the end of the positive loop of the input signal. When hysteresis is added, as in the Schmitt trigger, the first, positive-going zero crossing makes the output switch to logic 1. But the output switches back to logic 0 only when the input signal goes sufficiently negative to exceed the hysteresis offset voltage. Thus instead of producing glitches at its leading and trailing edges, as in (d), the output wave is clean, as we see in (e). The price we pay for this improvement is that the exact moment when the input wave crosses the axis appears to be less well defined.

12-3 THE 555 TIMER

One of the most widely used medium-scale integrated circuits is the 555 timer, whose circuit incorporates two comparators. The block diagram of the 555 is shown in Fig. 12-9. The reference voltage against which comparator 1 tests the threshold voltage is $(2R/3R)V_{CC}$. Thus, if the threshold voltage exceeds $(2/3)V_{CC}$, the flip-flop resets, $\overline{Q}$ goes to logic 1 (*high*), transistor Q_1 saturates, and the output stage delivers a *low* output at pin 3. This condition will persist until the flip-flop is set again (that is, Q goes to logic 1).

Comparator 2 is constructed so that as long as the voltage at pin 2 (trigger) *exceeds* $(R/3R)V_{CC}$, its output is zero. When the trigger input drops below this value, comparator 2 causes the flip-flop to set. In turn, this causes Q_1 to cut off and also causes the output stage to deliver a *high* output at pin 3.

A negative pulse applied to pin 4 (Reset) causes transistor Q_2 to override the flip-flop and makes Q_1 saturate and causes the output to go low. In most appli-

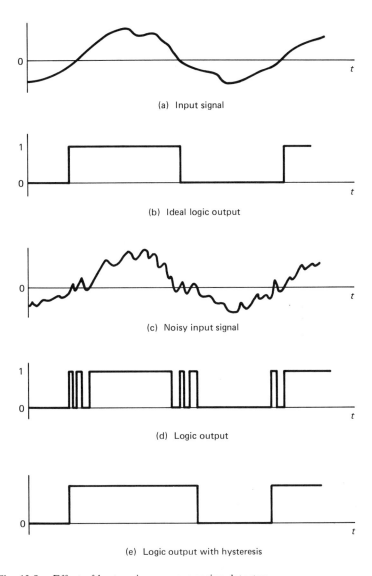

(a) Input signal

(b) Ideal logic output

(c) Noisy input signal

(d) Logic output

(e) Logic output with hysteresis

Fig. 12-8 Effect of hysteresis on zero-crossing detector.

cations, pin 4 is tied to pin 8 (V_{CC}) so that Q_2 is cut off and the reset function is not used.

12-3.1. The 555 in monostable operation

In Fig. 12-10, the 555 is connected to provide *monostable* operation. That is, upon the application of a negative pulse to the trigger input, the flip-flop is set and Q_1 is cut off, removing the short circuit that had existed across the capacitor C. The capacitor now charges toward V_{CC}. When the capacitor voltage v_c reaches $(2/3) V_{CC}$, the flip-flop resets, saturating Q_1 and thus discharging C. This is the initial state

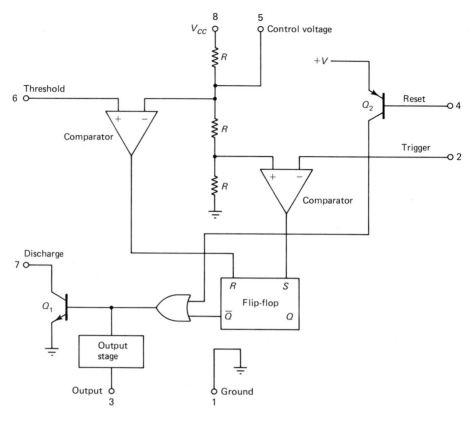

Fig. 12-9 Block diagram of the 555 timer.

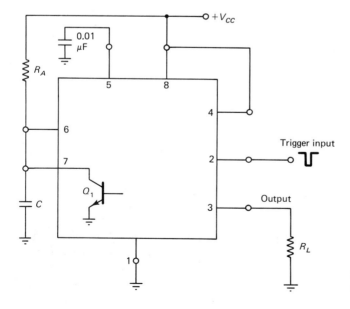

Fig. 12-10 The 555 connected for mono-stable operation.

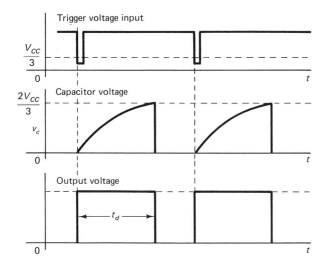

Fig. 12-11 Wave forms in monostable operation.

that prevailed before the trigger pulse occurred. During the time that the flip-flop is set and the capacitor C is charging, the output is high. The 0.01-μF capacitor stabilizes the internal connection of comparator 1 against switching transients and helps to prevent false triggering. The wave forms produced in normal monostable operation are shown in Fig. 12-11.

The time delay t_d, shown in Fig. 12-11, represents the time after the trigger pulse when the output returns to its low state. Clearly this depends on the time constant $\tau = R_A C$. During charging, the capacitor voltage V_c is given by

$$v_c = V_{CC}(1 - e^{-t/\tau})$$

When $t = t_d$, $v_c = 2V_{CC}/3$. It is left to you to show that $t_d = \tau \ln 3 \approx 1.1\tau$.

In Fig. 12-12, the monostable circuit of Fig. 12-10 is modified so that the resistor R_A is replaced by a current-mirror circuit. This causes the capacitor C to be charged by a constant current, thus causing the charging curve of v_c versus time to become linear. Since this linear function crosses the $2V_{CC}/3$ voltage level at a steeper angle than the exponential curve, the precise moment when C discharges is better defined. The time delay t_d for this circuit is inversely proportional to R.

12-3.2 The 555 in astable operation

The 555 timer can be made to operate as a free-running, or *astable*, circuit. Or, in other words, it can be used as an oscillator. This is accomplished by omitting the external trigger connection and tying the ungrounded side of capacitor C to the trigger input. The discharge of the capacitor must be slowed down to provide a satisfactory ratio of high time to low time in the output wave form. In the circuit shown in Fig. 12-13, the resistor R_E provides this lengthened discharge time. Note that capacitor C charges through the series combination of R_A and R_B but discharges through R_B alone. It is left to you to use these facts to determine the relationship between the frequency of oscillation and R_A, R_B, and C.

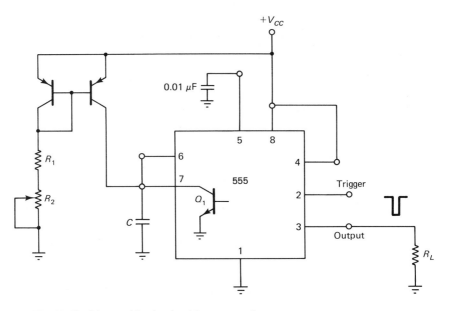

Fig. 12-12 Monostable circuit with current mirror.

As in Fig. 12-12, the circuit of Fig. 12-13 can be modified to include a current-mirror circuit in place of R_A.

By operating with V_{CC} equal to 5 V, the 555 can deliver output voltage pulses that are suitable for use with TTL logic gates. This makes the 555 useful as a low-frequency clock to drive counters, shift registers, and the like.

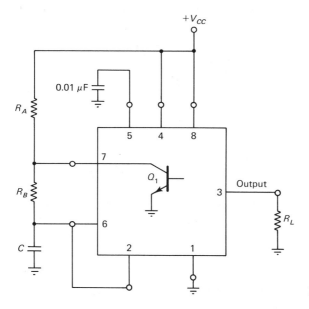

Fig. 12-13 Astable circuit.

Other Integrated-Circuit Devices   Chap. 12

12-4 DIGITAL-TO-ANALOG CONVERTERS

A *digital-to-analog converter*, or *D/A converter*, accepts several binary bits and produces an analog output whose level is related to the binary input information. Thus the D/A converter is a vital link between the analog world of quantities like temperature or velocity and the digital realm where such quantities may be processed, as in a digital computer. There are two widely used types of D/A converters—the weighted-current-source type and the *R-2R* type.

12-4.1 The weighted-current-source D/A converter

A simplified schematic diagram of a weighted-current-source D/A converter is shown in Fig. 12-14. Each current source is a *controlled* source in the sense that it is either on or off according to the logic level of its binary input. Thus, for an input of 1010, the current I_T would be $8I + 2I = 10I$. Assuming that the input current to the operational amplifier is zero, Kirchhoff's current law applied to the inverting input of the operational amplifier yields

$$I_T + v_o/R_F = 0$$

$$-v_o/R_F = 10I$$

$$v_o = -10IR_F \qquad (12\text{-}3)$$

You should recall that, in effect, Eq. (12-3) is derived using the same principles that were used to find the asymptotic gain of a feedback amplifier. Hence the inverting input terminal is a virtual ground.

Notice that this circuit converts the sum of the weighted *currents* into an output *voltage*. In fact, if we make I equal to 100 μA and R_F equal to 10 KΩ, we find that a binary input of 1010 yields an output of -10 V. Another unity-gain inverting amplifier can be added to restore the polarity of the output.

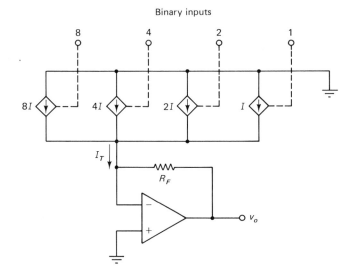

Fig. 12-14 Weighted-current-source D/A converter.

It is interesting to see how the weighted currents are precisely set and how they are switched into or out of the circuit. Figure 12-15 shows some of these details.

Transistor Q_1 acts to regulate the current through R_3 to maintain the potential of node B at a constant value of 1.2 V with respect to ground. Since the bases of transistors Q_2, Q_3, Q_4, Q_5, and Q_6 are all tied to node B, whereas their emitter-resistors are returned to the regulated output line driven by amplifier A_2, the emitter currents of Q_3, Q_4, Q_5, and Q_6 will be determined by the emitter current of Q_2. Because of the relative values of the emitter-resistors, the respective currents are 1, $\frac{1}{2}$, $\frac{1}{4}$, or $\frac{1}{8}$ times that of Q_2. To achieve high precision in weighting the emitter currents, transistors having closely matched characteristics are needed. This is readily accomplished using IC construction, as we have seen before. In addition, precision trimming of emitter-resistors is required. Laser trimming is sometimes used for this purpose.

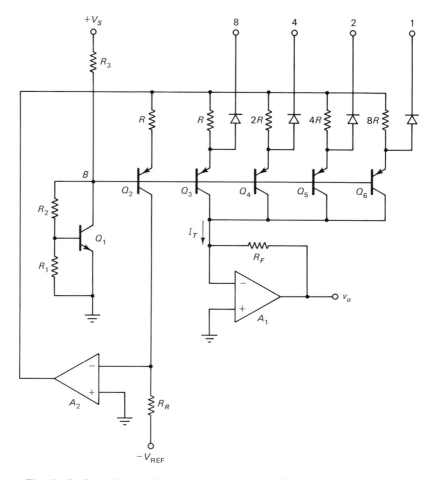

Fig. 12-15 Some details of weighted-current-source D/A converter.

The D/A converter of Fig. 12-15 is designed to be used with TTL logic. Thus if a binary input line is brought to a logic-0 level, the gate that produces this effect is capable of sinking the current that passes through the emitter resistor of the appropriate current-source transistor. In this way, that transistor is deprived of its emitter current and therefore delivers no current at its collector. If an input line is brought to a logic-1 level—for example, near $+5$ V—the diode in that line is switched off and the transistor in question delivers its collector current. Note that setting the bases to 1.2 V causes the emitters to operate at a potential of about 1.9 V when the transistors are conducting. Thus input-line voltage levels greater than 1.9 V leave the transistors on, while levels less than 1.9 V turn them off.

12-4.2. *R-2R* ladder types of D/A converters

The weighted-current-source D/A converter has the disadvantage of requiring a wide range of high-precision resistors. For example, if an eight-bit converter were constructed along the lines of the one shown in Fig. 12-14 using a resistor R of 1 KΩ in the most-significant-bit current source, a resistor of 128 KΩ would be required in the least-significant-bit source. The R-2R resistance ladder structure overcomes this problem.

Instead of the binary-weighted current sources that are used in the circuit of Fig. 12-15, we can design a D/A converter that uses equal current sources. When these current sources are connected in an R-2R ladder, as shown in Fig. 12-16, current weighting occurs so that proper conversion is achieved. To see why this happens, consider the simplified diagram shown in Fig. 12-17.

Notice that the resistance to ground at node A due only to the two parallel 2R resistors is R. Similarly, the resistance at node B due to the 2R resistor and all resistance looking to the left is also equal to R. The same is true at nodes C and D. Therefore, the resistance looking left from the inverting (negative) terminal of the operational amplifier is always equal to R, since switching the ideal current sources in or out of the network has no effect upon its resistance.

Using the superposition principle, we can consider the effect of one source at a time and sum the results to find the total effect. For the MSB source, the short-circuit current with node D grounded is simply I.

For the source that is connected to node C, the current that is delivered to the right through R, with node D short-circuited to ground, must be $I/2$.

The source connected to B sees resistance equal to R looking to the left but sees $R + 2R\|R$ looking to the right, again with D short-circuited to ground. This means that there is a current split at node B such that a current equal to $\frac{3}{8}I$ flows to the right. At the short-circuited node D, the current flowing through R is equal to $\frac{1}{4}I$.

Similarly, the source connected to A causes a current of $\frac{1}{8}I$ to flow through the hypothetical short circuit at node D. We have thus justified the binary weights that are indicated on the circuit diagram of Fig. 12-17. All that remains for us to do is to determine v_o in terms of the binary-weighted short-circuit currents that we have calculated. This is easily done when we recognize that a Norton equivalent circuit for each of these cases merely puts a current source equal to the appropriate

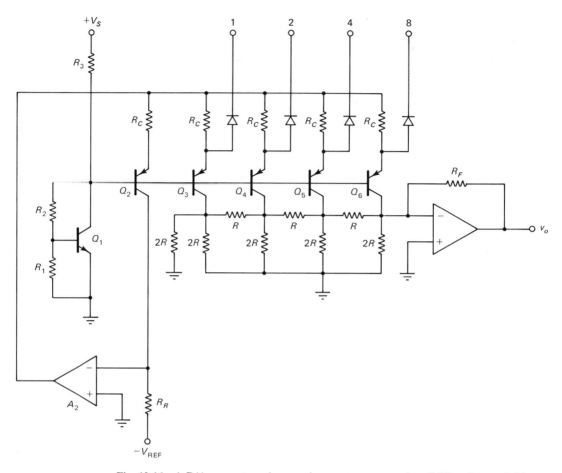

Fig. 12-16 A D/A converter using equal current sources and an R-$2R$ resistance ladder.

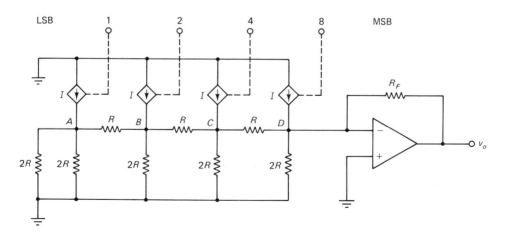

Fig. 12-17 Simplified circuit of D/A converter.

short-circuit current in parallel with the resistance seen looking to the left into node D. This resistance is, of course, simply R. The two Norton-circuit elements are connected between node D and ground with the current source directed toward node D.

If we visualize these Norton equivalent circuits as having been converted to Thevenin equivalents, then we have voltage sources equal to RI, $RI/2$, $RI/4$, or $RI/8$ in series with R. Again, the Thevenin circuits are connected between ground and node D, with the positive terminal of the voltage source connected to node D. It is clear that v_o must be either $-R_FI$, $-R_FI/2$, $-R_FI/4$, or $-R_FI/8$.

A simple form of D/A converter using the $R-2R$ ladder structure can be constructed using a single input voltage instead of several identical current sources. The analysis of that form of converter is left as an exercise.

12-5 ANALOG-TO-DIGITAL CONVERTERS

We are accustomed to thinking of most physical data—for example, temperatures, speeds, pressures, or voltages—as belonging to the analog family. A deeper look at some physical phenomena might make us take a different view—could we measure light intensity by counting photons per second or current by counting electrons per second? We shall leave such questions to physicists or philosophers and simply recognize that most quantities in which we are interested occur in analog form. Since we often want to process such quantities digitally—for example, by using a microprocessor—it is clear that we must first convert such analog data to digital form; hence the need exists for analog-to-digital (or A/D) converters.

12-5.1 The parallel converter

The simplest form of A/D converter is the parallel, or flash, converter. A simple example of a parallel converter is shown in Fig. 12-18. The reference voltage V_{REF} is applied to a four-element voltage divider having equal resistors of value R. Thus the inputs to the positive terminals of comparators C_1, C_2, and C_3 are $V_{REF}/4$, $V_{REF}/2$, and $3V_{REF}/4$, respectively. The analog input voltage v_a is applied to the inverting (negative) terminals of the comparators. If $V_{REF}/4 < v_a < V_{REF}/2$, then the output of C_1 would be logic 1, whereas the outputs of the other two comparators would be logic 0. Similarly, if $V_{REF}/2 < v_a < 3V_{REF}/4$, the outputs of both C_1 and C_2 would be logic 1, whereas that of C_3 would be logic 0. Clearly, the maximum detectable digital value of v_a is $3V_{REF}/4$, whereas the minimum is zero.

The logic-level outputs of the comparators are fed to a logic circuit consisting of gates (such as NANDs and NORs) so arranged that they drive two output lines. These output lines could be brought out to serve as the digital data output to some external device. Or, as in this example, they provide the parallel inputs to a two-bit data register.

If the analog voltage is varying with time, it may be necessary to sample it quickly and store its value using a device called a sample-and-hold. Or, the comparators can simply be strobed at the instant when we want to convert the input data.

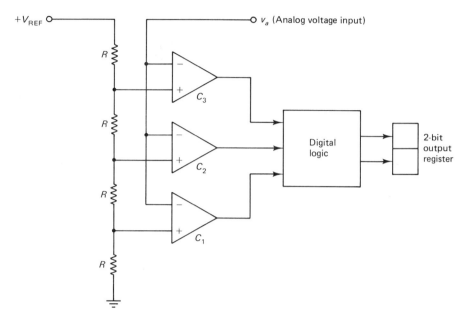

+V_{REF} o — o v_a (Analog voltage input)

R C_3

R C_2

R C_1

R

Digital logic

2-bit output register

Fig. 12-18 Simple parallel A/D converter.

Note that we are assuming that the digital result of the conversion is to be stored in the register as a binary fraction of the full-scale value V_{REF}. That is, register contents are to be interpreted as having the form $.r_1 r_2$, where the period is a *binary* point having a function similar to that of a decimal point. The binary digits r_1 and r_2 may be either 0 or 1. Thus .01 represents $E_{REF}/4$, .10 represents $E_{REF}/2$, while .11 represents $3E_{REF}/4$.

If we wish to improve the resolution of the converter to develop three bits of output data, we must divide E_{REF} by 8 instead of 4. We will then need eight resistors and seven comparators. In general, to obtain n bits of output data, we require $2^n - 1$ comparators. Clearly, as the resolution is made finer, the amount of hardware required grows rapidly. For that reason, parallel converters are not often used except in cases where very fast conversion is required. Since the comparators work simultaneously, the time required for data conversion would be the sum of the time required for one comparator to operate plus the time required for logic levels to propagate through the gates and into the register. This total time can be made very small. It is for this reason that converters of this type are sometimes called flash converters.

12-5.2 The dual-slope A/D converter

In contrast to the parallel converter, the dual-slope converter is extremely slow. However, its simplicity makes it useful for applications where speed is not important. Such applications include panel meters, DVMs and the like. Clearly, if slow conversion means that several milliseconds are required to produce a readout on a panel meter, no one will observe the delay.

The dual-slope A/D converter is shown in Fig. 12-19. The generation of a digital output that represents the level of the analog voltage v_a is a two-step process. To begin the first step, the counter is cleared and the control logic places switch S in position ①. This connects v_a, which is assumed to have a positive value, to the input of the integrator. At the same time, clock pulses are sent to the counter. The control logic acts in such a way that when a predetermined count is registered in the counter, the first measurement phase ends. At this point, the integration of the voltage v_a, which is assumed to be constant, places some negative voltage at the + terminal of the comparator.

The second phase of the measurement begins with the resetting of the counter and the switching of S to position ② to send $-V_{REF}$ to the integrator. This time, integration proceeds until the output of the integrator reaches zero. At that point, the control logic halts the process, and the quantity registered by the counter yields a value for v_a.

To understand how this process works, we should look at Fig. 12-20, which shows the output of the integrator as a function of time.

We see that time T_1 is fixed, since, in the first phase of operation, the counter totals clock pulses until a certain, predetermined count is reached. An easy way to do this is to allow the binary counter to count from 0 until every bit becomes a 1. Then the next clock pulse resets each bit to 0 but produces a carryout (overflow) from the most significant bit. Thus the counter is automatically reset for the second phase of measurement, and the overflow signal can be used to reset S.

At the end of the second phase, the counter contains a number proportional to T_2, which is the time required to integrate V_{REF} so that the integrator output

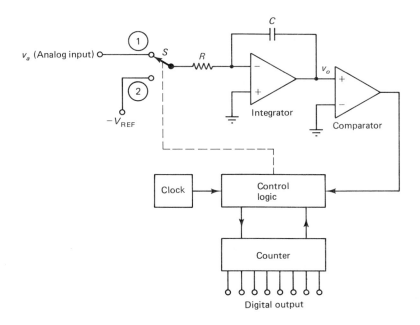

Fig. 12-19   Dual-slope A/D converter.

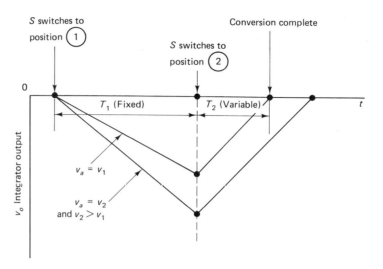

Fig. 12-20   Dual-slope integrator output.

returns to 0. It is clear that the time T_2 must depend on the value of integrator output at the beginning of the second phase.

Note that the slope of the integrator output function is variable in the first phase and fixed in the second. If we denote the maximum (negative) output of the integrator by v_o, then $v_o = K_1 v_a T_1$, where K_1 is a proportionality constant. Then $v_o/T_2 = K_2$, where K_2 represents the constant slope in the second phase. Hence we have $T_2 = (K_1 T_1/K_2)v_a$. Since K_1, K_2, and T_1 are all constants, we see that the counter contents that indicate T_2 can be made to register v_a by suitable adjustment of these constants.

Clearly, the total time required for conversion is $T_1 + T_2$, which can amount to a much longer time than was required for the parallel converter. However, we note that the dual-slope converter incorporates only one comparator and one integrator. The resolution is limited only by the clock-pulse repetition rate and the size of the counter register.

In general, all A/D converters except the parallel converter make use of some form of integration process to achieve conversion. This integration may be accomplished by an analog integrator, as in the case of the dual-slope converter. Or, the output counter contents can be converted by a D/A converter and the resulting analog (staircase) signal compared with the input signal. In all forms of integrating converters, time is required to perform the integration. As a general rule, for a given resolution, we trade off simplicity of construction against conversion time.

12-5.3 The successive-approximation A/D converter

One widely used type of A/D converter, whose design involves a compromise between complexity and conversion time, is the successive-approximation converter, shown schematically in Fig. 12-21.

As its name implies, the successive-approximation converter homes in on a digital approximation of the analog input voltage, whose resolution becomes finer

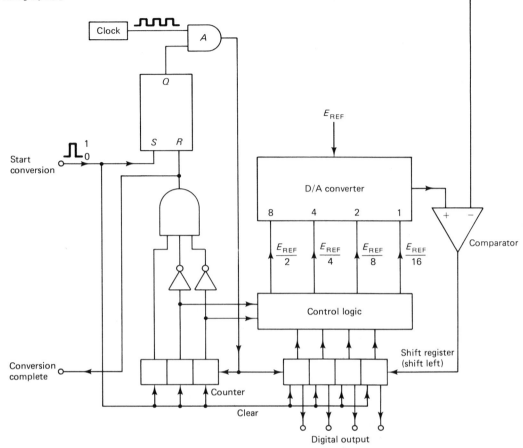

v_a (Analog input)

Clock

A

Q

S   R

E_{REF}

Start
conversion

1
0

D/A converter

8   4   2   1

$\dfrac{E_{REF}}{2}$ $\dfrac{E_{REF}}{4}$ $\dfrac{E_{REF}}{8}$ $\dfrac{E_{REF}}{16}$

+   −

Comparator

Control logic

Conversion
complete

Shift register
(shift left)

Counter

Clear

Digital output

Fig. 12-21   Successive-approximation A/D converter.

at each step of the process. This is accomplished by performing a sequence of
comparisons between the analog input voltage and selected fractions of the reference voltage. At each step, the reference-voltage fraction that is chosen depends
upon the results of preceding comparisons.

To begin the process, the logic inputs to the D/A converter are set so that
its output is $E_{REF}/2$. If $v_a > E_{REF}/2$, the comparator produces a logic-1 output,
which is left-shifted into the rightmost bit of the shift register. If $v_a < E_{REF}/2$, a
logic 0 is placed in this cell of the shift register.

The next comparison depends upon whether v_a was found to be greater than
or less than $E_{REF}/2$. If $v_a > E_{REF}/2$, the next comparison would be made to
determine whether $v_a > 3E_{REF}/2$. For the case where $v_a < E_{REF}/2$, the next
comparison would be with $E_{REF}/4$. After four comparisons, the digital representation of v_a will have been determined within $\pm E_{REF}/16$.

The circuit of Fig. 12-21 makes use of a D/A converter such as the one shown
in Fig. 12-16 to produce the appropriate fractions of E_{REF} at the comparator input.

The counter and the shift register provide information to the control logic about the number of comparisons performed and their results.

The circuit also includes an *SR* flip-flop. This permits a short pulse (start-conversion pulse) to set the flip-flop, thereby enabling AND gate *A*, which sends clock pulses to the counter and the shift register. The start-conversion pulse also clears the counter and shift register so that they initially contain only 0s. Operation proceeds until four comparisons have been performed, at which point the counter contains the binary number 100. AND gate *B* then delivers a 1 to the *R* terminal of the flip-flop, which resets it, bringing *Q* to 0 and shutting off further clock pulses. At the same time the logic level of the conversion complete line shifts to logic 1, signaling external equipment that data are available on the shift-register output lines.

Since four clock pulses are required to perform the conversion, it would seem that to get faster conversion we need only to increase the clock-pulse frequency. This is true up to the point where the time required for logic signals to propagate through gates, the time needed to shift one bit of data in the shift register, and the time required for comparator operation set a lower limit on the allowable time between clock pulses. Despite these limitations, it is possible, using a successive-approximation converter, to convert ten bits of data in 1 μs.

12-6 THE BALANCED MODULATOR

Balanced modulators accept as inputs two signals, which are multiplied together to produce an output. If one signal is a high-frequency carrier and the other is an audio signal, the result is an amplitude-modulated (AM) signal. The fact that the modulator is balanced allows the suppression of the carrier component in the output. This is a necessary requirement in single-sideband systems. Among other applications, the balanced modulator can be used as a phase detector. We shall take up this application in more detail when we consider phase-locked loops in a later section.

Shown in Fig. 12-22 is the MC1496 balanced modulator. Diode D1 represents a diode-connected transistor that, together with an external resistor connected to pin 5, forms a current mirror with transistors Q_7 and Q_8. These two transistors set the maximum currents that can be passed by Q_5 and Q_6. Connections to the emitters of Q_5 and Q_6 may be made externally through pins 2 and 3. These pins may be tied directly together, which has the effect of connecting the emitters of Q_5 and Q_6 to a single, double-amplitude current source. With such a connection in place, the possibility exists that Q_5 might pass the entire double-amplitude current, whereas Q_6 passes no current, or vice versa. Placing a finite resistance between pins 2 and 3 would allow Q_5, for instance, to pass not only enough current to satisfy the current demand of Q_7 but also an additional component that, added to the current passed by Q_6, will satisfy the current demand of Q_8. As we recall from our study of differential amplifiers in Chap. 3, the gain of a differential amplifier is proportional to the total current passed by the two connected emitters. Thus the gain of the Q_1-Q_2 pair is determined by the current passed by Q_5, whereas that of the Q_3-

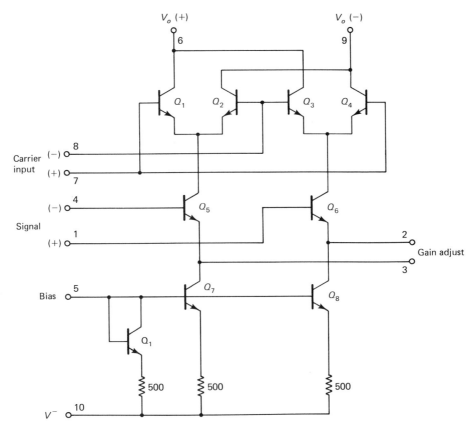

Fig. 12-22 The MC1496 balanced modulator.

Q_4 pair depends on the current passed by Q_6. We can therefore conclude that the maximum gain of each differential amplifier can be controlled by the value of the resistor connected between pins 2 and 3, zero resistance yielding maximum gain.

Figure 12-23 shows a stripped-down form of the balanced-modulator circuit with input voltages and collector currents indicated.

From Eq. (3-10), we found that a differential amplifier may be analyzed in terms of its differential transconductance g_{md}, which is directly proportional to the total current passed by the joined emitters. For the present analysis, we shall recall that this total emitter current depends, in turn, on the bias voltage applied between base and emitter of the emitter-current-source transistor, which is Q_5 or Q_6 in this circuit. Since the relationship between emitter-base voltage and collector current is not linear (exponential, in fact), we shall simply write $g_m(v_{s1})$ to denote that the transconductance of the differential pair $Q_1 - Q_2$ is a function of v_{s1}. Note that v_{s1} and v_{s2} denote signal voltages with respect to ground rather than the emitter-base voltages of Q_5 and Q_6. However, each of these emitter-base voltages must be a component of v_{s1} or v_{s2}. Hence we can infer that g_{md} is indeed a function of v_{s1} or v_{s2}.

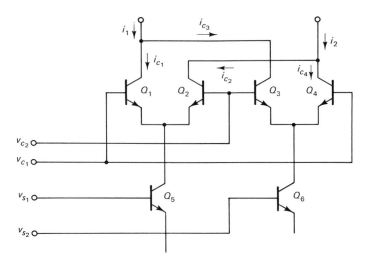

Fig. 12-23 Simplified balanced-modulator circuit.

From our knowledge of differential amplifiers, we can write

$$i_{c1} = g_m(v_{s1})\,(v_{c1} - v_{c2})$$

$$i_{c2} = g_m(v_{s1})\,(v_{c1} - v_{c2})$$

$$i_{c3} = g_m(v_{s2})\,(v_{c2} - v_{c1})$$

$$i_{c4} = g_m(v_{s2})\,(v_{c2} - v_{c1})$$

Note that all the variables are small-signal components. The currents i_1 and i_2 are given by

$$i_1 = i_{c1} + i_{c3} = [g_m(v_{s1}) - g_m(v_{s2})]\,(v_{c1} - v_{c2})$$

$$i_2 = i_{c2} + i_{c4} = [g_m(v_{s2}) - g_m(v_{s1})]\,(v_{c1} - v_{c2})$$

In actual application, equal resistors, which we can denote by R, would be connected from pin 6 and pin 9 to $+V_{CC}$ (see Fig. 12-22) and the output voltage signal would be taken between pin 6 and pin 9. The signal voltage at pin 6 would be $v_6 = -i_1 R$, whereas that at pin 9 would be $v_9 = -i_2 R$. The output voltage v_o would therefore be $v_6 - v_9$, or $(i_2 - i_1)R$. Hence we can write

$$v_o = 2R[g_m(v_{s2}) - g_m(v_{s1})]\,(v_{c1} - v_{c2}) \tag{12-3}$$

Since g_m is a nonlinear function of v_{s1}, we can write g_m as a power series

$$g_m(v_{s1}) = a_0 + a_1 v_{s1} + a_1 v_{s1}{}^2 + \cdots \tag{12-4}$$

The plus and minus markings applied to pins 1 and 4 of Fig. 12-22 imply that the modulating signal input is connected between these pins. That is equivalent to connecting equal, out-of-phase voltages from each of these pins to ground. Therefore, we can consider that $v_{s1} = v_s$ and that $v_{s2} = -v_s$. Then Eq. (12-4) takes

Other Integrated-Circuit Devices   Chap. 12

two forms:

$$g_m(v_{s1}) = a_0 + a_1 v_s + a_2 v_s^2$$

$$\tag{12-5}$$

$$g_m(v_{s2}) = a_0 - a_1 v_s + a_2 v_s^2$$

In Eqs. (12-5), we are neglecting terms of higher power than the second. Equation (12-3) now becomes

$$v_o = 2R(2a_1 v_s)(v_{c1} - v_{c2})$$

We recognize that the carrier input voltages v_{c1} and v_{c2} are equal and opposite in phase. Hence we can replace v_{c1} by v_c and v_{c2} by $-v_c$. Then v_o can be written

$$v_o = 2R(2a_1 v_s)(2v_c) = 8Ra_1 v_s v_c$$

Since a_1 and R are constants, we see that v_o can be taken simply to be

$$v_o = Kv_s v_o \tag{12-6}$$

We have thus established that the balanced modulator circuit is simply a multiplier. We should note, however, that the result just obtained depends upon an exact cancellation of the a_o and a_2 terms that appear in Eq. (12-5). The fact that it is possible to fabricate closely matched transistors in an integrated circuit helps to achieve this result. Nevertheless, some external circuit balancing may still be required.

12-6.1 Use of the balanced modulator in single-sideband transmission

It remains for us to see how the circuit works as a modulator. Suppose that the modulating signal (typically having an audio frequency) is given by $v_s = V_s \sin \omega_s t$, whereas the carrier signal is given by $v_c = V_c \sin \omega_c t$. From Eq. (12-6), we get

$$v_o = KV_s V_c \sin \omega_s t \sin \omega_c t$$

Using trigonometric identities, we can write

$$v_o = \frac{KV_s V_c}{2}[\cos(\omega_c - \omega_s)t - \cos(\omega_c + \omega_s)t] \tag{12-7}$$

Equation (12-7) shows that, using this type of modulator, among the frequencies that appear in the output are $\omega_c + \omega_s$ and $\omega_c - \omega_s$. These are the normal sideband frequencies that arise in amplitude modulation (AM). The frequency ω_c, the carrier frequency, does not appear. Also, the separation between the two sideband frequencies is $2\omega_s$. These facts make it fairly easy to filter out one sideband and transmit only the remaining one, making the balanced modulator attractive for single-sideband transmission.

12-6.2 The balanced modulator as a phase detector

Suppose that the signals applied to pins 1–4 and 7–8 have the same frequency but are out of phase with each other by $\theta + \pi/2$. We can then write $v_c = V_1 \sin \omega t$ and $v_s = V_2 \sin(\omega t + \theta + \pi/2)$. The phase displacement is taken as $\theta + \pi/2$

instead of θ to indicate that the two input signals are in phase quadrature (90° phase difference) and that one signal has an additional, variable phase displacement equal to θ. We can therefore write $v_s = V_2 \cos(\omega t + \theta)$. Then, from Eq. (12-6), we get $v_o = K V_1 V_2 \sin \omega t \cos(\omega t + \theta)$. The use of trigonometric identities leads to

$$v_o = \tfrac{1}{2} \sin 2 \omega t \cos \theta + \tfrac{1}{2} \cos 2 \omega t \sin \theta + \tfrac{1}{2} \sin \theta \qquad (12\text{-}8)$$

Equation (12-8) suggests that if the signal v_o were applied to a low-pass filter having a cutoff frequency low enough to reject the frequency 2ω, the output of the filter would yield only the quantity $\tfrac{1}{2} \sin \theta$. For small values of θ, $\sin \theta \approx \theta$. Thus we see that the balanced modulator circuit provides the basis for the construction of a linear phase detector.

12-7 THE PHASE-LOCKED LOOP (PLL)

Using a linear phase detector together with a low-pass filter, an amplifier, and a voltage-controlled oscillator, we can form the basic *phase-locked loop*, or PLL, system having the block diagram shown in Fig. 12-24.

You will readily recognize that the system of Fig. 12-24 embodies feedback. However, it differs from other feedback systems that we have encountered in that a *voltage* is transmitted from the low-pass filter through the amplifier to the voltage-controlled oscillator (VCO), while a *frequency* is fed from the VCO to the phase comparator. The voltage referred to is, of course, proportional to the phase error, ϕ, between the two quadrature frequency inputs, f_1, from an external source, and f_0, from the VCO. The feedback, as one might expect, is negative. That is, the amplifier output voltage is applied to the VCO in such a direction that the phase error, ϕ, is reduced. You should recall that the instantaneous phase angle ϕ associated with a sinusoidal function is given by $\phi = (\omega_2 - \omega_0)t$. The angular frequencies ω_1 and ω_0 are equal to $2\pi f_1$ and $2\pi f_0$, respectively. Clearly, reducing ϕ to 0 will bring f_0 to the point where it equals f_1. Such a condition is called *phase locking*. The PLL will maintain lock only for a restricted range of f_1 above and below f_0. The total range of values of f_1 within which lock can be maintained is called the *lock range*. In general, a PLL out of lock cannot achieve lock for a range of f_1 as great as the lock range. This (usually) narrower range wherein lock can be achieved is called the *capture range*.

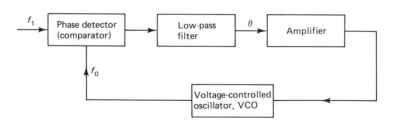

Fig. 12-24 Basic PLL configuration.

The output of the PLL is usually taken from either the amplifier output or the VCO output according to the application. This is explained shortly.

Shown in Fig. 12-25 is the block diagram of the NE565 PLL. It is not essential that the VCO of a PLL produce a sinusoidal output. In fact, in Fig. 12-25 the VCO is working as a voltage-controlled multivibrator. This explains the triangular wave form at pin 9 and the square wave at pin 4.

The output of the VCO is not connected internally to the phase detector. Instead, both the VCO output and the normally associated input to the phase detector are brought out to pins 4 and 5, respectively. The reasons for this will shortly become apparent.

12-7.1 The PLL applied to frequency-shift keying

Fig. 12-26 shows a PLL circuit configuration to be used for *frequency-shift keying* (*FSK*). FSK is used for data transmission in which a carrier *frequency* is switched between two values to represent a logic 1 or a logic 0. Such a scheme might be used in connection with the design of a modem. The circuit of Fig. 12-26 transforms the FSK input to TTL output.

In a typical FSK system, the two frequencies of 1070 Hz and 1270 Hz denote logic 0 and logic 1. Also, R_1 and C_1 adjust the free-running frequency of the VCO to yield a slightly positive output at pin 7 when the input frequency is 1070 Hz.

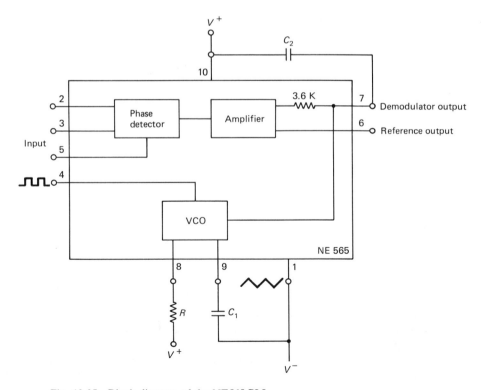

Fig. 12-25 Block diagram of the NE565 PLL.

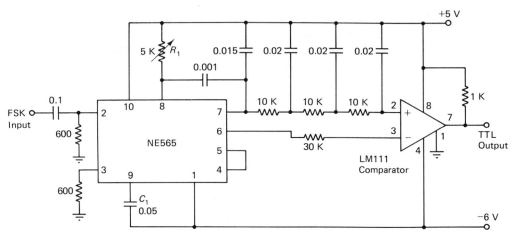

Fig. 12-26 FSK-TTL converter.

Note that pins 4 and 5 are connected so that the VCO drives the phase detector directly. When the FSK input frequency becomes 1270 Hz, pin 7 becomes more positive, driving pin 2 of the LM 111 comparator through the three-stage RC low-pass filter. This causes the TTL output of the comparator to go to the logic-1 level.

The RC filter used in this circuit has a 3-db cutoff frequency about equal to the keying frequency. For a bit rate of 300 baud, the keying frequency is 150 Hz. (*One* keying cycle corresponds to a sequence of a 1 followed by a 0, or *two* bits.)

If we assume that the output resistance of the NE565 at pin 7, equal to about 3600 Ω, is low enough, we can neglect the effect of the 0.015-μF capacitor insofar as it affects the low-pass filter. Then, assuming a high resistance looking into pin 2 of the LM 111 comparator, we can determine the transfer function of the RC filter. In terms of the Laplace variable S, this turns out to be

$$H(S) = \frac{1}{1 + 6RCS + 5(RCS)^2 + (RCS)^3}$$

For sinusoidal input, we let $S = j\omega$, so that for $|H(j\omega)|$ we obtain

$$|H(j\omega)| = \frac{1}{\sqrt{[1 - 5(\omega RC)]^2 + [6\,\omega RC - (\omega RC)^3]^2}}$$

For convenience we can let $\omega RC = \alpha$. We then write

$$|H(\alpha)| = \frac{1}{\sqrt{(1 - 5\alpha^2)^2 + (6\alpha - \alpha^3)^2}}$$

Clearly, when $\alpha = 0$ (DC operation), $|H(\alpha)| = 1$. To find the 3-db cutoff frequency analytically, we would have to equate the denominator to $\sqrt{2}$. This yields an intractable algebraic problem. However, with a programmable calculator, it is a fairly simple matter, using a trial-and-error procedure, to find that $|H(\alpha)| \approx 0.707$ when $\alpha \cong 0.195$. We then determine that $f_{3db} = 0.195/(2\pi \times 10^4 \times 2 \times 10^{-8})$ = 155 Hz. This result tells only part of the story. It can be shown that there are two other break frequencies, at 1217 Hz and at 2541 Hz. The response roll-off is down about 18 db at the 1217-Hz breakpoint, after which the roll-off slope becomes −40 db/decade.

The determination of filter transfer functions in order to optimize performance of a PLL is beyond the scope of this book and will not be discussed further.

12-7.2 The PLL as a frequency multiplier

As we have noted earlier, the output of the VCO is not connected directly to the input of the phase detector within the integrated circuit structure. Rather, the VCO output and the phase detector input are made available for external connection. The advantage of this arrangement is seen in the frequency-multiplier circuit shown in Fig. 12-27.

In this circuit, the VCO is set up to have a free-running frequency f_1, which may be a multiple of the reference frequency f_0. The VCO output, brought out on pin 4, is fed through the buffer transistor Q to a counter that acts as a divide-by-N circuit. (Suitable counters for this purpose are listed in Table 11-5, although that list is by no means comprehensive.) The output of the counter will be a pulse sequence whose repetition rate is f_1/N. This output is fed into pin 5, which applies it to the phase detector. When lock is achieved, $f_1/N = f_0$. It follows that f_1 must equal Nf_0. Thus the circuit allows the generation of a high frequency that is an integer multiple of a low reference frequency. It is possible, for example, to use a cascade connection of four four-bit binary counters to achieve divide-by-$(16)^4$ operation or, in other words, to make N equal 65,536. Then, using a 60-Hz low-frequency reference source, we can generate a frequency of $60 \times 65,536 = 3.93$ MHz.

Phase-locked loops have a number of other applications. An important application is to the demodulation of FM signals. We can appreciate the basic principle involved by supposing that the FM signal is applied to the reference-frequency input. As the instantaneous frequency of this signal shifts, assuming that lock is

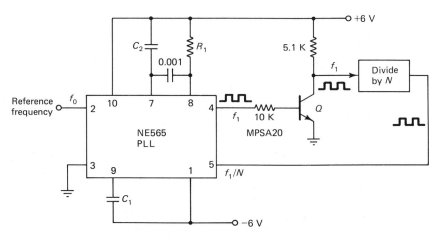

Fig. 12-27 Frequency multiplier using PLL.

achieved, the VCO will track the signal frequency. As it does so, the instantaneous *level* of the VCO control-voltage input will shift. Thus the modulating signal can be recovered from the variations of this VCO-input voltage.

REFERENCES

1. Signetics Corporation, *Analog Data Manual 1981* (Sunnyvale, Calif.: Signetics Corporation, 1980).
2. Daniel H. Sheingold, ed., *Analog-Digital Conversion Notes* (Norwood, Mass.: Analog Devices, Inc., 1977).
3. Intersil, Inc., *Data Acquisition and Conversion Handbook* (Cupertino, Calif.: Intersil, Inc., 1980).
4. Garth Nash, *Phase-Locked Loop Design Fundamentals, Application Note AN-535* (Phoenix, Ariz.: Motorola Semiconductor Products, Inc., 1970).

PROBLEMS

12-1. The voltages v_1 and v_2 are connected to a comparator as shown. Assume $v_o = +5$ V when $v_1 > v_2$ and $v_o = 0$ when $v_1 < v_2$. Sketch v_o as a function of time.

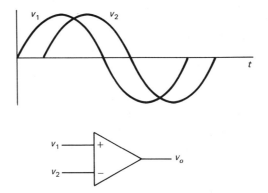

12-2. The Schmitt trigger circuit of Fig. 12-5 uses $R_1 = 1$ K Ω. The circuit is intended to drive *TTL* devices. Hence, $V_{CC} = +5$ V. A 10-mV (peak) noise voltage is superimposed on the analog input signal. Specify the maximum value of R_2 so that the circuit will be immune to this noise.

12-3. A 555 timer is to be used in monostable operation to generate a 1-ms output pulse. For the circuit of Fig. 12-10, specify the value of R_A if C is a 0.22-μF capacitor. If the input is triggered every 3 ms, sketch the output waveshape at pin 3. $V_{CC} = 10$ V.

12-4. The 555 timer of Prob. 12-3 is to be modified to use a current-mirror circuit as in Fig. 12-12. The potentiometer R_2 in series with R_1 is to be used to adjust the output pulse width from 0.5 ms to 2 ms. C is 0.22 μF. Triggering occurs every 5 ms. $V_{CC} = +10$ V. Specify approximate values for R_1 and R_2.

12-5. Using the astable circuit of Fig. 12-13, specify R_A and R_B to produce a 10-KHz output frequency with a 25% duty cycle (high, 25%; low, 75%). $C = 0.22$ μF. Sketch the output wave form if $V_{CC} = +5$ V.

12-6. It can be shown that for the regulator transistor Q_1 in Fig. 12-15, the change in collector voltage ΔV_C in response to a change in positive supply voltage ΔV_S is given by

$$\frac{\Delta V_C}{\Delta V_S} = \frac{1}{1 + R_3/(R_2 + R_1 \| h_{ie})\{1 + h_{fe}[R_1/(R_1 + h_{ie})]\}}$$

    (a) Neglecting the base current of Q_1 and assuming $V_{BE} = 0.7$ V, $V_B = 1.2$ V (collector terminal), $V_S = 10$ V, $R_1 = 7$ KΩ, and $I_C = 0.9$ mA, determine R_2 and R_3.

    (b) Assuming $h_{fe} = 100$ and $h_{ie} = 3$ KΩ, calculate the percent change in V_C for a 10% change in V_S.

12-7. A stripped-down circuit derived from the D/A converter of Fig. 12-15 is shown. Assuming $V_{BE} = 0.7$ V and the emitter current of Q_2 is 0.81 mA, determine values for R and R_R.

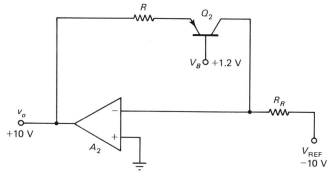

12-8. **(a)** Use the results of Prob. 7 to determine the current I_T in Fig. 12-15 when the binary input is 1010.

    (b) Determine the value of R_F so that a binary input of 1010 yields v_o equal to -10 V.

12-9. Part of the circuit of Fig. 12-17 is shown, in which only one current source is active and node D is short-circuited to ground. Prove that the short-circuit current I_x is equal to $I/2$.

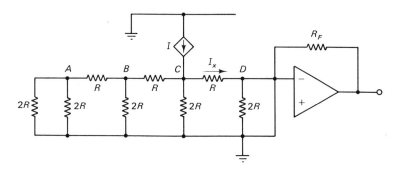

12-10. For current source I connected to node B in Fig. 12-17, show that the short-circuit current I_x (see Prob. 9) is $I/4$.

12-11. For the current source I connected to node A in Fig. 12-17, show that the short-circuit current I_x (see Prob. 9) is $I/8$.

12-12. For the circuit of Fig. 12-16, assume that the emitter current of Q_2 is 1 mA and that transistors Q_2, Q_3, Q_4, Q_5, and Q_6 are identical. Determine R_F so that for a binary input of 1111, v_o will be -9.375 V, where full-scale output voltage, if attainable, would be -10.00 V.

12-13. A D/A converter using an $R-2R$ ladder and a single, reference-voltage source of $+10$ V is shown.

    (a) Determine the currents in each of the ladder resistors.

    (b) Determine the value of R_F so that for a digital input of 1111, v_o will be -9.375 V.

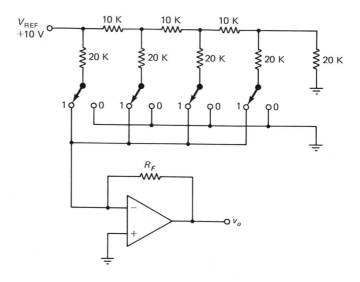

12-14. For the dual-slope A/D converter of Fig. 12-19, let $-V_{\mathrm{REF}} = -10$ V and assume that the counter is a two-digit BCD counter. (Maximum count is $99_{10} = 1001\ 1001$.) The clock-pulse rate is 1 KHz. If $C = 1\ \mu\mathrm{F}$, determine R so that when $v_a = 9.9$ V, v_o reaches its maximum value of 9.9 V at the end of the interval T_1.

12-15. Using the value for R determined in Prob. 14, find the total time $T_1 + T_2$ required to measure $v_a = 4.9$ V.

12-16. For the successive-approximation A/D converter of Fig. 12-21, the analog input voltage v_a is 5 V and E_{REF} is 16 V. What are the contents of the shift register at the end of three comparisons?

12-17. Design formulas for a phase-locked loop allow determination of the VCO free-running frequency f_o, lock range f_L, and capture range f_C. These formulas are

$$f_o = \frac{1.2}{4R_1C_1}\ \ Hz$$

$$f_L = \pm\frac{8f_o}{V_{CC}}$$

$$f_c = \pm\frac{1}{2\pi}\sqrt{\frac{2\pi f_L}{3.6 \times 10^3 C_2}}$$

R_1, R_2, and C_2 are shown in the diagram. For an FSK circuit, determine C_1 so that $f_o = (1070 + 1270)/2 = 1170$ Hz. For $V_{CC} = 6$ V, what is the value of f_L? Choose C_2 so that $f_C = \frac{1}{2} f_L$.

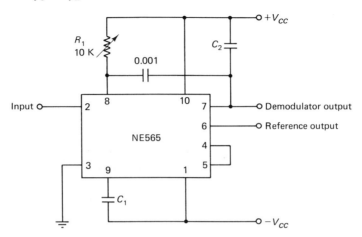

12-18. A designer wishes to use a PLL to generate a 15,525-Hz horizontal synchronizing frequency for a television picture raster. Three reference frequency sources are being considered—the 60-Hz line, a 1.009125-MHz crystal, and a 993,600-Hz crystal. Which sources are suitable? Discuss advantages and disadvantages of each.

APPENDIX A

CLASSIFICATION
OF FEEDBACK CIRCUITS

In the conventional analysis of feedback amplifiers, the circuits are classified as belonging to one of four types depending (1) on how the output signal is sensed and sampled, and (2) on how the feedback signal is injected at the input. It is easy to see that at the output, we can either place the sensing terminals *across* the output or *in series* with it (and the load). At the input, we can either put the feedback voltage *in series* with the source or inject a feedback current at an input node in superposition with the input current from the external source. The four possible connections are illustrated in Fig. A-1.

For the moment, we put aside the question of how to recognize, in a given circuit, which type of feedback is being used. Rather, we shall investigate input and output resistances of the four types. We note, before proceeding, that the dimensions of both A and β change according to which type of circuit is considered. Only in circuits (b) and (c) are they both dimensionless quantities.

For the current-series circuit of Fig. A-1(a), we draw the equivalent circuit shown in Fig. A-2. This circuit is highly simplified in the interest of getting at some important results with a minimum of complexity. For this reason, the β network is shown as having zero input resistance (short circuit) as well as zero output resistance. Moreover, this representation of the β network supposes that there is total isolation between its input and output sections. This is rarely true in practice since, typically, β networks are passive, often consisting only of resistive voltage dividers.

To find the input resistance with feedback R_{if}, we make use of Blackman's equation, Eq. (4-12). With the circuit dead, $R_{if}^o = R_i$. It is easy to determine that $T_{SC} = \beta A R_o/(R_o + R_L)$ and that $T_{OC} = 0$. We then obtain

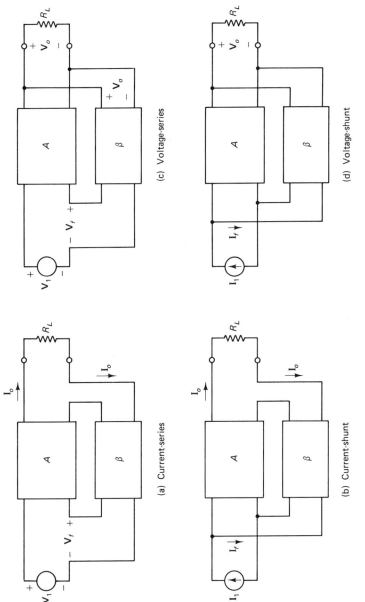

(a) Current-series

(b) Current-shunt

(c) Voltage-series

(d) Voltage-shunt

Fig. A-1   The four types of feedback circuit.

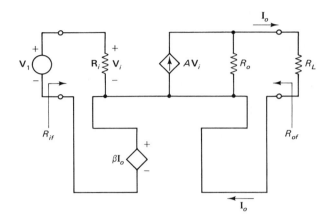

Fig. A-2 Equivalent current-series circuit.

$$R_{if} = R_i\left(1 + \frac{\beta A R_o}{R_o + R_L}\right) \tag{A-1}$$

To find R_{of}, it is much simpler to assume that R_L has been disconnected from the circuit. Under that condition, $R_{of}^0 = R_o$, $T_{OC} = 0$ and T_{SC} is simply equal to βA. Then Blackman's equation yields

$$R_{of} = R_o(1 + \beta A) \tag{A-2}$$

The equivalent circuit for that of Fig. A-1(b) is shown in Fig. A-3. To find R_{if} we note that $R_{if}^0 = R_i$. To find T_{OC} we remove the source I_1, replace AI_i with a source of A amperes and compute I_o to be $AR_o/(R_o + R_L)$. Then $I_i = -\beta AR_o/(R_o + R_L)$ and T_{OC} is $AR_o/(R_o + R_L)$. Because a short circuit at the input robs R_i of all current, $T_{SC} = 0$. Therefore, we have

$$R_{if} = \frac{R_i}{1 + \beta AR_o/(R_o + R_L)} \tag{A-3}$$

To find R_{of}, we remove R_L. Then $R_{of}^0 = R_o$. Because of the absence of R_L, T_{OC} will be 0. But with the output short-circuited, we find that $T_{SC} = A$. Therefore,

$$R_{of} = R_o (1 + \beta A) \tag{A-4}$$

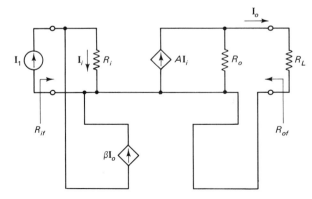

Fig. A-3 Equivalent current-shunt circuit.

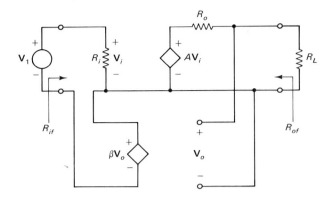

Fig. A-4 Equivalent voltage-series circuit.

For the voltage-series circuit of Fig. A-1(c), we draw the equivalent circuit of Fig. A-4. For this circuit R_{if} is found by calculating that $T_{SC} = \beta AR_L/(R_L + R_o)$, $T_{OC} = 0$, and $R_{if}^0 = R_i$. Hence we obtain

$$R_{if} = R_i \left(1 + \frac{\beta AR_L}{R_L + R_o} \right) \tag{A-5}$$

To find R_{of}, we remove R_L. Then $R_{of}^0 = R_o$, $T_{OC} = \beta A$, and $T_{SC} = 0$. Hence

$$R_{of} = \frac{R_o}{1 + A\beta} \tag{A-6}$$

The equivalent circuit for the voltage-shunt circuit of Fig. A-1(d) is shown in Fig. A-5. Clearly, $R_{if}^0 =$ is equal to R_i. $T_{OC} = \beta AR_L/(R_L + R_o)$, and $T_{SC} = 0$. Hence

$$R_{if} = \frac{R_i}{1 + \beta AR_L/(R_L + R_o)} \tag{A-7}$$

Removing R_L, we find $R_{of}^0 = R_o$, $T_{SC} = 0$, and $T_{OC} = \beta A$. This allows us to

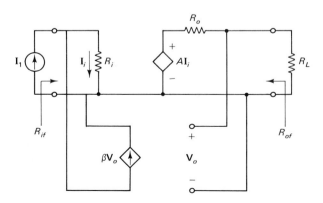

Fig. A-5 Equivalent voltage-shunt circuit.

Connection	R_{if}	R_{of}
Current-series	$R_i \left(1 + \dfrac{\beta A R_o}{R_o + R_L} \right)$	$R_o (1 + \beta A)$
Current-shunt	$\dfrac{R_i}{1 + \beta A R_o / (R_o + R_L)}$	$R_o (1 + \beta A)$
Voltage-series	$R_i \left(1 + \dfrac{\beta A R_L}{R_L + R_o} \right)$	$\dfrac{R_o}{1 + \beta A}$
Voltage-shunt	$\dfrac{R_i}{1 + \beta A R_L / (R_L + R_o)}$	$\dfrac{R_o}{1 + \beta A}$

write

$$R_{of} = \frac{R_o}{1 + \beta A} \tag{A-8}$$

We collect all the results we have found in Table A-1. As shown in the table, the input and output resistances may be reduced or magnified according to the form of feedback circuit used. Caution should be exercised in using these results since (1) the feedback (β) networks are highly idealized and (2) the load resistance R_L is omitted in determining the output resistances. Using the methods discussed in Chap. 4, we should have no difficulty taking into account more complex β networks. The effect of R_L is easily taken into account by simply calculating the modified output resistance $R_o = R_{of} \| R_L$.

In an actual circuit, the determination of which of the four circuit types we are dealing with is often tricky. Indeed, we may sometimes be able to analyze a given circuit using either one or another equivalent circuit. Such ambiguity does not arise when the return-ratio methods of Chap. 4 are used.

As an illustration of a simple feedback circuit whose analysis by conventional methods is often confusing, let us consider the emitter-follower shown in Fig. A-6.

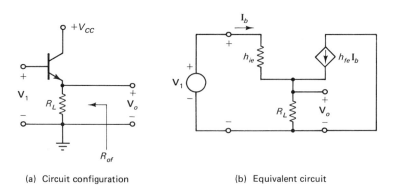

(a) Circuit configuration   (b) Equivalent circuit

Fig. A-6 Emitter-follower circuit.

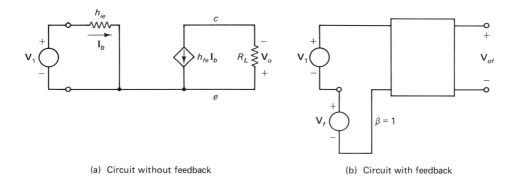

(a) Circuit without feedback   (b) Circuit with feedback

Fig. A-7   Equivalent circuits for emitter-follower.

In conventional analysis, we try to separate the network into its forward, amplifying section (A) and a feedback network (β). But how is this to be done?

If we take away the load resistor, leaving only an open circuit, the circuit becomes inoperative. Therefore, let us imagine that the emitter is grounded and that the load resistor R_L is "folded" around to lie in the collector side of the circuit. This yields the equivalent circuit shown in Fig. A-7(a).

Examining the input circuit of Fig. A-6(b), we see that all the voltage developed across the load resistor is fed back to the input. Hence $\beta = 1$. From the circuit without feedback shown in Fig. A-7, we find A_v to be $h_{fe}R_L/h_{ie}$. Using the relationship $A_{vf} = A_v/(1 + A_v\beta)$, we find A_{vf} to be $h_{fe}R_L/(h_{ie} + h_{fe}R_L)$, a result that can be easily verified using other, simpler methods. The return-ratio method developed in Chap. 4 produces this result directly, and without the clever manipulations required to justify the use of the circuit of Fig. A-7(a).

APPENDIX B

ANSWERS
TO SELECTED PROBLEMS

CHAPTER 1

2. -0.119 V.   **6.** $I_2/I_1 = 1.149$.   **9.** $V_{C1} = 4$ V, $V_{C2} = +14$ V.   **10.** $R_C = 10$ KΩ, Δi_B $= \pm 15$ μA.   **11. (a)** $I_B = 50$ μA, $V_{CE} = 7$ V; **(b)** $A_v = -250$.
15. $I_C = 2$ mA, $V_{BB} = 2.7$ V.   **17. (a)** $I_D = 4$ mA, $V_{DS} = 20$ V; **(b)** $A_v = -10$; **(c)** $v_o(t) = -1 \sin \omega t$.   **20. (a)** $R_S = 3.5$ KΩ, $R_D = 4.5$ KΩ; **(b)** $A_v \approx -9$.

CHAPTER 2

1. (a) $V_o = 0.3$ V, $I_C = 9.7$ mA; **(b)** $h_{FEmin} = 66.9$.   **2. (a)** 0.0392 mA; **(b)** 137 gates.
4. $v_n = 3.1$ V.   **5. (a)** NOR gate; **(b)** 38.3 KΩ.   **7.** 1.20 KΩ.   **13. (a)** $I_B = 74.8$ μA; **(b)** $I_{B2} = 1.43$ mA; **(c)** Q_3 is cut off; **(d)** $I_{B3} = 1.91$ mA; **(e)** $R = 280$ Ω.   **16.** 22.
17. (a) $V_{out} = 3.94$ V; **(b)** $V_A = 4.86$ V; **(c)** OR gate.   **19. (a)** $V_o = -1.555$ V; **(b)** $V_o = -0.7$ V.

CHAPTER 3

1. (a) 950 Ω; **(b)** 6.33; **(c)** 32.4 db.   **3. (a)** 4 mA; **(b)** 76.8; **(c)** 0.5×10^{-3}; **(d)** 1.54×10^5.   **5. (a)** 5.6 KΩ; **(b)** 7 V; **(c)** 37.8 KΩ; **(d)** 19.85.   **10.** 4.65 KΩ.
11. 2.2 V.   **13.** 10,300.   **14.** 1050 Ω.   **15.** 33.   **18.** 28,000.   **20. (a)** 8 KΩ; **(b)** 152.6.
23. (a) 198; **(b)** 8 nA.   **32. (a)** 6 KΩ; **(b)** 32.7.

CHAPTER 4

1. (a) 0.1; **(b)** 190. **3.** $A_v = 20 \times 10^4$, $\beta = 0.01$. **4. (a)** 0.02; **(b)** -33.3.
6. $K = 6$, $\beta = 0.0694$. **8.** 1.086×10^4. **9.** $K = 11$, $A_{vf} = 11$.
10. $T = 12.77$, $A_{vf} = 93.7$. **12. (a)** -100; **(b)** 4.15Ω; **(c)** 6.17Ω. **14. (a)** 0.971; **(b)** 73.9.
15. 0.985. **16.** $r_i = 203$ KΩ, $r_o = 10$ Ω. **18.** 21.55. **21.** 0.0939Ω. **25.** 244 Hz.
29. $A_{vf} = 47.6$, $r_i = 458$ KΩ, $r_o = 1147$ Ω. **32. (a)** 24.2; **(b)** -26.1; **(c)** 37.8 KΩ; **(d)** 272.1
Ω.

CHAPTER 5

1. $8 \times 10^4[1 + (f/f_o)^2]^{-3/2}$ $\underline{/-3 \tan^{-1} f/f_o}$. **5. (a)** 0.158; **(b)** 55 KHz. **6. (a)** $\frac{3}{8}$ A_v;
(b) $-150.7°$; **(c)** 63.2 KHz; **(d)** $-143°$. **7.** $R = 4$ KΩ, $C = 3979$ pF. **8. (a)** 100;
(b) -49.5, **(c)** 2680 Hz. **11. (a)** $\beta = 10^{-2}$; **(b)** $\omega = 10^5$ for $|T| = 0$ db.
12. (a) $\beta = 7.69 \times 10^{-6}$; **(b)** $|T| = 57.7$ db. **14. (a)** Break frequencies: ω_1 (two breaks),
$(3 - \sqrt{5})\,\omega_1/2$, $(3 + \sqrt{5})\,\omega_1/2$; **(b)** Ratio equals 6.854. **18. (a)** Approximately 100 MHz;
(b) $-218.7°$ **(c)** 33 MHz. **21. (a)** $R_2 = 1$ KΩ; **(b)** $R_1 = 4$ KΩ.
23. Reduce $|T|$ at $f = 0$ by 22.8 db.

CHAPTER 6

1. 10^6 rad/s, 6.005. **2.** $K = 7.58$, $f = 34.84$ Hz. **3.** $A = 4$, $f = 112$ KHz. **4.** 2.4 MΩ.
6. (c) $R = 1739$ Ω; **(d)** 3183 Hz. **10.** $C_1 = 51.6$ pF, $C_2 = 2837$ pF. **12. (b)** $\omega = 1/RC$;
(c) $g_m R = 3$. **16.** $R_D = 9.56$ KΩ, $C = 8156$ pF. **18. (a)** 2962 Hz, **(b)** 0.9667.
20. (a) $R_3 = 260\Omega$; **(b)** $f = 10,065$ Hz.

CHAPTER 7

1. (a) -450μ V; **(b)** 100 nA. **2.** 343 nA. **5.** 52 nA. **7.** -36 mV. **8.** 23,100.
9. (a) 4.2 V/μs; **(b)** 990 Hz; **(c)** slew rate sets limit. **12.** 14,788. **14.** 3. **15.** -15 V.
17. 0. **18.** 5. **22. (a)** 1.8 MHz; **(b)** 9 Hz. **25. (a)** 27.2 KHz; **(b)** 5305 pF;
(c) 37.7 KΩ; **(d)** 1/3. **28.** $R_1 = 1.4$ KΩ, $R_2 = 1.99$ KΩ, $R_3 = 1.99$ KΩ, $R_4 = 1.99$ KΩ.
31. (a) 1591 Hz; **(b)** 707 Ω.

CHAPTER 8

1. (a) 2.2 μs; **(b)** 318 KHz. **3.** $f_\beta = 7.99$ MHz, $f_T = 800$ MHz, $C_{b'e} = 35.3$ pF, $r_{b'e} = 520$
Ω, $r_{b'b} = 80$ Ω. **6.** 3.3 KΩ. **7.** $A_i = -50$, $f_1 = 2.11$ MHz. **10. (a)** 0.0446 mho; **(b)** 3
pF; **(c)** 56.8 MHz.
13. (a) 19.27 MHz; **(b)** 20-V supply should be suitable; **(c)** 384.

CHAPTER 9

1. (a) $Q_0 = 140$, $R = 11.37$ Ω, $L = 253.3$ μH; **(b)** 280.
4. 8160 Ω at resonance, 6371 Ω for $\delta = 0.01$. **5.** 46,960 Ω. **7. (a)** 26.526 KHz;

(b) 60; (c) 1326 Hz.   **10.** (a) 32, 6.22 KHz; (b) 20 KΩ.   **12.** (a) 56 db, 32.2 KHz;
(b) BW = 70.7 KHz, f_1 = 9.975 MHz, f_2 = 10.025 MHz.
14. R_1 = 247 Ω, R_2 = 53.05 KΩ.   **16.** (a) f_o = 2 KHz, BW = 100 Hz, (b) −5.93%.
19. R_2 (both amplifiers) = 75.07 KΩ, values of R_1 for each amplifier are 145 Ω and 213 Ω.
21. 3.98 MHz, 20.

CHAPTER 10

1. (a) 4 A pk-pk; (b) 80 W.   **3.** (a) 25 sin ωt mA; (b) 7.5 W; (c) 7.5 W.   **5.** (a) 2.12;
(b) 0.667 A, 12 V; (c) 8 W; (d) 0.667 A.   **9.** 141.7 mW, 188.9 mW.   **11.** 4.5 in².
13. (a) 0.1 A; (b) 60°C/W; (c) 1.25 W, 5.
15. V_{CC} = 20 V, $\hat{I}_C$ = 1 A, n_1/n_2 = $\sqrt{2}$ P_L = 10 W.   **18.** (a) 9.87 W, 41.1%; (b) 36 W.

CHAPTER 11

2. 16 ns.   **5.** (a) S = 1, R = 0, (b) 0 and 1.   **9.** 28.6 MHz.
11. $Q_B Q_A$ execute the sequence: 00, 01, 11, 10, 00, ... Repetition occurs after 4 pulses.
18. 4096.   **19.** 25 pins.

CHAPTER 12

2. 1 MΩ.   **3.** 4132 Ω.   **4.** R_1 = 3409 Ω, R_2 = 10,227 Ω.
6. (a) R_2 = 5 KΩ, R_3 = 8.8 KΩ; (b) 0.933%.   **8.** (a) 1.0125 mA; (b) 9876.5 Ω.
13. (a) 0.0625 mA, 0.125 mA, 0.25 mA, 0.5 mA; (b) 10 KΩ.   **15.** 148 ms.   **16.** 0010.
17. C_1 = 2564 pF, f_L = 1560 Hz, C_2 = 0.1133 μF.

Index